MIDDLETOWN

Robert S. Lynd was born in New Albany, Indiana, in 1892. He graduated from Princeton University in 1914, received a B.D. from Union Theological Seminary in 1923 and a Ph.D. from Columbia University in 1931. From 1914 to 1918 he was managing editor of *Publishers' Weekly*. He was director of the Small City Study under the auspices of the Institute of Social and Religious Research from 1923 to 1926, assistant director of the educational research division of the Commonwealth Fund from 1926 to 1927, and permanent secretary of the Social Science Research Council from 1927 to 1931. Since 1931 he has been Professor of Sociology at Columbia University Graduate School. He is author of *Knowledge for What* (1939), and has contributed many articles to social-science journals. He and his wife, Helen Merrell Lynd, collaborated on *Middletown* (1929) and *Middletown in Transition* (1937).

Helen Merrell Lynd was born in the Middle West and now lives in New York. After graduating from Wellesley, she took her doctorate at Columbia University, where she worked in history and psychology as well as in philosophy. Since 1928 she has taught Social Philosophy at Sarah Lawrence College. Her books include *England in the Eighteen-Eighties: Toward a Social Basis for Freedom* (1945) and *On Shame and the Search for Identity* (1958). She has also published articles in the *Journal of Philosophy, The American Scholar,* and other periodicals.

MIDDLETOWN

A STUDY IN AMERICAN CULTURE

Robert S. Lynd & Helen Merrell Lynd

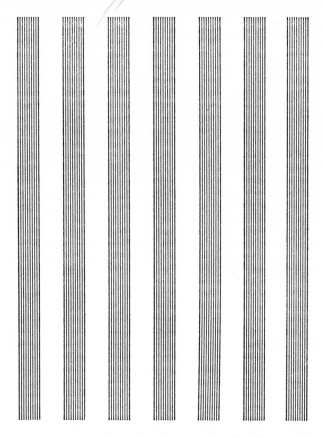

FOREWORD BY CLARK WISSLER

A Harvest Book

Harcourt, Brace & World, Inc. · New York

FOREWORD

On every hand we hear the admonition, "The study of society must be made objective." When one asks what is meant by this, he is referred to the natural and the biological sciences. But while the average man has little difficulty in comprehending what is meant by objective in the study of electricity, bees, etc., he finds himself at a loss to visualize the objects of study in a social inquiry. There is nothing strange in this, because the professionals in social science are still far from confident that they have their hands upon the social reality. True, many attempts have been made to find the basic factors in society, but these factors have been sought, for the most part, in the laboratories of biology and psychology, which is not unlike groping behind the scenes and digging under the stage, disregarding the comedies, tragedies, and dramas in plain sight. On the other hand, experience with social phenomena is bringing us nearer and nearer to a realization that we must deal directly with life itself, that the realities of social science are what people do. Seemingly in full realization of this, the authors of this book have patiently observed an American community and sketched out for us, in the large, the whole round of its activities. No one had ever subjected an American community to such a scrutiny; probably few would regard it as worth while. Rather have we been taught to set store by studies of the individual on one hand, and on the other, on the gathering of intimate statistics as to wages, living conditions, etc., for groups in our national population at large, as coal miners, teamsters, working girls, etc. The first of these seems to have been ordered upon the theory that maladjustments of individuals might be dealt with effectively if one knew a true sample of personal histories, and, in the main, studies of this kind have justified their making. The second seems to rest on the assumption that occupational groups present collective problems which can be dealt with on a national level, the maladjustments in this case arising in the failure of these groups to articulate properly with other groups. Here again insight has been achieved by statistical and analytical studies of wide scope. There remains, however, the obvious condition that the masses of individuals concerned live and function in communities, and that the picture will not be complete until these communities also are made objects of study. Whatever else a social phenomenon is, it is a community affair. The communities that collectively are American are also objective, they are realities, and if, as we are told, we can never know society until it is subjected to objective methods, then here is one place to begin.

So this volume needs no defense; it is put forth for what it is, a pioneer attempt to deal with a sample American community after the manner of social anthropology. To most people, anthropology is a mass of curious information about savages, and this is so far true, in that most of its observations are on the less civilized. What is not realized is that anthropology deals with the communities of mankind, takes the community, or tribe, as the biological and social unit, and in its studies seeks to arrive at a perspective of society by comparing and contrasting these communities; and whatever may be the deficiencies of anthropology, it achieves a large measure of objectivity, because anthropologists are by the nature of the case "outsiders." To study ourselves as through the eye of an outsider is the basic difficulty in social science, and may be insurmountable, but the authors of this volume have made a serious attempt, by approaching an American community as an anthropologist does a primitive tribe. It is in this that the contribution lies, an experiment not only in method, but in a new field, the social anthropology of contemporary life.

Finally, irrespective of the interests of social science, this volume is a contribution to history, not the usual kind of history, but the kind that is coming more and more into demand, a cross-section of the activities of a community today as projected from the background of yesterday, and the authors are to be commended for their foresight in revealing the Middletown of 1890 as a genesis of the Middletown of today, not as its contrast. Every reader of these pages will realize more clearly than before the changes each decade has brought and the imperfect way in which our communities, of which this is a sample, have met the new conditions under which they must function, and incomplete though this record is, its perusal should enlighten the conscientious citizen and serve as a suggestion as to what information is needed by those who attempt to direct the affairs of an American town.

CLARK WISSLER.

1929, American Museum of Natural History.

CONTENTS

IV. Using Leisure

V. Engaging in Religious Practices

VI. Engaging in Community Activities

Appendix

PREFACE

Behind the tightly filed record sheets, schedules, question-naires, tables, and maps from which this report has been assembled lie days and nights of patient observation, interviewing, reading old records, and checking and re-checking data on the part of the staff of field assistants:

Dr. Faith Moors Williams
Miss Dorothea Davis
Miss Frances Flournoy

Their tireless coöperation and ready ingenuity made the study possible. Dr. Williams, in particular, contributed to the planning of the study as it developed and supervised a large part of the statistical handling of the data as well as coöperating in the actual field work. She was directly responsible for collecting the data on cost of living and on the income of working class families.

No one can be more aware than the writers of the short-comings of the report—the lack of adequate data at certain points and frequent unevenness of method. Furthermore, the field work was completed in 1925; the point of view of the investigators has developed during the subsequent years, and the treatment would be at many points more adequate were the investigation to be undertaken now.

Invaluable counsel has been received from Professor Clark Wissler, Professor L. C. Marshall, Professor William F. Ogburn, Mr. Lawrence K. Frank, Dr. Joseph Chassell, and Dr. Gardner Murphy. None of the inadequacies of the study, however, lies at their door.

To the Institute of Social and Religious Research, which financed the investigation, and to its technical staff, which has been generous in criticism and suggestion, the investigation owes its support.

A final acknowledgment should be made to the patient subject of this picture, the people of Middletown, without whose generous friendship the marrow of the study would be lacking.

R. S. L.
H. M. L.

New York, June, 1928

INTRODUCTION

Chapter I

NATURE OF THE INVESTIGATION

The aim of the field investigation recorded in the following pages was to study synchronously the interwoven trends that are the life of a small American city. A typical city, strictly speaking, does not exist, but the city studied was selected as having many features common to a wide group of communities. Neither field work nor report has attempted to prove any thesis; the aim has been, rather, to record observed phenomena, thereby raising questions and suggesting possible fresh points of departure in the study of group behavior.

The stubborn resistance which "social problems" offer may be related in part to the common habit of piecemeal attack upon them. Students of human behavior are recognizing increasingly, however, that "the different aspects of civilization interlock and intertwine, presenting—in a word—a continuum." [1] The present investigation, accordingly, set out to approach the life of the people in the city selected as a unit complex of interwoven trends of behavior.

Two major difficulties present themselves at the outset of such a total-situation study of a contemporary civilization: *first,* the danger, never wholly avoidable, of not being completely objective in viewing a culture in which one's life is imbedded, of falling into the old error of starting out, despite oneself, with emotionally weighted presuppositions and consequently failing ever to get outside the field one set out so bravely to objectify and study; and, *second,* granted that no one phase of living can be adequately understood without a study of all the rest, how is one to set about the investigation of anything as multifarious as the gross-total thing that is Schenectady, Akron, Dallas, or Keokuk?

A clew to the securing both of the maximum objectivity and of some kind of orderly procedure in such a maze may be found in the approach of the cultural anthropologist. There are, after

[1] A. A. Goldenweiser, *Early Civilization* (New York; Knopf, 1919), p. 31.

all, despite infinite variations in detail, not so many major kinds
of things that people do. Whether in an Arunta village in Central Australia or in our own seemingly intricate institutional
life of corporations, dividends, coming-out parties, prayer meetings, freshmen, and Congress, human behavior appears to consist in variations upon a few major lines of activity: getting
the material necessities for food, clothing, shelter; mating; initiating the young into the group habits of thought and behavior; and so on. This study, accordingly, proceeds on the
assumption that all the things people do in this American city
may be viewed as falling under one or another of the following
six main-trunk activities:

> Getting a living.
> Making a home.
> Training the young.
> Using leisure in various forms of play, art, and so on.
> Engaging in religious practices.
> Engaging in community activities.

This particular grouping of activities is used with no idea
of its exclusive merit but simply as a methodological expedient.[2]
By viewing the institutional life of this city as simply the form
which human behavior under this particular set of conditions
has come to assume, it is hoped that the study has been lifted on
to an impersonal plane that will save it from the otherwise
inevitable charge at certain points of seeming to deal in personalities or to criticize the local life. For, after all, having one's
accustomed ways scrutinized by an outsider may be discon-
certing at best. Like Aunt Polly in Donald Ogden Stewart's
Aunt Polly's Story of Mankind, many of us are prone to view
the process of evolution as the ascent from the nasty amoeba
to Uncle Frederick triumphantly standing at the top of the
long and tortuous course in a Prince Albert with one gloved
hand resting upon the First National Bank and the other upon

[2] W. H. R. Rivers in his *Social Organization* (New York; Knopf, 1924)
sets forth a sixfold classification of social groupings identical with the six
types of activity employed here. Clark Wissler presents a ninefold culture
scheme, in *Man and Culture* (New York; Crowell, 1923), Chs. V and XII.
Frederick J. Teggart criticizes Wissler's use of a universal culture pattern,
but himself implicitly recognizes certain activities as common to men
everywhere, in *Theory of History* (New Haven; Yale University Press,
1925), p. 171.

the Presbyterian church. To many of us who might be quite willing to discuss dispassionately the quaintly patterned ways of behaving that make up the customs of uncivilized peoples, it is distinctly distasteful to turn with equal candor to the life of which we are a local ornament. Yet nothing can be more enlightening than to gain precisely that degree of objectivity and perspective with which we view "savage" peoples. Even though such a venture in contemporary anthropology may be somewhat hazy and distorted, the very trial may yield a degree of detachment indispensable for clearer vision.

It is a commonplace to say that an outstanding characteristic of the ways of living of any people at any given time is that they are in process of change, the rate and direction of change depending upon proximity to strong centers of cultural diffusion, the appearance of new inventions, migration, and other factors which alter the process. We are coming to realize, moreover, that we today are probably living in one of the eras of greatest rapidity of change in the history of human institutions. New tools and techniques are being developed with stupendous celerity, while in the wake of these technical developments increasingly frequent and strong culture waves sweep over us from without, drenching us with the material and non-material habits of other centers. In the face of such a situation it would be a serious defect to omit this developmental aspect from a study of contemporary life.[3]

The further device has, therefore, been adopted in this investigation, wherever the data available permitted, of using as a groundwork for the observed behavior of today the reconstructed and in so far as possible equally objectively observed behavior of 1890. The year 1890 was selected as the base-line against which to project the culture of today because of greater availability of data from that year onward and because not until the end of 1886 was natural gas struck in the city under study and the boom begun which was to transform the placid county-seat during the nineties into a manufacturing city. This narrow strip of thirty-five years comprehends for hundreds of

[3] Cf. Rivers' closing sentence in *The History of Melanesian Society* (Cambridge; Cambridge University Press, 1914) : "It is because we can only hope to understand the present of any society through a knowledge of its past that such historical studies as those of which this book is an example are necessary steps toward the construction of a science of social psychology."

American communities the industrial revolution that has descended upon villages and towns, metamorphosing them into a thing of Rotary Clubs, central trade councils, and Chamber of Commerce contests for "bigger and better" cities.

Had time and available funds permitted, it would obviously have been desirable to plot more points in observed trends between 1890 and the present. But the procedure followed enables us to view the city of today against the background of the city of a generation ago out of which it has grown and by which it is conditioned, to see the present situation as the most recent point in a moving trend.

To sum up, then: the following pages aim to present a dynamic, functional [4] study of the contemporary life of this specific American community in the light of the trends of changing behavior observable in it during the last thirty-five years.

So comprehensive an approach necessarily involves the use of data of widely varying degrees of overtness and statistical adequacy. Some types of behavior in the city studied lie open to observation over the whole period since 1890; in other cases only slight wisps of evidence are obtainable. Much folk talk, for instance—the rattle of conversation that goes on around a luncheon table, on street corners, or while waiting for a basket ball game to commence—is here presented, not because it offers scientifically valid evidence, but because it affords indispensable insights into the moods and habits of thought of the city. In the attempt to combine these various types of data into a total-situation picture, omissions and faults in proportion will appear. But two saving facts must be borne in mind: no effort is being made to prove any thesis with the data presented, and every effort is made throughout to warn where the ice is thin.

Since the field work aimed at the integration of diverse regions of behavior rather than at the discovery of new material in a narrowly isolated field, it will be easy to say of much of the specific data presented, "We knew that already." Underlying the study, however, is the assumption that by the presentation of these phenomena, familiar though some of them may be, in their inter-relatedness in a specific situation, fresh light may be thrown upon old problems and so give rise to further investigation.

[4] "Function" as here used denotes a major life-activity or something contributing to the performance of a major life-activity.

Chapter II

THE CITY SELECTED

The city will be called Middletown. A community as small as thirty-odd thousand affords at best about as much privacy as Irvin Cobb's celebrated goldfish enjoyed, and it has not seemed desirable to increase this high visibility in the discussion of local conditions by singling out the city by its actual name.

There were no ulterior motives in the selection of Middletown. It was not consulted about the project, and no organization or person in the city contributed anything to the cost of the investigation. Two main considerations guided the selection of a location for the study: (1) that the city be as representative as possible of contemporary American life, and (2) that it be at the same time compact and homogeneous enough to be manageable in such a total-situation study.

In line with the first of these considerations the following characteristics were considered desirable: (1) A temperate climate.[1] (2) A sufficiently rapid rate of growth to insure the presence of a plentiful assortment of the growing pains accompanying contemporary social change. (3) An industrial culture with modern high-speed machine production. (4) The absence of dominance of the city's industry by a single plant, i.e., not a one-industry town. (5) A substantial local artistic life to balance its industrial activity; also a largely self-contained artistic life, e.g., not that of a college town in which the college imports the community's music and lectures. (6) The absence of any outstanding peculiarities or acute local problems which would mark it off from the mid-channel sort of American community. After further consideration, a seventh qualification was added: the city should, if possible, be in that common-denominator of

[1] The relation of climate to the elaborate equilibrium of activities that make up living is suggested by the late James J. Hill's motto to which he is said absolutely to have adhered: "You can't interest me in any proposition in any place where it doesn't snow," or, more picturesquely, "No man on whom the snow does not fall ever amounts to a tinker's dam." (Quoted in J. Russell Smith's *North America*, New York; Harcourt, Brace and Company, 1925, p. 8.)

America, the Middle West.[2] Two streams of colonists met in this middle region of the United States: "The Yankees from New England and New York came by way of the Erie Canal into northern Ohio. . . . The southern stream of colonists, having passed through the Cumberland Gap into Kentucky, went down the Ohio River."[3] With the first of these came also a foreign-born stock, largely from Great Britain, Ireland, and Germany.

In order to secure a certain amount of compactness and homogeneity, the following characteristics were sought: (1) A city of the 25,000–50,000 group. This meant selection from among a possible 143 cities, according to the 1920 Census. A city of this size, it was felt, would be large enough to have put on long trousers and to take itself seriously, and yet small enough to be studied from many aspects as a unit. (2) A city as nearly self-contained as is possible in this era of rapid and pervasive inter-communication, not a satellite city. (3) A small Negro and foreign-born population. In a difficult study of this sort it seemed a distinct advantage to deal with a homogeneous, native-born population, even though such a population is unusual in an American industrial city. Thus, instead of being forced to handle two major variables, racial change and cultural change, the field staff was enabled to concentrate upon cultural change. The study thus became one of the interplay of a relatively constant native American stock and its changing environment. As such it may possibly afford a base-line group against which the process of social change in the type of community that includes different racial backgrounds may be studied by future workers.

Middletown, selected in the light of these considerations from a number of cities visited, is in the East-North-Central group of states that includes Ohio, Indiana, Illinois, Michigan, and Wisconsin. The mean annual temperature is 50.8° F. The highest recorded temperature is 102° F. in July and the lowest

2 "The 'Middle West,' the prairie country, has been the center of active social philanthropies and political progressivism. It has formed the solid element in our diffuse national life and heterogeneous populations. . . . It has been the middle in every sense of the word and in every movement. Like every mean, it has held things together and given unity and stability of movement." John Dewey, "The American Intellectual Frontier" (*The New Republic*, May 10, 1922).

3 Smith, *op. cit.*, pp. 296-7.

—24° F. in January, but such extremes are ordinarily of short duration, and weather below zero is extremely rare. The city was in 1885 an agricultural county-seat of some 6,000 persons; by 1890 the population had passed 11,000, and in 1920 it had topped 35,000. This growth has accompanied its evolution into an aggressive industrial city. There is no single controlling industrial plant; three plants on June 30, 1923, had between 1,000 and 2,000 on the payroll, and eight others from 300 to 1,000; glass, metal, and automobile industries predominate. The census of 1890 showed slightly less than 5 per cent. of the city's population to be foreign-born [4] and less than 4 per cent. Negroes, as against approximately 2 per cent. foreign-born in 1920 and nearly 6 per cent. Negroes; over 81 per cent. of the population in 1890 and nearly 85 per cent. in 1920 was native white of native parentage. In the main this study confines itself to the white population and more particularly to the native whites, who compose 92 per cent. of the total population.

The nearest big city, a city under 350,000, is sixty miles away, nearly a two-hour trip by train, with no through hard-surface road for motoring at the time the study was made. It is a long half-day train trip to a larger city. Since the eighties Middletown has been known all over the state as "a good music town." Its civic and women's clubs are strong, and practically none of the local artistic life was in 1924 in any way traceable to the, until then, weak normal school on the outskirts.

The very middle-of-the-road quality about Middletown would have made it unsuitable for a different kind of investigation. Had this study sought simply to observe the institution of the home under extreme urban conditions, the recreational life of industrial workers, or any one of dozens of other special "social problems," a far more spectacular city than Middletown might readily have been found. But although it was its characteristic rather than its exceptional features which led to the selection of Middletown, no claim is made that it is a "typical" city, and the findings of this study can, naturally, only with caution be applied to other cities or to American life in general.

[4] The census of 1890 shows 62.1 per cent. of the foreign-born in the state to have been of German-speaking stock and 24.5 British and Irish. Belgian glass workers were prominent among Middletown's foreign-born population in the nineties.

Chapter III

THE HISTORICAL SETTING

Two major experiences in Middletown antedate 1890, the date taken as the horizon of this study: the pioneer life of the earlier part of the century, and the gas boom of the end of the eighties which ushered in Middletown's industrial revolution. Both are within the memory of men who still walk the streets of the city.

The first permanent settlement in this county occurred in 1820, and county government was granted in 1827. The memory of one of the oldest citizens, a leading local physician throughout the nineties, reaches back to the eighteen-forties. Within the lifetime of this one man local transportation has changed from virtually the "hoof and sail" methods in use in the time of Homer; grain has ceased to be cut in the state by thrusting the sickle into the ripened grain as in the days of Ruth and threshing done by trampling out by horses on the threshing-floor or by flail; getting a living and making a home have ceased to be conducted under one roof by the majority of the American people; education has ceased to be a luxury accessible only to the few; in his own field of medicine the X-ray, anaesthetics, asepsis, and other developments have tended to make the healing art a science; electricity, the telephone, telegraph, and radio have appeared; and the theory of evolution has shaken the theological cosmogony that had reigned for centuries.[1]

This local physician whose lifetime so nearly spans that of

[1] That this stupendous change within a single lifetime was a phenomenon of the whole country, not merely of a backwoods section, is indicated by the recollections of a man born a year earlier than this physician, under the shadow of Boston State House: Henry Adams writes, ". . . on looking back, fifty years later, at his own figure in 1854, and pondering on the needs of the twentieth century, he wondered whether, on the whole, the boy of 1854 stood nearer to the thought of 1904, or to that of the year 1 . . . —in essentials like religion, ethics, philosophy; in history, literature, art; in the concepts of all science, except perhaps mathematics, the American boy of 1854 stood nearer the year 1 than to the year 1900." *Education of Henry Adams* (Boston; Houghton Mifflin, 1918), p. 53.

Middletown, the tenth of a family of eleven, was named, with the characteristic political fervor of the time, General William Harrison K———.[2] The log farmhouse of his father was ceiled inside without plaster, the walls bare save for three prized pictures of Washington, Jackson, and Clay. All meals were cooked before the great kitchen fireplace, corn pones and "cracklings" and bread being baked in the glare of a large curved reflector set before the open fire. At night the rooms were lighted by the open fire and by tallow dips; there was great excitement later when the first candle mold appeared in the neighborhood. Standard time was unknown; few owned watches, and sun time was good enough during the day, while early and late candle lighting served to distinguish the periods at night. When the fire went out on the family hearth the boy ran to a neighbor's to bring home fire between two boards; it was not until later that the first box of little sticks tipped with sulphur startled the neighborhood.

The homely wisdom of pioneer life prescribed that children be passed through a hole in the trunk of a hollow tree to cure "short growth"; hogs must be slaughtered at certain times of the moon or the bacon would shrink; babies must be weaned at certain times of the zodiac; the "madstone," "a small bone from the heart of a deer," was a valuable antidote for hydrophobia or snake-bite; certain persons "blew the fire out of a burn," arrested hemorrhage or cured erysipelas by uttering mysterious charms; a pan of water under the bed was used to check night sweats; bleeding was the sovereign remedy for fits, loss of consciousness, fever, and many other ills; and "in eruptive fevers, especially measles, where the eruption was delayed, a tea made of sheep's dung, popularly known as 'nanny tea,' was a household remedy."

Social calls were unknown, but all-day visits were the rule, a family going to visit either by horseback, the children seated behind the grown-ups, or in chairs set in the springless farm

[2] The boy grew up not in the county in which Middletown is situated but in a near-by county. His boyhood environment described here was not that of the rude pioneer villages of the state but of the open country; but the facts that life in the diminutive Middletown of 1840 did not differ markedly in fundamentals from that of the open country around it and that some people in Middletown today grew up under open country conditions not unlike those described are the reasons for the inclusion of this material here.

wagon. Social intercourse performed a highly important service; there were no daily papers in the region, and much news traveled by word of mouth. Nobody came to the home around mealtime who was not urged to take his place at the table —preachers being particularly welcome. Men would talk together for hours on the Providential portent of the great Comet of 1843, or of the time ten years before when the "stars fell." Men and women went miles and spent days in order to hear champions argue disputed political or religious points. People "got religion" and were "awakened to sin" at camp meetings under the vivid exhortation of baptizing preachers. The "Word" wove its influence closely about everyday acts.

Forty years later, in 1885, before gas and wealth spouted from the earth, bringing in their wake a helter-skelter industrial development, Middletown, a placid county-seat of some 6,000 souls, still retained some of the simplicity of this early pioneer life. "On the streets . . . on fair days lawyers, doctors, the officials of the county courts, and the merchants walked about in their shirt sleeves. The house painter went along with his ladder on his shoulder. In the stillness there could be heard the hammers of the carpenters building a new house for the son of a merchant who had married the daughter of a blacksmith." [8] Men in their prime who had grown up under pioneer conditions now controlled the affairs of Middletown. They were occupied with such momentous matters as offering "$200 for the scalp or body of any person in the city caught setting fire to the property of another," or passing regulations in response to complaints about neighborhood cows running through the streets and destroying lawns, or with badly bungling the job of laying the first town sewer.

The thin edge of industry was beginning to appear, though few people thought of the place then as anything but an agricultural county-seat: a bagging plant employed from a hundred to a hundred and fifty people, making bags from the flax grown in the surrounding countryside; a clay tile yard employed some fifteen; a roller-skate "company" in an old barn up an alley, perhaps eight; a feather-duster "factory," five or six; a small foundry, half a dozen; and a planing mill and two flour

[8] From Sherwood Anderson's description of the even tenor of life in these Middletowns of the '80's in his *Poor White*.

mills, a few more. It was still for Middletown the age of wood, and a new industry meant a hardwood skewer shop, a barrel-heading shop, or a small wooden pump works.

Such modest ventures in manufacturing as the community exhibited were the tentative responses of small local capital to the thing that was happening to the whole Middle West. The Federal Census reveals a steady movement westward of the center of manufacturing; in 1880 it was still in Pennsylvania, but by 1890 it had pushed on until it was eight and one-half miles west of Canton, Ohio. Dry-goods clerks were beginning to spend their evenings perfecting little models of washing-machines, mechanical hair-clippers, can-openers, various power-driven devices. The proprietor of a small Middletown restaurant who led a town band in the evening and "was always neglecting business to tinker around at things" saw a crude cash-register in a saloon in a neighboring city while on a trip there with his band, conceived the idea of a self-adding register, and set to work in the hope of making his fortune. The annual total of patents registered in Washington, which had remained practically constant during the decade of the seventies, jumped in 1890 to roughly double the 1880 figure.

In the state in which Middletown is located, the number of wage-earners increased from 69,508 in 1880 to 110,590 in 1890, and by 1900 was to total 155,956. The capital invested in manufacturing plants in the state doubled between 1880 and 1890 and was almost to double again by 1900.

The quiet life of the town drowsing about its courthouse square with its wooden pump—and iron dippers, punctually renewed every Fourth of July—was beginning to stir to these outside influences. A small Business and Manufacturing Association was formed about 1886 for "the promotion of any and all undertakings calculated to advance the interests, improvements and general welfare of the city."

And then in the fall of '86 came gas.

In 1876 a company boring for coal twelve miles north of the town had plugged up the hole and abandoned the project after boring 600 feet: all they "struck" was a foul odor and a roaring sound deep in the bowels of the earth, and rumor had it that they had invaded "his Satanic Majesty's domain." Nine years later, when natural gas was discovered at other points in the Middle West, the incident of the plugged-up hole north of town was recalled. In October, 1886, there was great local excitement over the plans "to bore for gas or oil or both." In November we read, "The persons employed to bore for oil have this morning 'struck' gas, and everybody is on the way to see for themselves." The roar of the escaping gas is said to have been audible for two miles and the flame when it was "lit up" could be seen in Middletown a dozen miles away.

The boom was on.

The laying of a pipe line to bring the gas into the county-seat began immediately, and new wells were sunk. By the following April a local well was producing 5,000,000 feet daily. New wells multiplied on every hand. In January, 1891, the local paper exclaimed, "We have a new gas well which really does eclipse all others in the [gas] belt. Daily output is nearly 15,000,000 feet, and they worked over thirty hours trying to anchor the flow." No wonder the little town went wild!

Meanwhile, from the spring of '87 on through '91 and '92, the "boomers" were arriving:

"Four vestibule, one dining-room and one baggage special train from Buffalo with 134 of its capitalists came in last night to see for themselves what gas can do and are much pleased. . . . Taken in carriages to all the factories and sites. . . . Grand manufacturing exhibition at the Rink, and a beautiful display of four open street cars." "A trainload of 1,200 from Cincinnati." "Quite a number of New York City capitalists and newspaper men came in from the East last night; three and one-half pages of the —— Hotel register were covered with their signatures." "American Association for the Advancement of Science visits the city and witnesses the wonders of natural gas; 300 scientists and men of affairs in the party."

Real estate was being turned over with dizzy rapidity. In 1888 a man tried to buy an eight-acre chunk of farm land on the outskirts of town but, shying at the price of $1,600, took only a sixty-day option. Before the sixty-day option expired the eight acres changed hands five times, the final price being $3,200.

Nothing short of the sky seemed an adequate limit to the citizens of Middletown. A contemporary parody runs—

"Tell me not in mournful numbers
That the town is full of gloom,
For the man's a crank who slumbers
In these bursting days of boom."

Optimists predicted a population of 50,000 in five years and even the pessimists allowed only ten years. The general senti- ment was that the gas supply was inexhaustible. Some called it "The City of Eternal Gas." The Introduction to the Middle- town City Directory announced confidently, "Every forty acres will supply a gas well, and 576 wells can be drilled within . . . [the] corporate limits and suburbs." "The mathematical deduction would be," chanted a "boom book," "that the continu- ance of this supply would be, at least, one hundred times as long as at Pittsburgh, which would be 700 years." Great flam- beaus burned recklessly day and night in the streets and at the wells. When the pipe lines were laid, consumers were charged by the fixture rather than by any system of exact measure- ment. It was cheaper to leave the gas on and to throw open doors and windows than to expend a match in relighting it.[4]

With the boomers came new industries lured by free fuel and free building sites. The earlier Business and Manufactur- ing Association awakened to new life in February, 1887, as the "Board of Trade," and concerted efforts were made to "sell the town" to industrial capital. Glass came first. Next were the iron mills—a bridge company, a nail works. A diary for 1888-9 buzzes with rumors of the coming of these new plants:

"Report that another glass factory is coming immediately." "Work progressing on the pulp mill and rubber factory." "A nail works wants to come here from ——." "Considerable talk about a Palace Stock Car Factory." "A boot and shoe factory is coming; building commenced this afternoon."

[4] "For the past six months" (the latter half of 1887), according to the State Geologist, "there has been an average waste of about 100,000,000 cubic feet of gas per day in [this state]. This is worth $10,000. . . . The volume of gas wasted in the last six months is . . worth $1,500,000." (*Sixteenth Annual Report of the [State] Department of Geology and Natural History:* 1888, p. 202.)

The value of the natural gas produced (not including that wasted) in the state in 1886 was $300,000; it doubled in '87 and again in '88; by 1890 it was two and one-third million dollars, in '95 passed five million, and in 1900 reached its high point of seven and one-quarter millions.

By the summer of 1890 the local paper speaks of the thriving little "gasopolis" with pardonable pride:

"Two and one-half years ago when natural gas was first discovered [Middletown] was a county-seat of 7,000 inhabitants. . . . It has grown since that time to a busy manufacturing city of 12,000. . . . Over forty factories have located here during that time. . . . There has been $1,500,000 invested in Middletown manufacturing enterprises employing 3,000 men. . . . Over thirty gas wells have been drilled in and around the city, every one of which is good. . . ." [5]

The first boom of '87 and '88 was the spontaneous, unorganized rush to a new El Dorado. When the earlier boom was renewed in '91 it was engineered by the Eastern land syndicate and carried forward by the local boosters' association, the Citizens' Enterprise Company, organized in August, 1891. The last-named organization raised a $200,000 fund to lure new industries with free sites and capital. [6]

Several years later, as abruptly as it had come, the gas departed. By the turn of the century or shortly thereafter, natural gas for manufacturing purposes was virtually a thing of the past in Middletown. But the city had grown by then to 20,000, and, while industry after industry moved away, a substantial foundation had been laid for the industrial life of the city of today.

And yet it is easy, peering back at the little city of 1890 through the spectacles of the present, to see in the dust and clatter of its new industrialism a developed industrial culture that did not exist. Crop reports were still printed on the front page of the leading paper in 1890, and the paper carried a daily

[5] These figures should all be deflated a little, for among the "over forty factories" were many that were operating on a "shoe-string" or less, and plant after plant failed to weather the first year. The air was full of new inventions, and these infant industries plunged courageously into manufacturing anything and everything for a frequently, as yet, vague market. Thus one local industry manufactured a wooden clothes washing-machine, a fire-kindler proclaimed "surely one of the grandest inventions of the age," and a patent can-opener.

[6] Bidding was keen for new industries among cities in the gas belt. In return for specific aid from local capital, the new company would frequently pledge itself to grow at a desirable rate, e.g., a Brass and Novelty Company from Rochester, N. Y., contracted "to employ fifty men the first six months, one hundred at the end of the first year, and 150 at the end of the second year."

column of agricultural suggestions headed "Farm and Garden." Local retail stores were overgrown country stores swaggering under such names as "The Temple of Economy" and "The Beehive Bazaar." The young Goliath, Industry, was still a neighborly sort of fellow. The agricultural predominance in the county-seat was gone, but the diffusion of the new industrial type of culture was as yet largely superficial—only skin-deep.

This, then, suggests the background of the city which is the subject of this field investigation.

I. GETTING A LIVING

Chapter IV

THE DOMINANCE OF GETTING A LIVING

A stranger unfamiliar with the ways of Middletown, dropped down into the city, as was the field staff in January, 1924, would be a lonely person. He would find people intently engaged day after day in some largely routinized, specialized occupation. Only the infants, the totteringly old, and a fringe of women would seem to be available to answer his endless questions.

In a word—

    43 people out of every 100 in Middletown are primarily occupied with getting the living of the entire group.

    23 of every 100 are engaged in making the homes of the bulk of the city.

    19 of every 100 are receiving day after day the training required of the young.

    15 of every 100, the remainder, are chiefly those under six years, and the very old.

Not only do those engaged in getting the living of the group predominate numerically, but as the study progressed it became more and more apparent that the money medium of exchange and the cluster of activities associated with its acquisition drastically condition the other activities of the people. Rivers begins his study of the Todas with an account of the ritual of the buffalo dairy, because "the ideas borrowed from the ritual of the dairy so pervade the whole of Toda ceremonial." [1] A similar situation leads to the treatment of the activities of Middletown concerned with getting a living first among the six groups of activities to be described. The extent of the dominance of this sector in the lives of the people will appear as the study progresses.

[1] W. H. R. Rivers, *The Todas* (New York; Macmillian, 1906), p. 16, also p. 38: "The lives of the people are largely devoted to their buffaloes. . . . The ordinary operations of the dairy have become a religious ritual and ceremonies of a religious character accompany nearly every important incident in the lives of the buffaloes."

At first glance it is difficult to see any semblance of pattern in the workaday life of a community exhibiting a crazy-quilt array of nearly four hundred ways of getting its living—such diverse things as being abstractors, accountants, auditors, bank cashiers, bank tellers, bookkeepers, cashiers, checkers, core makers, crane operators, craters, crushers, cupola tenders, dye-workers, efficiency engineers, electricians, electrical engineers, embalmers, entomologists, estimating engineers, illuminating engineers, linotypists, mechanical engineers, metallurgists, meteorologists, riggers, riveters, rivet makers, and so on indefinitely. On closer scrutiny, however, this welter may be resolved into two kinds of activities. The people who engage in them will be referred to throughout the report as the Working Class and the Business Class.[2] Members of the first group, by and large, address their activities in getting their living primarily to *things,* utilizing material tools in the making of things and the performance of services, while the members of the second group address their activities predominantly to *people* in the selling or promotion of things, services, and ideas. This second group supplies to Middletown the multitude of non-material institutional activities such as "credit," "legal contract," "education," "sale for a price," "management," and "city government" by which Middletown people negotiate with each other in converting the narrowly specialized product of their workaday lives into "a comfortable evening at home," "a Sunday afternoon out in the car," "fire protection," "a new go-cart for the baby," and all the other things that constitute living in Middletown. If the Federal Census distribution of those gainfully employed in Middletown in 1920 is reclassified according to this grouping we find that there are two and one-half times as many in the working class as in the business class—seventy-one in each 100 as against twenty-nine.[3]

[2] Other terms which might be utilized to differentiate these two groups by their vocational activities are: people who address their activities to things and people who address their activities to persons; those who work with their hands and those who work with their tongues; those who make things and those who sell or promote things and ideas; those who use material tools and those who use various non-material institutional devices.

[3] See Table I for the basis of this distribution.

Four of the twenty-nine in each 100 grouped with the business class belong to a group of users of highly-skilled techniques—architects, surgeons, chemists, and so on—who, though addressing their activities in getting a living more to things than to people, are not here grouped with the

No such classification is entirely satisfactory. The aerial photographer inevitably sacrifices minor contours as he ascends high enough to view a total terrain. Within these two major groups there is an infinite number of gradations—all the way from the roughest day laborer to the foreman, the foundry molder, and the linotype operator in the one group, and from the retail clerk and cashier to the factory owner and professional man in the other. There is naturally, too, a twilight belt in which some members of the two groups overlap or merge.

Were a minute structural diagram the aim of this study, it would be necessary to decipher in much greater detail the multitude of overlapping groupings observable in Middletown. Since what is sought, however, is an understanding of the major functional characteristics of this changing culture, it is important that significant outlines be not lost in detail, and the groups in the city which exhibit the dominant characteristics most clearly must, therefore, form the foci of the report. While an effort will be made to make clear at certain points variant behavior within these two groups, it is after all this division into working class and business class that constitutes the outstanding cleavage in Middletown. The mere fact of being born upon one or the other side of the watershed roughly formed by these

working class because all their other activities would place them with the business class. It should be borne in mind throughout that the term business class, as here used, includes these and other professional workers. Since it is the business interests of the city that dominate and give their tone, in the main, to the lawyer, chemist, architect, engineer, teacher, and even to some extent preacher and doctor, such a grouping by and large accurately represents the facts.

Careful consideration was given to the applicability for the purposes of this study of the conventional tripartite division into Lower Class, Middle Class, and Upper Class. This was rejected, however, for the following reasons: (1) Since the dominance of the local getting-a-living activities impresses upon the group a pattern of social stratification based primarily upon vocational activity, it seemed advisable to utilize terms that hold this vocational cleavage to the fore. (2) In so far as the traditional threefold classification might be applied to Middletown today, the city would have to be regarded as having only a lower and a middle class; eight or nine households might conceivably be considered as an upper class, but these families are not a group apart but are merged in the life of the mass of businessfolk. R. H. Gretton, while pointing out the difficulty of separating out any group in present-day industrial society as "Middle Class," defines it as precisely that group here called the business class: "The Middle Class is that portion of the community to which money is the primary condition and the primary instrument of life. . . . It . . . includes merchant and capitalist manufacturer . . . [and the] professional class." *The English Middle Class* (London; Bell, 1917), pp. 1-13.

two groups is the most significant single cultural factor tending to influence what one does all day long throughout one's life; whom one marries; when one gets up in the morning; whether one belongs to the Holy Roller or Presbyterian church; or drives a Ford or a Buick; whether or not one's daughter makes the desirable high school Violet Club; or one's wife meets with the Sew We Do Club or with the Art Students' League; whether one belongs to the Odd Fellows or to the Masonic Shrine; whether one sits about evenings with one's necktie off; and so on indefinitely throughout the daily comings and goings of a Middletown man, woman, or child.

Wherever throughout the report either Middletown or any group within the city is referred to as a unit, such a mode of expression must be regarded as simply a shorthand symbol. Any discussions of characteristics of groups are of necessity approximations only and the fact that the behavior of individuals is the basis of social behavior must never be lost sight of.

Chapter V

WHO EARN MIDDLETOWN'S LIVING?

Who are the forty-three people out of every 100 in Middletown who specialize day after day in getting its living?

Four out of five of them are males. Today as in 1890 a healthy adult male, whether married or unmarried, loses caste sharply by not engaging with the rest of the group in the traditional male activity of getting a living.

Among the women, however, no such constancy of tradition is apparent. "What has become of the useful maiden aunt?" asks a current newspaper advertisement of a women's magazine, showing a picture of a woman in her late thirties dressed in sober black, and bearing the date "Anno Domini 1900." "She isn't darning anybody's stockings," it adds succinctly, "not even her own. She is a draftsman or an author, a photographer or a real estate agent. . . . She is the new phenomenon in everyday life." Thirty-five years ago when the daughter of a prominent family became the first woman court reporter in the city, an old friend of her mother's protested that such work would "un-sex" her. The State Factory Inspector in 1900 shook his head over the spectacle of the new influx of women into industry:

"It is a sad comment on our civilization when young women prefer to be employed where they are compelled to mingle with partially clad men, doing the work of men and boys, for little more than they would receive for doing the work usually allotted to women in the home. . . . [One fears] the loss of all maidenly modesty and those qualities which are so highly prized by the true man. . . ."[1]

Throughout the entire state one woman in every ten, ten years old and over, was classified by the 1890 Census as occupied at getting a living, as over against one in six in 1920,

[1] *Eighth Biennial Report of* [*the State*] *Dept. of Statistics,* 1899-1900, pp. 212-13.

while at the latter date nearly one in four in Middletown was so occupied. This fact, coupled with the recent rise of two business women's luncheon clubs, one of them with the brisk motto, "Better business women for a better business world," would surprise the Middletown editor who so confidently proclaimed in 1891 that "it is true that qualities inherent in the nature of women impede their progress as wage-earners. . . . Women are uniformly timid and are under a disadvantage in the struggle for a livelihood."

The general attitude reflected in such characteristic school graduation essays of the 1890 period as "Woman Is Most Perfect When Most Womanly" and "Cooking, the Highest Art of Woman" contrasts sharply with the idea of getting one's own living current among the Middletown high school girls of today: 89 per cent. of 446 girls in the three upper classes in 1924 stated that they were planning to work after graduation, and 2 per cent. more were "undecided"; only 3 per cent. said definitely that they did not expect to work.[2]

But the married woman in business or industry finds herself much less readily accepted than her unmarried sister. As late as 1875 the Supreme Court of the state held that a wife's earnings were the property of her husband, and even today there is a widespread tendency to adhere to the view of a generation ago that the employment of married women involves an "ethical" problem.[3] Wives who do not themselves work may grumble that married women who work displace men and lower wages, and that they neglect their children or avoid the responsibility of child-bearing, while through their free and easy as-

[2] Six per cent. did not answer. See Appendix on Method regarding this questionnaire given to the three upper years of the high school. Large allowance must be made for subsequent changes of mind. It is not "the thing" today for a girl to admit that she plans to marry and be dependent, though the point is, of course, precisely that such an attitude has come to prevail so strongly since 1890. It is noteworthy that Middletown offers relatively few positions of instrinsic interest to a girl of the business group who has graduated from high school; this operates after Commencement to deflate considerably the zeal for working.

It should be borne in mind that many girls of this age not in high school are already actually working.

[3] As, for example, in Carroll D. Wright's *The Industrial Evolution of the United States* (New York; Scribner, 1901), p. 3.

Here, as in the case of child-rearing and of the institution of marriage discussed in later chapters, the relatively slower rate of secularization of the home and family than of business and industry is apparent.

sociation with men in the factory they encourage divorce. Many husbands, in their turn, oppose their wives' working as a reflection upon their ability as "good providers." These objections are, however, in the main, back-eddies in a current moving in the other direction. The Federal Census for 1920 showed that approximately twenty-eight women in every hundred women gainfully employed in Middletown were married, and among those employed in "manufacturing and mechanical industries," thirty-three in every hundred.[4]

These married women workers, according to the Census distribution, go largely into working class occupations. Only one of forty business class women interviewed had worked for money during the previous five years (1920-24), and she in work of a semi-artistic nature. Of the fifty-five wives out of a sample of 124 working class families[5] who had worked at some time during the previous five years (1920-24), twenty-four pointed to their husbands' unemployment as a major reason for their working, six to money needed for their children's education, five to debt, four spoke of "always needing extra money," or "It takes the work of two to keep a family nowadays," three of needing to help out with "so many children"; the other answers were scattered: "Just decided I'd like to try factory work. I was tired of housekeeping and had a baby old enough [five months] to be left"; "I needed clothes"; "I wanted spending

[4] A check by the Industrial Secretary of the local Y.W.C.A. on 889 female employees in twenty-four factories, retail stores, banks, and public utilities in 1924 showed 6 per cent. divorced, 4 per cent. widowed, 38 per cent. married, and 52 per cent. single.

The percentage of women workers who are married has more than doubled since 1890 in the state in which Middletown is located. It is significant of the trend that the pre-war unwritten rule in certain local plants that a women automatically loses her job when she marries is disappearing. Cf. Mary N. Winslow, *Married Women in Industry* (Washington, D. C.; Bulletin of the Women's Bureau, No. 38, 1924), p. 6 ff., for qualification of this trend toward equal acceptance of married women in industry.

[5] See Appendix on Method for detailed account of the selection of these and of the forty business class families interviewed, the methods of interviewing, and the occupational distribution of both groups of families. It is important to bear in mind in consideration of all data based upon these two groups of families that, whereas the 124 working class families represent what is believed to be a fair sample of the various levels of Middletown's working class in the dominant manufacturing and mechanical industries, the forty business class families include a somewhat larger proportion of the prosperous and influential than would be characteristic of the entire group. Only families with children of school age were included in either group.

money of my own"; "Other women could and I felt like I ought to"; and "The mister was sick and I had to."

The cases of a few representative women will make more specific the complex of factors involved in the wife's working:

In one family, characteristic of a large number of those in which the mother works, a woman of forty-five, mother of four children aged eighteen, sixteen, fifteen, and twelve, had worked fifteen months during the previous five years at two different factories. At the first she worked ten hours a day for $15.15 a week, stopping work because of a lay-off; at the second nine and a half hours a day for approximately the same wages, stopping because her health "gave out." She went into factory work because "We always seemed to have a doctor's bill around. The mister had an operation and I wanted to help pay that bill. Then he got back to work and was laid off again. He was out of work nine months last year. The children needed clothes and I had to do it." But although the mother did what she could at home after her day at the factory and washed and ironed on Sundays, the oldest daughter had to leave high school and give up going to the Girl Reserves to look after the children. "I made a big mistake in leaving them. The youngest got to running away from home with other girls. *Then* was the time I should have been home with her."

Another type of situation, less frequent than the above, appears in a family of five—a woman of forty-six, her husband of forty-nine, a farmer prior to 1920 and now employed fairly steadily at semi-skilled machine shop work, and their three boys of nineteen, thirteen, and ten. The oldest boy is in the small local college and the mother works continuously at factory work in order that all three boys may go through high school and college, "so that they can get along easier than their father." In a recent stretch of family unemployment the boy borrowed $125 to keep on at his schooling, both parents going on his note. The family manages by all buckling to the common job: husband and boys have taken over much of the housework; the boy of ten has dinner ready when the family gets home at noon.

In some more prosperous families securing a higher standard of living as well as education for her children leads the mother to work. One mother of two high school boys, a woman of forty-two, the wife of a pipe-fitter, goes outside her home to do cleaning in one of the city's public institutions six days a week. "I began to work during the war," she said, "when every one else did; we had to meet payments on our house and everything else was getting so high. The mister objected at first, but now he don't mind. I'd rather keep on working so my boys can play football and basketball and have spending money their father can't give them. We've built our own home, a nice brown and white bungalow, by a building and loan like every one else does. We have it almost all paid off and it's worth about $6,000. No, I don't lose out with my neighbors because I work; some of them have jobs and those who don't envy us who do. I have felt better since I worked than ever before in my life. I get up at five-thirty. My husband takes his dinner and the boys buy theirs uptown and I cook supper. We have an electric washing machine, electric iron, and vacuum sweeper. I don't even have to ask my husband any more because I buy these things with my own money. I bought an icebox last year—a big one that holds 125 pounds; most of the time I don't fill it, but we have our folks visit us from back East and then I do. We own a $1,200 Studebaker with a nice California top, semi-enclosed. Last summer we all spent our vacation going back to Pennsylvania—taking in Niagara Falls on the way. The two boys want to go to college, and I want them to. I graduated from high school myself, but I feel if I can't give my boys a little more all my work will have been useless."

This increasing employment of married women, which at the last Census involved nearly a thousand wives from the upwards of nine thousand families then in Middletown, must be viewed as a process of readjustment jammed in among the other changes occurring in the home and other sectors of Middletown life. Fifty-six per cent. of the 124 working class wives interviewed had not worked for money during the five years 1920-24, while 75 per cent. of 102 of their mothers on whom data was secured had not worked for money during their entire married lives. These figures undoubtedly dwarf the extent of the shift, as the interviews took place in most cases during the day and therefore included few women continuously employed away from home at the time. Of the twenty-five mothers of the 1890 period who worked for pay, all but one worked either at home, e.g., taking in washing, or at work such as sewing and cleaning that took them away from home only occasionally, while thirty of the fifty-five present-day wives who had worked had worked in factories or other places necessitat-

ing absence from home all day and every day.[6] These women,
two-thirds of them reared in a farm or village environment in
which family life centered about the wife and mother in the
home, must now attempt to integrate with these early habits the
diverse business of being a wife and mother in a city culture
where from time to time their best energies are expended for
eight and a half to ten hours a day in extraneous work away
from home.

Thus, from one point of view the section of Middletown's
population that gets its living by working for money is becoming larger; to a greater extent than thirty-five years ago women
share this activity with men.

From the point of view of age, however, the section of the
population which gets the city's living is somewhat narrower.
In general, it appears that male members of the working class
start to work from fourteen to eighteen, reach their prime in
the twenties, and begin to fail in their late forties, whereas the
young males of the business class tend to continue their schooling longer, start to work from eighteen to twenty-two, reach
their prime in their thirties, and begin to fail somewhat later
than the working class in cases where ripening years do not
actually bring increased prestige. The whole working population tends to start to work from two to five years later than
in 1890.[7] The compulsory school laws in force today make
school attendance obligatory for all children until fourteen,
allow working before sixteen only under very restricted condi-

[6] These figures may not be used in comparing urban conditions in the
nineties and today, since fifty-seven of the 113 working class wives for
whom data were secured on place of childhood residence had lived on farms,
fifteen in villages, and only forty-one in towns and cities. They do, however,
reflect the actual magnitude of the readjustment these women in Middletown today are having to make.

Of the forty wives of the business class interviewed, seven had been
brought up on farms.

[7] Whereas the working class constitutes 71 per cent. of those who get the
city's living, 52 per cent. of 309 boys in the three upper years of the high
school answering a questionnaire on vocational choices were sons of working class fathers.

Cf. Chs. XI and XIII for the number of years spent in school by members
of the business and of the working class and for the fresh problems this
in-between generation is creating in the home.

tions which tend to encourage remaining in school, and retain some control of working conditions until eighteen. In 1890, when there were no compulsory school attendance laws and two boys and a dozen girls constituted a year's graduates from the high school, an abundance of boys available for factory work was a civic asset. The press of this early period reported a deputation from the national Flint Glassmakers' Union as looking over Middletown before locating a coöperative glass factory "to make sure there will be no trouble in securing sufficient juvenile help," and the editor lamented that "Boys are not as plentiful as blackberries." Forty-one per cent. of the 425 employees of the leading Middletown glass plant in 1892 were "boys," according to the Report of the State Statistician for 1891-92. State laws forbade the employment of boys under twelve years of age or for longer than ten hours a day, but "they had little practical effect, because no special officers were designated to administer them." [8] A state report in 1897 spoke of the "dwarfed and undeveloped appearance" of many of the boys in the factories in Middletown's section of the state, "who had been engaged in the factories from the age of ten years or younger."

But although the labor of children in their early teens or younger has ceased in Middletown, youth plays a more prominent rôle than ever before in getting the city's living; in fact, among the numerically dominant group, the working class, the relative positions of the young and the old would appear to be shifting. "When tradition is a matter of the spoken word, the advantage is all on the side of age. The elder is in the saddle." [9] Much the same condition holds when tradition is a matter of elaborate learned skills of hand and eye. But machine production is shifting traditional skills from the spoken word and the fingers of the master craftsman of the Middletown of the nineties to the cams and levers of the increasingly versatile machine. And in modern machine production it is speed and endurance that are at a premium. [10] A boy of nineteen may, after a

[8] W. A. P. Rawles, *Centralizing Tendencies in the Administration of [the State]*, (New York; Columbia University Press, 1903), p. 315.

[9] Goldenweiser, *op. cit.*, pp. 407-8.

[10] Cf. the discussion of the displacement of skill by the machine in Chapter VI below.

few weeks of experience on a machine, turn out an amount of
work greater than that of his father of forty-five.[11]

An analysis of the ages of all male workers, exclusive of
office personnel, in three leading Middletown plants shows that
two of the three (plants II and III in the Table), both modern
machine shops, have respectively 19 and 27 per cent. of their
male workers in the twenty to twenty-four age group, although
only 12 per cent. of the city's male population aged fifteen years
and over in 1920 fell in this age group. At the other end, although 27 per cent. of the male population of fifteen years and
over was aged forty-five to sixty-four, only 17 and 12 per cent.
respectively of the male workers in these shops are of this age;
in the group aged sixty-five and older were 7 per cent. of the
male population, but only 1 and 2 per cent. respectively of the
workers in these shops. Plant I, on the other hand, which dates
back to the end of the eighties, is not of the predominant machine shop type, and has the reputation of being "one of the few
places in town that tries to look out for its older workers,"
follows closely the population distribution.[12]

For the state as a whole, which has undergone a heavy industrial development since 1890, the percentage of increase between 1890 and 1920 in the male population of the fifteen to
twenty-four age group engaged in manufacturing and mechanical industries was roughly seventeen times greater than the
percentage of increase in the total male population of this age;

[11] It is not uncommon for a father to be laid off during slack times while
the son continues at work.

Cf. regarding earnings of workers under twenty-one: *Minors in Automobile and Metal-Manufacturing Industries in Michigan* (Washington, D. C.;
Children's Bureau Publication No. 126, 1923), p. 12 ff.

[12] See Table II. Plant II is a high-speed modern machine shop owned
by one of the great automotive corporations of the country and known
locally as "hard-boiled": it has half again as high a percentage of young
men twenty to twenty-four years old as Plant I, one-fifth higher twenty-
five to forty-four, and two-fifths lower forty-five to sixty-four. (Plant II,
however, has been running only about fifteen years and only five years under
the present ownership.) Plant III, another machine shop plant producing
metal products, draws 16 per cent. of its total personnel (both male and
female) from the sixteen to nineteen year group, and 30 per cent. from the
twenty to twenty-four group, or nearly half of its entire personnel (46 per
cent.) from that section of the population under twenty-five years, while
only 11 per cent. of all its workers are forty-five or over, despite the fact
that it has been operated in Middletown since the nineties. Both of these
plants are typical of current conditions, although it should be borne in mind
that these large industrial plants represent the trend in its most advanced
form.

in the twenty-five to forty-four group, four and one-fourth times greater; in the forty-five to sixty-four group two and one-half times greater; in the sixty-five and over age group only one and two-thirds times.[18]

Like many another trend in Middletown this increasing demand for young workers appears in a different light to the two groups concerned in it. To managerial members of the business group, it naturally appears largely as a problem of production:

The head of a leading machine shop: "I think there's less opportunity for older men in industry now than there used to be. The principal change I've seen in the plant here has been the speeding up of machines and the eliminating of the human factor by machinery. The company has no definite policy of firing men when they reach a certain age nor of hiring men under a certain age, but in general we find that when a man reaches fifty he is slipping down in production."

The general manager of another prominent machine shop: "Only about 25 per cent. of our workers are over forty. Speed and specialization tend to bring us younger men. We do not have an age line at which we fire men."

The personnel manager of another outstanding machine shop: "In production work forty to forty-five is the age limit because of the speed needed in the work. Men over forty are hired as sweepers and for similar jobs. We have no set age for discharging men."

The manager of another large plant in which 75 per cent. of the men are under forty-five: "We have a good many routine jobs a man can do if he is still strong. We try to find a place for these older men even when they are as old as fifty-five if there is no danger in their working near machinery."

The superintendent of a small foundry: "Molders are working up to sixty-five in Middletown at present. After a man reaches forty to forty-five he begins to slow down, but these older, experienced men are often valuable about the shop. But that's not true in the machine shops. There a man is harnessed to a machine and he *can't* slow down. If he does, his machine runs away with him."

[18] No detailed study was made of the types of work performed by aging industrial workers. It is usually not a question of their total superannuation but rather of their slipping down the scale to work of a sort carrying less prestige or less pay, e.g., sweeping up about the plant, as indicated in the quotations cited.

The superintendent of a foundry: Fifty per cent. of the men now employed by us are forty or over, but the company has decided to adopt a policy of firing every employee as he reaches sixty, because it takes a man over sixty so long to recover from accidents and the State law requires us to pay compensation during the entire period of recovery." [14]

The superintendent of another major plant: "The age dead line is creeping down on those men—I'd say that by forty-five they are through."

The old-age dead line as it looks from some apparently characteristic homes of Middletown working class families is suggested by the answers of certain of the 124 wives to the staff interviewers' question: "What seems to be the future of your husband's job?" These answers exhibit a more obvious pessimism, perhaps, because it was a time of local unemployment and many of the working class families were preoccupied with the immediate urgency of keeping a job or getting a new one. [15]

(*Husband a laborer, age forty.*) "Whenever you get old they are done with you. The only thing a man can do is to keep as young as he can and save as much as he can."

(*Husband a foreman, age fifty-six.*) "Good future if he's not getting too old. The [plant] is getting greedier and pushing more every year."

(*Husband a molder, age fifty-one.*) "He often wonders what he'll do when he gets a little older. He hopes and prays they'll get the State old-age pension through pretty soon."

(*Husband a machine tender, age thirty-nine.*) "The company is pretty apt to look after him [i.e., not lay him off]. But when he gets older, then I don't know."

(*Husband a machinist, age forty-four.*) "They keep men there until they die."

(*Husband a machinist, age forty-six.*) "I worry about what we'll do when he gets older and isn't wanted at the factories and I am unable to go to work. We can't expect our children to support us and we can't seem to save any money for that time."

[14] This shows how action aiming to "solve" one "social problem" frequently aggravates another. At certain of the Middletown plants men over fifty are barred from the "mutual aid" insurance plan for a similar reason.
[15] Cf. Ch. VII below on promotion for further answers to this question.

(*Husband a pattern maker, age forty.*) "He is forty and in about ten years now will be on the shelf. A pattern maker really isn't much wanted after forty-five. They always put in the young men. What will we do? Well, that is just what I don't know. We are not saving a penny, but we are saving our boys." (Both boys attend the small local college.)

As noted above, approximately half of the working class wives interviewed were farm-bred. If roughly the same proportion holds for their husbands, perhaps a third of the men in Middletown are having to shift from a world where the physical decline is gradual and even the very old are useful to a new environment in which "economic superannuation takes place abruptly and earlier in life and stands like a specter before the industrial worker." [16] In a culture in which economic authority is so pervasive, the maladjustment of habits occasioned by loss of vocational and financial dominance by the elders may be expected to have extensive repercussions throughout the rest of the living of the group.[17]

Meanwhile, among the business class of Middletown, to a somewhat greater extent than among the working class, advancing age still appears to mean increasing or stable earning power and social prestige. Among some members of the lower ranks of the business class, however, such as retail salespeople and clerical workers, old age is increasingly precarious as aggressive outside chain stores or new owners are more and more dominating local retailing methods. For instance, outside Jewish capital recently took over one of the men's clothing stores and inaugurated a strict sales rule docking any clerk who spends three-quarters of an hour with a customer and fails to make a sale. This rule has resulted in the dropping of at least one of the older, slower clerks.[18] And even in the professions,

[16] Abraham Epstein, *Facing Old Age* (New York; Knopf, 1922), p. 3.

[17] Cf. in this connection James Mickel Williams' *Our Rural Heritage* (New York; Knopf, 1925), p. 67.

Cf. Ch. XI for discussion of the decline of parental dominance over children, particularly in their late teens.

[18] Old age is not generally considered a "social problem," though signs of social strain connected with it are increasingly common—e.g., the relation between smaller houses shorn of "spare bed rooms" and the apparently diminishing tendency of married children to take elderly parents into their homes, noted elsewhere. The attitude toward old age appears to be going through the same cycle traveled by child labor, accident compensation, woman suffrage, factory inspection, and other social changes: for years there

such as teaching and the ministry, the demand for youth is making itself felt more than a generation ago.

Racially, those getting Middletown's living are very largely native-born white Americans. Negroes, both in 1890 and today, have totaled only about 5 per cent. of the earners, virtually all of them being in the working class.

A further factor, that of apparent variations in "intelligence" among the people getting the city's living, cannot be ignored. Whatever the extent of modification of native endowment by varying environmental conditions in the traits measured by "intelligence tests," these tests do by and large seem to reflect differences in the equipment with which, at any given time, children must grapple with their world. A cross-section of the white population was secured in the form of scores (Intelligence Quotients) of all white first-grade (1A and 1B) children in the public schools, according to the Terman Revision of the Binet-Simon Intelligence Tests, administered by the professional school psychologist.[19] Five of the twelve schools draw their children from both business and working class to such an extent as not to be clearly classifiable as predominantly one or the other; three schools with a total of

was no "problem," the situation was aggravated by other social changes, became more acute, until two factions emerged, the one "for" doing something publicly about it and the other "against" doing something about it—until one side won and the new measure became taken for granted, or changed institutional factors rendered the issue obsolete. Limitation of child labor, factory inspection, workmen's compensation, and tax-supported employment offices were opposed step by step by one group in Middletown as "socialistic" and making competition with other centers more difficult; they were pushed with equal persistence by another group. Provision for old age is just reaching the stage in Middletown of occasional questionings of the adequacy under machine production in urban surroundings of the traditionally assumed benefits of the threat of old age as an incentive to saving, and also of the adequacy of the poor house as the wisest instrument for caring for the aged needy. In 1925 the Middletown Eagles, a working class lodge, actively backed the state lodge in introducing an old-age pension bill into the Legislature, while a business group opposed and defeated the bill.

[19] The only white first-grade pupils not included were eighty-three children in the Catholic parochial school and four in the city's one private school. Scores for these children were not available.

It is not intended here to underwrite current extravagant claims for intelligence tests. These scores are presented simply in the belief that they do represent, however inadequately, certain variations which must be taken into account in the study of this community.

97 first-grade children and four schools with 290 first-grade children may, however, be fairly accurately classified as markedly business class and working class, respectively. Their scores compare as follows with those of the total 667 first-grade children in the city:

	Percentage Distribution of 97 Children Chiefly of Business Class Parents	Percentage Distribution of 290 Children Chiefly of Working Class Parents	Percentage Distribution of 667 Children in All Schools
"Near Genius" (I. Q. 140 +)	0.0	0.0	0.2
"Very Superior" (I. Q. 120-139)	7.2	1.0	3.7
"Superior" (I. Q. 110-119)	18.6	5.5	9.7
"Normal or Average" (I. Q. 90-109)	60.8	51.0	52.6
"Dull, rarely feeble-minded" (I. Q. 80-89)	11.3	25.5	21.0
"Borderline, often feeble-minded" (I. Q. 70-79)	2.1	10.7	8.1
"Moron" (I. Q. 50-69)	0.0	5.9	3.0
"Imbecile" (I. Q. 25-49)	0.0	0.4	1.2
"Idiot" (I. Q. Below 25)	0.0	0.0	0.5
	100.0	100.0	100.0

The tendency to diverge suggested by the first two columns above should be borne in mind throughout the entire range of earning a living, making a home, leisure time, training the young, religious, and community activities to follow.

A final point worth noting in regard to those who get Middletown's living is the constant process whereby Middletown tends to recruit its population from the outlying smaller communities about it and itself in turn to lose certain of its young potential leaders to larger cities. A check of the Middletown residents in the graduating classes of the high school for the years 1916-19, for the number who in May, 1925, still lived in Middletown and the number living elsewhere, revealed the fact that roughly half of the 135 boys and a third of the 221 girls residing in Middletown when they graduated from high school in the four years 1916-19 had not returned to the city to live five years later. For example, of eighteen boys in the class of 1916, eight were nine years later in Middletown, one elsewhere in the state, four in other of the East-North-Central group, and five elsewhere in the United States. These migrants undoubtedly contain a fairly high percentage of the more energetic young men of the type who go off to college. On the other hand, of the seven boys graduating from the high school of a small neighboring town in 1908, every one has left the little town: two going to Middletown, where one is president of the young business men's Dynamo Club of the Chamber of Commerce, two to another near-by city, and three others to cities still more remote.

Nobody knows exactly what such a depletion from above and enrichment from below means to the life of a city. It is, however, pertinent to bear it in mind as a possible factor influencing the energy and quality of all the activities of Middletown, notably the degree of resistance to social change. One student of American life has remarked that "frequently the loss of even the best tenth will cut down by 50 per cent. the effective support the community gives to higher interests." [20]

[20] E. A. Ross, *The Social Trend* (New York; Century, 1922), p. 45.

Chapter VI

WHAT MIDDLETOWN DOES TO GET ITS LIVING

Little connection appears between most of the nearly four hundred routinized activities in which these men and women are engrossed day after day in their specialized places of work and the food, sex, and shelter needs of human beings. A few of these workers buy and sell quantities of food, clothing, and fuel made by other specialized workers in other communities, and a few others spend their days in making houses for other members of the group. Only to a negligible extent does Middletown make the food it eats and the clothing it wears. Instead, it makes hundreds of thousands of glass bottles or scores of thousands of insulators or automobile engine parts. The annual output of a single plant, employing a thousand of the city's total of 17,000 who get its living, aggregates $12,000,-000; everything this plant makes is promptly shipped away, and perhaps one-tenth of 1 per cent. of it ever returns as a few obscure parts hidden in some of the automobiles Middletown drives.

And this gap between the things the people do to get a living and the actual needs of living is widening. Radical changes in the activities of the working class in the predominant industries of Middletown during the last four decades have driven the individual workman ever farther from his farm and village background of the eighties.[1] Inventions and technology continue rapidly to supplant muscle and the cunning hand of the master craftsman by batteries of tireless iron men doing narrowly specialized things over and over and merely "operated"

[1] In the pages that follow more space is devoted to the activities of the working class in getting a living than to those of the business class. A number of considerations prompted this treatment: (1) the heavy numerical preponderance of the working class; (2) the more sweeping changes in the types of work performed by them, including that from hand to machine labor, with all the social dislocations involved therein; (3) the fact that the working man's life is buttressed or assailed at more points exclusively by his job; he is supported by fewer kinds of social and other ties, while his life

or "tended" in their orderly clangorous repetitive processes by the human worker. The newness of the iron man in Middletown is reflected in the fact that as recently as 1900 the local press reported only seventy-five "machinists" in the city.[2]

The coming of machine brains and brawn is vividly revealed by the tool-using processes in a local glass plant in 1890 and today. Then, the blowing of glass jars was almost entirely a hand skill. The furnace in which the glass was melted held eight to fourteen "pots" of molten glass, each pot being the focus of the activities of a "shop" or crew of two highly-skilled "blowers" and three boy assistants—one "gatherer," one "taking-out boy," and one "carry-in boy." Through a narrow, unprotected ring-hole the gatherer collected a small gob of glass on a long iron blow-pipe from the blistering interior of the pot. Moving back to a tub six feet from the furnace, he quickly smoothed the ball of glass in a "block" and passed it to one of the blowers. The latter swung to the rhythm of his work at a distance of eight feet from the furnace, setting the pipe to his lips, swinging it up until the glowing ball on the other end was above the level of his lips, blowing, lowering the balloon of glass into the jaws of a waiting mold shut with a foot treadle, blowing a third time until a thin bright "blow-over" of molten glass oozed over the mouth of the mold and he could twist his pipe free from the thin glass without hurting the part of the jar inside. It took as a rule three deep breaths to a jar and he averaged about twenty-five seconds to a jar—something under 100 dozen quart jars a day. At his side on a stool

is more frequently and drastically disrupted by such purely job occurrences as lay-offs and accidents; and (4) the greater reticence of the business class in talking to strangers about certain intimate matters, e.g., their hopes and fears about their work and the details of their financial status, which forced the research staff to content itself with data on the working class alone at certain points. More detailed study of the work of the business class, however, would obviously have been desirable.

[2] The sweeping industrial changes since the end of the eighties are suggested by the fact that the evolution of the iron and steel industry, second in the 1919 census in value of product, did not begin on a large scale until about 1887; the automobile industry, third in value of product in 1919, was non-existent in the late eighties; the motion picture industry was non-existent; the chemical and electrical industries were in their infancy. "Ninety per cent. of the total growth of the electrical industry as a whole has occurred during the past twenty-five years." (*Electrical World,* Fiftieth Anniversary Number, Vol. 84, No. 12, September 20, 1924, p. 640.) "In the early eighties there was no college in this country where one could take a course in electrical engineering." (*Ibid.* p. 566.)

by the molds sat a third worker, the taking-out boy, who took the red-hot jar from the mold and placed it on a tray for a fourth worker, who carried it several yards to the annealing oven. The three boys in the shop kept the two blowers going. It was hot, steady work with only a couple of five-minute halts from 7:00 to 10:00, when a fifteen-minute "tempo" occurred, during which one of the boys might be sent across the street with a row of buckets strung on a long pole for beer; then at it again from 10:15 to 12:00 with one or two more five-minute breaks; lunch, 12:00 to 1:00; work, 1:00 to 3:00; another fifteen-minute "tempo"; and then the last leg till 5:00, each of the two afternoon "spells" with one or two brief breaks. On hot days in early summer the pauses might have to come every half hour; in midsummer the plants closed down entirely. It is important to note that the speed and rhythm of the work were set by the human organism, not by a machine. And with all the repetition of movement involved, the remark of an old glass-blower should be borne in mind, that "you never learn all there is to glass-blowing, as there's always some new twist occurring to you."

The annealing ovens of 1890 were stacked full of hot jars by a "layer-in boy" using a long fork, the oven sealed, heated to 1000° F., and then left for seventy-two hours to cool. The jars were then removed, again by hand, the irregular "blow-over" of glass about the lip of each jar chipped off by hand and then ground smooth by girls standing in front of revolving iron plates covered with wet sand, washed by other girls, and then carried to the packing room. One twelve horse-power engine furnished all the mechanical power for this entire plant at the end of the eighties.

Today this entire process, save the last step of transporting to the packing room, occurs without the intervention of the human hand. The development of the Owens and other bottle-blowing machines shortly after 1900 "eliminated all skill and labor" and rendered a hand process that had come down largely unchanged from the early Egyptians as obsolete as the stone ax. Batteries of these Briareus-like machines revolve endlessly day and night, summer and winter, in this factory today, dipping in turn as they pass a ghostly finger at the end of each of their ten or fifteen arms into the slowly revolving pot of molten glass. Enough glass to make a jar is drawn up, molded roughly in a blank mold, which mold withdraws while a second or finishing mold rises, closes about the red-hot glass, compressed air forces the glass into the crannies of the mold—and a jar is made. As giant Briareus circles endlessly round and round, each arm in turn drops its red-hot burden on to an automatic belt conveyor which winds its way to the "lehr"— the modern annealing oven. Into the electrically operated lehr the jars march, single file, and take their places on the slowly moving floor, to journey for three hours under the care of pyrometers and emerge eighty feet farther on, cool enough to be handled with the naked hand.

In 1890, 1,600 dozen quart jars could be turned out in a one-shift day from an eight-pot furnace manned by twenty-one men and twenty-four boys. Today one furnace manned by three ten-arm machines and a human crew of eight men turns out 6,600 dozen quart jars in a one-shift day of the same length.

Similar, though perhaps not so spectacular, changes have occurred throughout the other departments of the plant.[3] It is in these other departments that one observes particularly the speeding-up process of the iron man. In the room in which zinc caps for the fruit jars are punched, the punches were operated by a foot-treadle in 1890 and, at the will of the operator, about seventy caps could be punched in a minute. Today the power-driven machines hit a pace hour after hour of 188 punches a minute, or at the rate of two caps a punch, 366 caps a minute. The noise necessitates shouting close to the ear if one is to be heard.

Even more characteristic perhaps of the machine age in Middletown is the high-speed specialization of the several automobile part plants. Having in this new automobile industry little previous heritage of plant and machinery, the planning and control of the engineer have had a clear field for the introduction of machine production methods.[4] Specialization in

[3] In 1890 a man with a two-wheel truck trundled six dozen jars at a time to the packing room, whereas today one man with an electric tow motor hauls 150 dozen jars. Freight cars today are loaded by a conveyor that extends into the car. Automatic mixers operated by electricity have replaced men with shovels in the batch room where the sand and other ingredients are mixed for the furnace.

[4] Cf. Veblen's comment regarding the significance for the industrial rise of modern Germany of "the break with an earlier and traditional situation in trade and industry [which] left German enterprise hampered with fewer

tool processes is proceeding rapidly.[5] Here, for instance, is what a typical worker in a local plant making automobile parts does to get a living. The worker is drilling metal joint rings for the front of a well-known automobile. He stands all day in front of his multiple drill-press, undrilled rings being brought constantly to his elbow and his product carted away. Three times each minute, nine hours a day, he does the following set of things: He picks up two joint rings with his left hand, inserts them in his iron man with both hands, turns two levers waist-high with both hands to close jigs, pushes up shoulder-high lever with right hand, takes hold of levers at waist line with both hands and turns the rotating table with its four jigs halfway round, pulls shoulder-high lever with right hand, pulls shoulder-high lever with left hand, removes the two rings which have been drilled with right hand, while he picks up two new rings with his left hand, inserts the two new rings into the drill press with both hands—and does the process all over again.[6]

In still a third type of plant, a foundry, changes have involved chiefly the process of making the sand molds and the methods of conveying the materials about the plant. All the castings were made in a leading Middletown foundry in the nineties by highly-skilled hand molders, "bench molders," whereas, despite strong union opposition, 60 per cent. of the castings today are made by machine molders who need only a fortnight or so of training. In the transporting of materials a magno-electric crane and three men today pile two and one-half times as much scrap into a furnace as twelve men could

conventional restrictions and less obsolescent equipment and organization on its hands than the corresponding agencies of retardation in any of the contemporary English-speaking countries." *Imperial Germany* (New York; Huebsch, 1918), p. 186.

[5] "From the beginning, division of labor and inventions were stimulated by each other. Directly a job was divided up into simple elements, it became comparatively easy to devise a tool or machine for doing that particular work, and profitably if the market were large enough to keep the machinery in full use. Conversely, the increased effective use of machinery tended to greater and greater division of labor; and this interaction of machinery and the division of labor acted in turn on the development of the factory and the organization of big business, owing to the economic advantage of the continuous employment of specialists and specialized machinery." P. Sargant Florence, *Economics of Fatigue and Unrest* (New York; Holt, 1924), p. 34.

[6] In estimating his production, on which his pay is based, the company allows 10 per cent. of this worker's time for taking care of his machine, and 10 per cent. for rest and delay.

thirty years ago in the same time. "The men in the nineties carried boxes of scrap on their heads; it was heavy work and it was hard to keep a man on the job." Today the crane transports from one end of the plant to the other in a minute and a half materials it took a man a half day of fifteen journeys to wheel in 1890. As Frederick W. Taylor remarked, for such work "the Gorilla type are no more needed."

Other Middletown tool-using activities record similar changes. The diary of an employee in a leading confectionery shop in 1893 gleefully announces the installation of a machine to turn the ice cream freezers. Two men with a compressed air pick and chisel outfit known as a "paving breaker" can today accomplish as much work as fifteen laborers under the old pick-and-shovel-spit-on-your-hands régime. One Middletown industry, manufacturing annually enough woven wire fence to enclose half the United States, which began when a farmer's inventive son who knew the labor of rail-splitting devised a hand-power wire fence weaver, today produces with a single power machine sixteen times as much fence as the early hand machine.

If the working class in Middletown does not make the material necessities of its everyday life, the activities of the business class appear at many points even more remote. As the population has forsaken the less vicarious life of the farm or village and as industrial tools have become increasingly elaborated, there has been a noticeable swelling in the number and complexity of the institutional rituals by which the specialized products of the individual worker are converted into the biological and social essentials of living. It is by carrying on these institutional rituals that the business group gets its living.

In the main these are elaborations of similar, if simpler, devices of the nineties, not exhibiting such spectacular changes as that involved, for instance, in the shift from hand to machine processes. Thus the "general manager" of the glass factory of a generation ago has been succeeded by a "production manager," a "sales manager," an "advertising manager," a "personnel manager," and an "office manager." The whole business structure is dominated by the necessity for keeping costly machines busy. As the business man moves intently to and fro "hiring," "laying off," "arranging a new line of credit," "putting on a bargain sale," "increasing turnover," "meeting competition," "advertising," and otherwise operating his appro-

priate set of rituals under the rules prescribed by "business,"
he seems subject to almost as many restrictions as the machine
dictates to the worker who manipulates its levers.

Chief among these devices for converting the actual products
of labor into the necessities and satisfactions of life is the ex-
changing or arranging for exchange of money for usable things
in stores, banks, and offices. Retail selling remains much the
same kind of thing that it was a generation ago, though, to be
sure, the pace has quickened since the middle of the eighties,
when a leading retailer recorded placidly in his diary at the end
of the day, "Quiet in the way of trade. Farmers are busy and
kept at home," or "We have had a fair trade today—sold
twenty screen doors." But here, too, specialization is apparent.
The Busy Bee Bazaar and the Temple of Economy on Main
Street are being displaced by brisk, competing men's wear,
women's wear, electrical, gift, leather-goods, and other "specialty" shops. A swarm of chain stores is pressing hard upon
the small independent retailer, who had things far more his
own way in the nineties; during an apparently characteristic ten
months from April, 1924, through January, 1925, three Middletown clothing stores and one shoe store were taken over by
selling agencies having at least one store in another city, and
four new chains entered the city with one or more branches.[7]
Trade papers, new to Middletown since the nineties, hammer
away at the local retailer about "increasing turnover," while
selling promotion men sent out by manufacturers' associations
worry him at his civic club luncheons by telling him that his
"clerks sell only 15 per cent. of their time," and "salaries ought
to be paid on a sliding scale based on individual sales." [8]

The allied business institution of "credit" is coming rapidly
to pervade and underlie more and more of the whole institutional structure within which Middletown earns its living.
Middletown in the early eighties may almost be compared to
an English provincial town in the middle of the eighteenth century when "there was little capital laid down in fixed plant and

---

[7] These four chains included one clothing store, two food stores, and one
ten-cent store. The last of these did not open a store at the time but bought
a central location necessitating the removal of two established retail busi-
nesses.

[8] Cf. Ch. VII for a description of the trend in the number of hours worked
in retail stores away from the leisurely open-all-day-and-all-evening plan
of thirty-five years ago.

the machinery of finance and credit was very slight." When the fathers of the present generation in Middletown wanted to buy a piece of land they were likely to save up the money and "pay cash" for it, and it was a matter of pride to be able to say, "I always pay cash for the things I buy." "In 1890," says a local banker who was then a rising young business man, "you had to have cash to buy. I wanted to buy a $750 lot and had only $350 in cash. The man wanted cash and there was no place in town where I could raise the money, so I lost the place." [9] A store on Main Street was usually owned lock, stock, and barrel by the man in spectacles who sold a customer three yards of calico or a pound of ten-penny nails. This man, when he bought five bolts of calico from the man whose workers made it, might also have paid cash for the goods. A great many private citizens kept their surplus money in a trunk or hidden away about the home, though some might take it to Mr. —— at the Middletown County Bank or even buy government bonds. People dreaded "being in debt," but a man who owned a house or a business might in an emergency borrow small sums of the local banks.

Today Middletown lives by a credit economy that is available in some form to nearly every family in the community. The rise and spread of the dollar-down-and-so-much-per plan extends credit for virtually everything—homes, $200 over-stuffed living-room suites, electric washing machines, automobiles, fur coats, diamond rings—to persons of whom frequently little is known as to their intention or ability to pay. [10] Likewise, the building of a house by the local carpenter today is increasingly ceasing to be the simple act of tool-using in return for the prompt payment of a sum of money. The contractor is extensively financed by the banker, and this more and more frequently involves such machinery as "discounting second-mort-

[9] Cf. Ch. IX for a discussion of the facilities for home financing today.

[10] This sudden expansion of the miraculous ability to make things belong to one immediately under the installment payment plan has telescoped the future into the present. It would be interesting to study the extent to which this emphasis upon the immediately possessed is altering Middletown's habits as touching all manner of things involving the future, e.g., the increasing unwillingness today, noted elsewhere, of young working class boys to learn more than is necessary to operate a single machine so as to earn immediate big pay, regardless of the future and of how this early specialization may affect their chances to become foremen. Elsewhere will be noted the frequent loss of homes today—with resulting disorganization of many kinds—by people who attempt to purchase "on time" with inadequate resources.

gage notes." A veteran official of a local building and loan company summed up the present-day optimistic reliance upon credit for all things great and small: "People don't think anything nowadays of borrowing sums they'd never have thought of borrowing in the old days. They will assume an obligation for $2,000 today as calmly as they would have borrowed $300 or $400 in 1890."

As the study progresses, the tendency of this sensitive institution of credit to serve as a repressive agent tending to standardize widening sectors of the habits of the business class—to vote the Republican ticket, to adopt golf as their recreation, and to refrain from "queer," i.e., atypical, behavior—will be noted.

Advertising has grown rapidly since 1890, when the local press first began to urge that "advertising is to a business what fertilizer is to a farm." Local grocers rely less upon the mild advertising of giving the children a bag of candy or cookies when they pay the monthly bill and now scratch their heads over "copy" for the press. The first electric sign was hung out before a local store in 1900 and the press pronounced the "effect . . . dazzling." Today all sorts of advertising devices are tried: a local drug chain hires an airplane to blazon "Hook's Drugs" on the sky; a shoe store conducting a sale offers one dollar each to the first twenty-five women appearing at the store on Monday morning; semi-annual "dollar days" and "suburban days" are conducted by the press, Ad. Club and Merchants' Association; and an airplane drops a thousand coupons in the town's "trading area," each coupon good for from $0.25 to $5.00 worth of trade in some local store. The advertising carried in the leading daily paper is six times that in the leading daily of 1890.[11]

In response to these elaborations of the business system, the law, concerned in large part with facilitating its operation and maintaining the sanctions of "Private Property," "Free Competition," and "Individual Initiative" upon which it rests, has likewise grown in complexity, notably through the addition of "corporation law."[12] It is characteristic of the trend in the

[11] See Table XXII for distribution of this advertising by kind.

[12] "No invention of modern times, not even that of negotiable paper, has so changed the face of commerce and delighted lawyers with a variety of new and intricate problems as the creation of incorporated joint-stock companies. America, though she came latest into the field, has developed

folk-ways of this city and its overwhelming preoccupation with business that just as the high school "professor" has been surpassed in salary and prestige by the vocational teacher, and the dominance of the professional man has been largely usurped by the business man, so the prestige of the judge in the legal profession has yielded to that of the corporation lawyer.

Yet, pervasive as these elaborations in ways of doing business are, externally and in terms of the actual activities involved they exhibit less tendency to change than do those of the working class. There is little here comparable to the shift of the glass-blower learning to tend an Owens machine, the teamster learning to run a motor truck, or the compositor a Mergenthaler linotype. The merchant, the retail clerk, the banker, the lawyer still do, with some exceptions, essentially the same things that they did in 1890; they sit or stand at desks or tables dealing with people in face-to-face relations or through writing or advertising copy at a distance, arranging for them to exchange money for things or to carry on other standardized relations with each other according to increasingly complicated developments of the old devices. It is not so much a question of business men doing different kinds of things in Middletown as compared with a generation ago as of their doing the same things or specialized segments of the same things more intensively.[18]

With these nearly four hundred kinds of work available in Middletown and many more in adjacent localities, and with presumably a wide range of human aptitudes and propensities represented in the city's population, an intricate matching of jobs and personalities must contrive somehow to solve itself before young Middletown starts to work. Most of the city's boys and girls "stumble on" or "fall into" the particular jobs that become literally their life work. The pioneer tradition that "you can't keep a good man down" and the religious tradition of free rational choice in finding one's "calling" have helped to foster a *laissez-faire* attitude toward matching the individual and the job.

these on a grander scale and with a more refined skill than the countries of the Old World." James Bryce, *The American Commonwealth* (New York; Macmillan, 1897), Vol. II, p. 655.

[18] The work of teachers, doctors, and ministers is treated in subsequent chapters.

But the complacency of even a generation ago in this regard is less general in Middletown today. Living in all its aspects leans upon money more than ever before, and the conviction is growing that chance should not be given such a free rein in so crucial a matter as finding one's livelihood. Not satisfied with the vocations chance has dealt them, many parents want to do something more for their children, but, particularly among the working class, they are frequently at a loss as to how to go about it. Such ideas as working class parents have for their children's future are largely negative: "I hope they won't have to work as hard as their father"; or, *"He* don't want the girls to go into no factory if he can help it." For many, the magic symbol of education takes the place of any definite plans for vocation: "We want them to have a good education so they can get along easier than their father"; and, "If they don't have a good education, they'll never know anything but hard work."

If a child of a business class family shows particular fitness or desire for a certain vocation he is usually encouraged to go on with it, provided it will yield money or social position comparable to or better than his father's. In the absence of marked individual aptitude, family preferences of varying degrees of relevance and the easy attraction of the familiar carry great weight, and latent abilities may go undiscovered. In this group, also, parental uncertainty frequently takes refuge in exposing the children to as much education as possible.

Religious groups supplement such counsel as parents are able to give by setting forth the claims of the life of a minister or missionary. But the most marked, though as yet incipient, effort to guide the young among the maze of occupations appears in the public schools. Aside from the occasional suggestion of individual teachers, this consists in two recently established devices: a solitary "vocational guidance" class in the junior high school which seeks to introduce pupils collectively to the training for and opportunities in various trades and professions, and a series of "chapel" talks given to the high school seniors by local men each April and May on the "opportunities" in Business, in the Ministry, in Law, in Education, and similar more obvious "callings." These last are usually general talks urging the graduates to be "concerned" about the work they enter, to have "vision" and "determination." Their tone is indicated by the concluding words of one speaker: "This is a

great old world; America is one of the most wonderful coun‧
tries on the face of the earth, and we must prepare our lives
to serve our country—prepare for worth-while occupations."
The vocational work in the high school also furnishes a trial-
and-error sifting among certain tool-using occupations pursued
chiefly by the working class.[14]

But such new sorting devices affect relatively few of Middletown's children. For the most part they "go to work" by taking advantage of some vacancy they or their families happen to
hear about and spend the rest of their lives doing that thing
with what satisfaction it may or may not happen to afford. The
pressure of the customary in confining the choice of Middletown's youth to the obvious jobs available in the city can
scarcely be over-emphasized. As the zero hour for starting to
work draws near, the tendency to conform grows more apparent. The vocational plans of a random sample of 225 young
boys aged eight to fourteen revealed a range of wished-for
occupations that included "running a museum," "being an
astronomer," "being a cartoonist," an architect, a musician, and
other uncommon preferences. It is probably safe to say, however, that few of these boys will eventually find their way into
work so far from the usual group habits. What will probably
become of them, even those who go on to high school, is suggested by Book's study of the vocational choices of high school
seniors in this state:

"Only sixteen different lines of work were chosen by our total
group of more than 6,000 seniors. Some of these occupations were
selected by so few seniors as to make them almost negligible. The
occupations selected most often by the boys were engineering (31
per cent.) and farming (24 per cent.); by the girls teaching (47
per cent.) and clerical work (34 per cent.). . . . We have boys
and girls coming into our high schools from all classes and occu-

[14] Of 120 pupils who have graduated from the Middletown high school
with a degree in electricity since the course was inaugurated, fifty-two
have gone into the electrical trade or to study electrical engineering in
college, while sixty-eight have gone into other lines. Of eighty-one who
have graduated in the drafting course, forty-five are draftsmen or in related
trades or in engineering schools, while thirty-six are in other trades. In the
printing course eleven out of fifty-six boys who have taken the training are
now printers.

pational groups. The high school is unconsciously directing them towards a few lines of work—the traditional professions." [15]

But, though boys naturally tend to gravitate towards the stock occupations understood and recognized by the community, owing to the spread of high school education and other factors they apparently enter the same line of work as their fathers somewhat less commonly than a generation ago. City directories show fewer persons having the same name engaged in the same work. An analysis of the vocational preferences of a sample of 309 Middletown high school sophomore, junior, and senior boys as compared with the occupations of their fathers shows a marked tendency among these children who have gone to high school, especially those of working class parents, to break away radically from the work of their fathers. Of the 309 fathers 52 per cent. are working class men, and 48 per cent. belong to the business and professional class. Only 20 per cent. of the 309 boys answered the question, "What are you planning to do to earn a living?" by indicating some manual activity of the working class sort—a third of the 20 per cent. listing industrial jobs, a quarter building trades jobs, and a third printers, barbers, railroad men, etc.; 18 per cent. want to go into business other than the professions; while 54 per cent. wish to enter learned or technological professions, fully half of them as engineers, architects, chemists, etc., a quarter as doctors, lawyers, and ministers, and the final quarter as teachers, musicians, and writers.[16] These totals greatly underestimate the actual shift away from the occupations of the fathers, particularly among the working class: e.g., of the sixty-two boys who want to enter manual livelihoods, only twenty-two want to enter tool-using occupations in the manufacturing and mechanical industries, although 121 of the fathers are so engaged; while five boys want to be wood workers, fourteen printers, and ten electricians, not one of the

[15] William F. Book, *The Intelligence of High School Seniors* (New York; Macmillan, 1922), pp. 139-142. The influence of high school vocational work is too recent to be adequately reflected in this study. As will be pointed out later, the vocational work, too, confines itself largely to preparing children for a few dominant tool skills.

[16] Seven per cent. were uncertain, and 1 per cent. fell in a miscellaneous group answering boxer, acrobat, etc.

fathers is either a wood worker or a printer, and only two are electricians. Fourteen sons want to be teachers, while only two of the fathers are teachers, the fathers of at least twelve of the would-be teachers being working men. Thirteen sons want to be doctors, while only seven fathers are doctors, and actually at least ten of these thirteen sons are not children of doctors and eight of them are working class children. Eighty-five boys want to be engineers, architects, chemists, etc., although only eleven of the fathers are so classified.

It is, of course, impossible to say how far these choices will carry over into the work actually done by these boys. The assignment of life occupations, for the most part so casual, with the emphasis in the selection more upon the work available than upon the subtleties of the individual, tends to fasten upon getting a living an instrumental rather than an inherently satisfying rôle. This tendency must never be lost sight of as we observe Middletown getting its living, making its homes, spending its leisure, and engaging in other vital activities.

Furthermore, eighty-five out of each hundred of those engaged in earning the city's living work for others and are closely directed by them, while only the remaining fifteen are either independent workers or persons whose work involves some considerable independence and ability to "manage" others.[17]

This whole complex of doing day after day fortuitously assigned things, chiefly at the behest of other people, has in the main to be strained through a pecuniary sieve before it assumes vital meaning. This helps to account for the importance of money in Middletown, and, as an outcome of this dislocation of energy expenditure from so many of the dynamic aspects of living, we are likely to find some compensatory adjustments in other regions of the city's life.

[17] Computed after Hookstadt from the U. S. Census occupational distribution for Middletown for 1920 (cf. Carl Hookstadt, "Reclassification of the U. S. 1920 Occupational Census by Industry," *Monthly Labor Review*, Vol. XVII, No. 1).

Chapter VII

THE LONG ARM OF THE JOB

As one prowls Middletown streets about six o'clock of a winter morning one notes two kinds of homes: the dark ones where people still sleep, and the ones with a light in the kitchen where the adults of the household may be seen moving about, starting the business of the day. For the seven out of every ten of those gainfully employed who constitute the working class, getting a living means being at work in the morning anywhere between six-fifteen and seven-thirty o'clock, chiefly seven. For the other three in each ten, the business class, being at work in the morning means seven-forty-five, eight or eight-thirty, or even nine o'clock, but chiefly eight-thirty. Of the sample of 112 working class housewives reporting on this point, forty-eight (two out of five) rise at or before five o'clock, seventy-nine (nearly three-fourths) by five-thirty, and 104 (over nine-tenths) are up at or before six. Among the group of forty business class housewives interviewed, none rises before six, only six at six, fourteen at any time before seven, and twenty-six rise at seven or later.

This gap between the rising hours of the two sections of the population touches the interlocked complex of Middletown life at many points. A prominent citizen speaking on the curtailing of family life by clubs, committees, and other organized activities urged the parents of the city to "Help solve the boy problem by making breakfast a time of leisurely family reunion." He did not realize that such a solution could apply to only about one-third of the city's families, since in the other two-thirds the father gets up in the dark in winter, eats hastily in the kitchen in the gray dawn, and is at work from an hour to two and a quarter hours before his children have to be at school. Or take another local "problem"—the deadlock between north and south sides of the city in the spring of 1925 over daylight saving time; the working class majority overwhelmed the measure before the city officials on the plea that in summer their small dwellings cool off slowly, often remaining warm until after midnight, and that they can ill spare an hour of cool early-morning sleep before they must get up to work. The business men, on the other hand, urged the need of daylight time because of golf and because standard time put local business at a two-hour disadvantage in dealing with Eastern business. Each group thought the other unreasonable.

The rising hours of business and working class differed less thirty-five years ago, as early rising was then somewhat more characteristic of the entire city. Nowadays one does not find doctors keeping seven to nine o'clock morning office hours as in 1890. During the eighties retail stores opened at seven or seven-thirty and closed at eight or nine, a thirteen-hour day.[1] About 1890 a six o'clock closing hour, except on Saturdays, was tried by a few merchants, and gradually the practice prevailed. Today stores open at eight or eight-thirty and close at five-thirty.

Ten hours a day, six days a week, was the standard rhythm of work for Middletown industrial workers in 1890.[2] In 1914, 73 per cent. of them, according to the Federal Census, worked sixty hours a week or longer. By 1919 only 33 per cent. worked sixty hours or longer, although another 35 per cent. worked from fifty-five to sixty hours a week. The coming of the now almost universal Saturday half-holiday is the outstanding shift in industrial hours of work since 1890.

Year in and year out, about 300 working men work all night and sleep during the day. Periodically, however, a force of 3,000–4,000 men is either shifted from day work or recruited afresh by leading plants to work at night, thus establishing con-

[1] In the leading men's clothing store in 1890 the hours were 7 A.M. to 10 P.M. on Monday, 7 A.M. to 9 P.M. Tuesday to Friday, and on Saturday 7 A.M. to midnight. Stores were frequently open parts of such holidays as Thanksgiving and Christmas.

In 1890 the Middletown jewelry clerks "organized a union . . . and waited upon their employers and made known their desire of being off duty at 7.30 each evening and allowed to attend all ball games." "We are glad to say," adds the press account, "that, rather than have trouble, the jewelry men have acceded to their demands."

[2] All four of the representative Middletown iron works and the four leading wood-working plants listed in the 1891 state *Biennial Report* had a ten-hour day. Among the glass workers, where there was a high degree of organization, two plants had a nine-hour day and two a ten-hour day.

tinuous day and night use of machinery.[3] These periods of night work continue usually five to six months, after which the workers are discharged or shifted to day work. The repercussion upon home, leisure time, community life, and other activities of these periodic dislocations of the rhythms of living, when anywhere from several hundred to three or four thousand heads of families "go on night shift," should be borne in mind; the normal relations between husband and wife, children's customary noisy play around home, family leisure-time activities, lodge life, jury duty, civic interest, and other concerns are deranged as by the tipping over of one in a long line of dominoes. "I work nights, judge, and sleep during the day, and I haven't been able to keep in touch with George," pled a father to the judge of the juvenile court in behalf of his son. The fact that, with few exceptions, this dislocating factor affects only the working class has direct bearing upon the differential concern of the two groups for such things as the civic welfare of "Magic Middletown."

Not only does the accident of membership in one or the other of the two main groups in the city determine the number of hours worked and the liability to night work, but it also determines to a considerable degree whether one is allowed to get a living uninterruptedly year after year or is subject to periodic partial or total debarments from these necessary activities.[4] The most prosperous two-thirds of the business group, at a rough estimate, now as in 1890, are virtually never subject to interruptions of this kind so long as they do good work, while the other third is somewhat subject to cessation of work, though to a less extent than the working class. When "times were very bad" in 1924 the leading department store laid off small groups of clerks alternate weeks without pay. During 1923 the office force of a leading machine shop plant dropped

[3] Only three times, for five or six months each, in the five years between January 1, 1920, and January 1, 1925, have "times" been sufficiently "good" in Middletown for this to happen generally throughout the major industries of the city. At other times night shifts are put on for short periods to meet the needs of individual plants.

[4] The institution of an annual vacation of one or two weeks with pay is another point at which the rhythms of work of working man and business man differ. Among the latter, vacations are today a well-nigh universal rule, but no working man gets vacations with pay, save an occasional foreman who may get a single week. Cf. discussion of the growth of the vacation habit since 1890 in Ch. XVIII.

at one time during the year to 79 per cent. of its peak number, while the wage-earners declined to 32 per cent. of the peak.[5]

Among the working class, however, the business device of the "shut-down" or "lay-off" is a recurrent phenomenon. If the number of working men employed in seven leading Middletown plants [6] on June 30, 1920, be taken as 100, the number allowed to get a living on December 31, 1921, was sixty-eight; on December 31, 1922, ninety-three; on June 30, 1923, 121; on December 31, 1923, 114; on June 30, 1924, seventy-seven; on December 31, 1924, sixty-one; on June 30, 1925, eighty-one.[7] The month-by-month record of one of these plants, a leading machine shop, during 1923, again taking the number employed on June 30, 1920, as 100, was:

January 61	May 117	September 57
February 75	June 92	October 48
March 93	July 66	November 43
April 110	August 63	December 46

In one leading plant 1,000 is regarded as the "normal force." When interviewed in the summer of 1924, about 250 men were actually getting a living at this plant, though the bosses "think of about 550 [of the normal 1,000] as our men." The other 450 are floaters picked up when needed. In another large plant the number of men employed on December 31, 1923, was 802,

[5] The relative seriousness of "bad times" to business and working class personnel is revealed in Willford I. King's *Employment Hours and Earnings in Prosperity and Depression* (New York; National Bureau of Economic Research, 1923), p. 53 ff., in the estimate for the continental United States of the percentage of maximum cyclical decline over the period of industrial strain from the beginning of 1920 through the first quarter of 1922 in the total hours actually worked, as follows:

	Enterprises having less than 21 employees.	Enterprises having 21-100 employees.	Enterprises having over 100 employees.	Total enterprises of all sizes.
Commerce and trade ...	1.27%	5.81%	9.94%	2.78%
Retail only	1.31	4.66	10.84	2.75
All factories...........	8.21	19.21	38.56	29.97
Metal and metal products only	17.89	52.10	52.65	50.25

In this connection the predominance of metal industries in Middletown should be borne in mind.

[6] These seven plants were used by a local bank as an index of local employment in its monthly summaries of local business.

[7] These intervals are uneven because the data were available only for the dates given.

and six months later, June 30, 1924, was 316, but only 205 of these men worked continuously throughout the entire six months with no lay-offs.

Of the sample of 165 working class families for whom data on steadiness of work was secured, 72 per cent. of the male heads of families lost no time at work in the twelve months of 1923 when "times were good," another 15 per cent. lost less than a month, and 13 per cent. lost a month or more; during the first nine months of 1924, throughout the last six of which "times were bad," only 38 per cent. of the 165 lost no time, another 19 per cent. lost less than a month, and 43 per cent. lost a month or more.[8] Among the forty families of business men interviewed, only one of the men had been unemployed at any time during the two years, 1923-24—and that was not due to a lay-off.[9]

It is difficult to say whether employment tends to be more or less regular in Middletown today than a generation ago. Sharper competition throughout markets that have become nation-wide, the rise of the new technique of cost-accounting, the resulting substantial overhead charges on expensive plant and machinery, and the imperturbability of machines in the doggiest of "dog-days" discourage today the easy custom of closing down the plant altogether which flourished among the flimsy factories and hand-workers of a generation ago. A characteristic summer news item in the Middletown *Times* for June 12, 1890, says: "Ninety per cent. of the glass houses in the U. S. A. close on Saturday until the first of September." Short shut-downs of two weeks or so at other times in the year were not uncommon. And yet, despite modern compulsions to maintain at least minimum production, and in fact because of such impersonal techniques as cost-accounting, lay-offs have become much more automatic than the reluctant personal decision of a sympathetic employer.[10] The sheer increase in the

---

[8] See Table III.

[9] See Appendix on Method for the way in which the families interviewed were selected.

[10] "On the transition to the machine technology . . . the individual workman has been falling into the position of an auxiliary factor, nearly into that of an article of supply, to be charged up as an item of operating expenses." Veblen, *The Nature of Peace* (New York; Huebsch, 1919), pp. 320-1.

Cf. J. L. and Barbara Hammond, *The Town Labourer, 1760-1832* (London; Longmans, 1920), especially Chs. I, II, and VI.

size of present-day plants [11] operates to make these periodic increases and curtailments in working force more obligatory when the need for them arises.[12]

As in the case of the lowering of the old-age deadline, described in Chapter V, the phenomenon of recurrent industrial unemployment assumes totally different aspects as it is viewed through the eyes of a business man or of a working man. For the dominant manufacturing group, the peremptory little figures on the cost sheets require that there shall always be on hand enough workers to take care of any fluctuations in business. The condition of there being more men than available jobs, though dreaded by the working man, is commonly called by his bosses "an easier labor market." [13] In March, 1924, when the long slump of unemployment was commencing and employers in other cities ran "want ads." in the Middletown papers offering work, two special delivery letters were laid by the plate of the president of the Middletown Advertising Club at one of its weekly luncheons, asking the Club to use its influence to suppress such advertisements because they tended to draw unemployed machinists from town. The president of the club agreed

---

[11] The Middletown press in 1890 hailed as "a gigantic concern" a new industry which was to have "when in full operation" 200 hands. The largest working group in the city in January, 1891, was 225. In 1923, eleven plants each employed more than 300—three of the eleven employed more than 1,000, while one of these three regards its "normal force" as over the 2,000 mark.

[12] See N. 5 above regarding the relative impact of "hard times" on enterprises of different sizes. According to King's evidence, whether owing to the fact that "the small employer keeps less accurate accounts," to the fact that "the small employer, being well acquainted with his employees, is so much interested in the welfare of the latter that his relationships with them are not governed primarily by purely business considerations," or to other factors, the bump tends to hit the big enterprises several times as hard as the little fellows.

[13] This business men's psychology is well illustrated by the following statement by one of the city's influential manufacturers: "In 1922 we were so rushed with orders we couldn't possibly fill them or get enough men here in town to carry on, so we had to import some men from Kentucky and West Virginia. Our men from our local district here, born and bred on the farms near here, knowing the use of machinery of some sort from their boyhood, reliable, steady, we call 'corn-feds.' These men we brought in from the mountains we called 'green peas.' We brought two train loads of them down. Some of them learned quickly, and some of them didn't. Most of them have drifted back by now. We figured it cost $75-$200 to train each one of them, and there was such a demand for labor about town that they didn't stay with us. They drifted about from shop to shop, and of course when the slump came we fired them and kept our old men."

that this was "something the Ad. Club certainly ought to back," and the representatives of all papers agreed to suppress the advertisements. The sentiment of the club was that it was important that plenty of skilled workers be kept in town.[14]

"People come to the house a great deal and tell me they can't get work," remarked the wife of a prominent business man. "Of course, I don't really believe that. I believe that any one who really tries can get work of some kind." This remark appears to sum up the philosophy of unemployment of many of the business class in Middletown. Others believe, as one outstanding business man put it, that "About the only thing that might be improved in the condition of working men today is unsteady employment. But that cannot possibly be helped. An employer cannot give employment to workmen if he cannot sell his goods."

To the working men, however, unemployment as a "problem" varies from a cloud the size of a man's hand when "times are good" to a black pall in a time of "easy labor market" that may overspread all the rest of their lives. It happened that times were not good during the late summer and fall of 1924 when the staff interviewed Middletown families. Over and over again, the wives interviewed answered the question, "What seems to be the future in your husband's job?" in terms of:

"He's to be laid off Saturday."

"He's just lucky if this job keeps up. He never knows from day to day whether his job will be there."

"He can't tell when he'll be laid off. One day he comes home thinking the work is over and then the next day he believes it will last a few weeks longer."

[14] To the bosses there is no "problem" in the abrupt posting, rarely more than a day in advance, of the announcement of a lay-off, or in the absence of any machinery for talking over with the men the reasons for the lay-off or its probable duration, or in the practice of "hiring at the gate."

During the depression of the summer of 1924 much press publicity was given to the announcement that a local plant would take on a thousand men the following Monday. The men crowded about the plant gates on the appointed day and a total of forty-eight were hired.

Cf. pp. 37-8 of Shelby M. Harrison's *Public Employment Offices* (New York; Russell Sage Foundation, 1924) for reasons why employers favor hiring at the gate. Cf. also Whiting Williams' *What's on the Worker's Mind* (New York; Scribner's, 1921), pp. 6-7, for the worker's view of this system.

For many of them the dread had become an actuality:

"I know people that have been out of work since June," one woman said in October, "and they're almost crazy because of it. Maybe if more people understood what it means something could be done about it."

"Not even the foreman knew the lay-off was coming," said one quiet, intelligent-looking woman who with her husband had been laid off in a leading plant the night before, at the close of the first week in December. "Last week the whole plant worked overtime every night on straight time pay. A petition asking for more wages was circulated by the men, but my husband and two others wouldn't sign it because they thought it was no time to ask for a raise with so many out of work. Now we're told the lay-off came because of the petition, because orders have stopped coming in. We can't figure that out. . . . What'll we do? I don't know, but we must not take the boy out of school if we can any way get along."

"He's awfully blue because his job is gone," said another wife in November. "He's trying to get work at ——. He hopes his old job will open up again in the spring."

Several of these women, all of them having husbands over thirty-five, said that their husbands had taken or would want work that paid less and had less future if it seemed likely to be "steady" and less subject to lay-offs. Steady work appeared to be generally valued by these older workers above high wages.[15]

The commonest working class solution of the problem of unemployment is to "get another job." Of the 182 sample workers for whom data was secured on this point, including 124 with children of school age, over a quarter (27 per cent.) had been with their present employers less than a year, over a third (38 per cent.) less than two years, and over half (55 per cent.) less than five years.[16] This "getting another

---

[15] Cf. Whiting Williams' statement: "If there is one thing I have learned on my labor travels it is that 'the job's the thing.' Wages are interesting, but the job is the axis on which the whole world turns for the workingman." (*Op. cit.*, p. 39.)

[16] See Table IV. These figures do not mean that these men had had no unemployment during these periods or had not in many cases worked temporarily in other plants until their regular jobs "opened up again."

Between January 1, 1919, and October 1, 1924, of 178 men for whom these data are tabulable, 46 per cent. had had one job, 20 per cent. two jobs, 22 per cent. three jobs, and the remaining 12 per cent. more than three.

job" frequently involves leaving the city: "In the summer we took to the Ford and went looking for work." "He has a job now over in —— [twenty-five miles away] and likes it so much he may stay on there." [17]

Failing in finding another chance to get a living, the whole family settles down to the siege.[18] Of 122 housewives, who gave information regarding readjustments occasioned by unemployment,[19] eighty-three reported unemployment during the preceding fifty-seven months. Sixty-eight, the great majority of those reporting unemployment, had made changes in their routine habits of living to meet the emergency.[20] Of these,

---

Here again brief fill-in jobs were not counted, provided a man returned to his old job when "work opened up."

It should be remembered that the interviewers had to rely upon the wife for these data in nearly every case, though every effort was made to see that she did not omit any pertinent data.

[17] This migratory tendency which modern industry invites and the Ford car enormously facilitates may be expected to have far-reaching influence throughout the rest of the workers' lives: e.g., the more frequent moving of working class families noted in Ch. IX and the decline in neighborliness and intimate friends among the wives noted in Ch. XIX. Cf. the statement by Roscoe Pound in *Criminal Justice in Cleveland:* "Some studies made during the war indicate that the moral implications of an increasingly migratory laboring population call for serious consideration. Our institutions presuppose a stable, home-owning, tax-paying population, of which each individual has and feels a personal interest in its legal and political institutions and bears his share in the conduct of them. Irregularity and discontinuity of employment and consequent migration from city to city, or back and forth between city and country, preclude the sort of society for which our institutions were shaped." (Cleveland; Cleveland Foundation, 1921; Part VIII), pp. 610-11.

[18] At least two factors make the incidence of unemployment more difficult for the worker today than formerly: (1) The decline of trade unions and of neighborhood spirit (cf. Ch. VIII). (2) The extension of the precarious habit of leaning the present upon the future by long-term commitments to pay for the purchase of a home (cf. Ch. IX), insurance, household appliances, education of the children, and so on. To take but the case of life insurance: in the sixteen years between 1910 and 1926 the number of individual policies in force with one national company in Middletown and a portion of the surrounding county increased from 3,800 to 23,000; this number should be reduced by approximately 40 per cent. to get the number of policy holders.

[19] These data are based upon the memory of the housewife; she had no opportunity to check up her recollection by talking to her husband. Undoubtedly certain minor lay-offs and times when work was reduced for short periods to three or four days a week were overlooked. These figures are therefore probably conservative throughout.

[20] This does not include cessation of saving and inroads upon accumulated savings. If this factor be included, it is probably safe to say that unemployment affected the behavior of the entire group.

47 cut on clothing;

43 cut on food;

27 of the wives worked for pay either at home or away from home;

14 of the 60 carrying some form of insurance got behind on payments;

6 moved to a cheaper home;

5 of the 20 having a telephone had it taken out;

4 of the 35 with children in high school took a child from school.[21]

Such comments as the following by some of these housewives reflect the derangement of established habits in "bad times":

As touching savings. "We had been saving to buy a home but lost all our savings paying rent while he was laid off." "We had to use up all our savings to keep going." "We lost both our auto and our house. We had paid $334.00 on the auto and had just a little over a hundred to pay. We had been paying on the house a little over a year." "My husband has just gone everywhere for work. We would have been out of debt now if he hadn't been out. It seems like a person just can't save. We started to buy a house a couple of years ago and his company would have paid the first payment, but the very next day he got his arm broke. I never plan nothing any more." "We haven't lost our life insurance yet. Last year we had to let a thousand-dollar policy go when he was out."

As touching shelter. "We don't know where the rent for this month is coming from. We're out of coal, too." "We have cut down all we can on food and the phone is the next thing to go. I am not strong enough to wash as I used to when he was laid off. He hates to see the phone go. It's the only way we hear from our children."

As touching food. "Now they have a new man in the grocery and we're afraid he won't allow us to charge things so long. We had a $60.00 grocery bill when he went back to work in 1922." "We get on the cheapest we can. Our living expenses are never more than $5.00 a week" [family of five]. "We have been buying no fresh milk this year, using only canned milk" [a family including two boys age seven and nine]. "We just live as close as we can all the time. I tell the children if they get a little candy for Christmas this year they'll be lucky; they haven't had anything

[21] These changes did not usually come singly, and the families involved in the above categories therefore overlap.

but clothes and things they absolutely need for the last two or three years." "We have cut our food all we can and have beans and potatoes two times a day with about $2.00 worth of meat scattered through the week. I don't know what we'll do if there isn't work soon." "Last winter our grocery bill ran eight or nine dollars a week. Now it is five or six dollars, partly because we trade at a cheaper place and partly because we're economizing."

As touching leisure time. "I haven't been able to afford a movie show since January" [ten months].

The forced choices during times of unemployment reveal sharply the things some of these working class people live by:

A woman who had just returned to the store a new winter coat because her husband had lost his job said she planned to cut down on "picture shows"—"but I'll never cut on gas! I'd go without a meal before I'd cut down on using the car."

Another woman said: "I'll give up my home last. A friend of mine belongs to several clubs and won't resign from any of them even though her husband has been laid off three months. She says she'll give up her home before her clubs."

One woman spoke for many others when she said: "We'll give up everything except our insurance. We just can't let that go." The head of a local insurance company reported that unemployment has relatively little effect upon insurance policies. Of the 100 working class families for whom income distribution on certain items was secured, all but seven reported money spent for life insurance in annual amounts ranging from $2.25 to $350.00.[22]

To Middletown as a whole in its corporate group capacity, unemployment as a "problem" virtually does not exist. At most it becomes a matter for privately supported charity to cope with. In the extreme bad times of the winter of 1921-22 when local unemployment overwhelmed these charitable agencies, a supplementary fund of $40,000 was raised by popular subscription to be distributed in doles. And yet it was in February of this winter, when local hardship was most acute, that the City Council voted to discontinue support of the highly successful

[22] See Table VI.

tax-supported free employment office launched during the War and in operation for two and one-half years.[23]

The mobility afforded by new modes of transportation combines with these periodic waves of employment, unemployment, and reëmployment to diminish the tendency for the workers in a given factory to live together immediately about the plant. Everybody in Middletown in 1890 got to work by walking, and workers tended to settle in the immediate neighborhood of a given factory; as a new factory was located on the outskirts of the community it formed a magnet drawing new dwellings close about it.[24] Today, when one gets about the city and the country surrounding it by bicycle, fifteen-minute street-car service, regular bus service, and five interurban lines, and approximately two out of every three families in the city own a passenger automobile,[25] decentralization of residence is apparent. A check of the residences of all workers in the shops of three local plants, a total of 2,171, showed that 28 per cent. lived within one-half mile of their places of work and 55 per cent. less than a mile away, while 45 per cent. lived a mile or more away; 20 per cent. of the men lived outside the city and from three to forty-five miles away,[26] and 14 per cent. of the women lived from three to nineteen miles away. Two of these

[23] The failure of the Council to vote the $1,500 needed for the upkeep of the office caused its abandonment. Both Chamber of Commerce and Trades Council had favored the employment office and a leading local paper called it "one of the best investments ever made by the city and county." One powerful councilman, connected with a leading industry, is said to have led the opposition to the bureau, and is quoted as declaring the office of "no assistance whatever to the manufacturers or to the laboring man. If a man wants work in this city he can get it without going through the employment office." The office was abandoned and part of the director's salary left unpaid, although the State Attorney General ruled that the city was liable for it.

This incident affords an instance of the way much of the group business is conducted in Middletown and of the relative inarticulateness and helplessness of the group in the face of a powerful minority. The weaker of the two Middletown dailies called in vain for a frank statement of the nature of the "nigger in the woodpile" from "certain persons" who have made "a protracted effort . . . to end the official existence of the bureau."

[24] E.g., the following from the local press in 1890: "Work has commenced on the Westside Glass Works. The location of this factory at Westside has caused a great demand for residences in that vicinity."

[25] Cf. Ch. XVIII for discussion of ownership and use of automobiles.

[26] Five men went back and forth together in an automobile from a city of the same size forty-five miles distant.

Distances up to three miles were figured "as the crow flies" and are therefore somewhat underestimated. (See Table V, N. 2.)

plants are old industries that have been in Middletown since gas boom days, while the third is a modern machine shop, located in Middletown more than a decade, of the sort that today dominates the city's industrial life. In the latter only 19 per cent. lived within one-half mile and only 43 per cent. within a mile, while 57 per cent. live over a mile away, and 29 per cent. of the males lived three to forty-five miles away, the number of women employed being negligible.[27]

This trend towards decentralization of workers' dwellings means that instead of a family's activities in getting a living, making a home, play, church-going, and so on, largely overlapping and bolstering each other, one's neighbors may work at shops at the other end of the city, while those with whom one works may have their homes and other interests anywhere from one to two-score miles distant.

Meanwhile, in season and out, regardless of such vicissitudes as unemployment, everybody who gets a living in Middletown is theoretically in process of "getting there"; the traditional social philosophy assumes that each person has a large degree of freedom to climb the ladder to ever wider responsibility, independence, and money income.[28] As a matter of fact, in six Middletown plants employing an average of 4,240 workers during the first six months of 1923 [29] there were ten vacancies for foremen over the period of twenty-one months from Jan-

[27] See Table V. These addresses represent the summer force. In the machine shop, the bulk of whose employees require little training, the winter force is heavily recruited from farmers. (Cf. King, op. cit., p. 91, for the reason agricultural labor flocks to the machine shops.) In response to a protest from local labor that they discriminated against city labor in the winter in favor of this cheaper-priced farm labor, Middletown manufacturers informed the Chamber of Commerce that they "consider [the] county a unit and not the city." The ease with which farmers can "when times are good" get work in machine shops and the general diffusion of Ford cars and surfaced roads is prompting some workers to return to small farms, preferably midway between Middletown and another small industrial city, where a garden can help out on food and work be drawn from either city. Cf. in Ch. IX the data on the housing mobility of the working class families interviewed.

[28] Thirty-four per cent. of 241 high school boys answered "true" to the extreme statement, "It is entirely the fault of a man himself if he does not succeed," while 16 per cent. more were "uncertain," and 49 per cent. thought the statement "false," the final 1 per cent. not answering. Forty-five per cent. of 315 girls thought the statement "true," while 9 per cent. were "uncertain," 44 per cent. marked it "false," and 2 per cent. did not answer.

[29] Average of total payrolls as of December 31, 1922, and June 30, 1923, less estimated averages of 600 foremen and office workers.

uary 1, 1923, to October 1, 1924.[30] This means that in a year and three-fourths there was a chance for one man in 424 to be promoted.[31] The total number of men estimated by the plants as of sufficient experience on January 1, 1923, to be eligible for consideration for promotion to foremanship was 531. Of this picked group one man in fifty-three got his chance in twenty-one months.[32]

The chance of promotion as it appears to the working class may be glimpsed from the answers of the wives in 105 of the 124 sample families to the question, "What seems to be the future in your husband's job?" It was a time of considerable local unemployment. Ten of the 105 husbands were already out of work, and "future" meant hope for the naked chance to begin getting a living again at anything; for twenty-two other wives future meant nothing beyond the possible date when "the mister" would be laid off—for two of them this future was no further off than "next Saturday"; to four others the future meant predominantly a fear of the old-age dead line; to eleven others a "good" future meant, "He'll probably have steady work"; nineteen others were hopeful in regard to their husbands' work and their chances in it;[33] while the remaining thirty-nine faced the future with no expressed hope of getting ahead. Of these thirty-nine, thirty-two, while not at the moment out of work or driven by an active fear of unemployment, voiced keen discouragement. Such answers as the following from this last group, to whom, with those unemployed or fearing a lay-off, the future shows no outlet toward greater

[30] One of the six plants reported vacancies over only eighteen months, from January, 1923, through June, 1924.

The condition of the Middletown labor market during these eighteen months can be seen in its setting from the index numbers of employment in seven leading plants given earlier in this chapter.

[31] The job of assistant foreman is not considered here, as it apparently counts for little. Promotion to a foremanship is the real step up.

[32] Not only were promotions infrequent, but during these twenty-one months a number of foremen were temporarily demoted to the ranks when forces were reduced, e.g., night shifts abandoned.

R. R. Lutz found that in the course of a year only one man in seventy-seven in a group of 618 eligible men in the metal trades in Cleveland had a chance of promotion. *The Metal Trades* (Cleveland; Cleveland Education Survey, 1916), p. 100.

[33] In one of these cases here counted as "hopeful" the wife said: "It's hard to say. There's not much opportunity for advancement but he is reading trade papers and studying his trade all the time to be able to take advantage of any opportunity that comes."

security or recognition, reflect an outlook on life that probably conditions profoundly all their other activities:

(*Husband a machinist, age thirty-eight.*) "Well, he's been doing the same thing over and over for fifteen years, hoping he'd get ahead, and he's never had a chance; so I don't suppose he ever will."

(*Husband a machinist, age twenty-six.*) "There's nothing ahead where he's at and there's nothing to do about it."

(*Husband a machine-tender, age forty-six.*) "There won't never be anything for him as long as he stays where he is and I don't know where else he can go."

Husband a foreman, age thirty-eight.) "He's been there nine years and there's no chance of promotion. The work is so hard he's always exhausted. He wants to get back on a farm. He's been lucky so far in not being laid off, but we're never sure."

(*Husband a factory laborer, age thirty.*) "He'll never get any better job. He'll be lucky if they keep him on this one."

And yet the chance of becoming a foreman, small as it is, would appear to be somewhat better than it was a generation ago. The experience of individual plants, cited below, suggests that foremen have increased more rapidly than the number of workmen. On the other hand, increasing technological complexity and the resulting tendency to insert college-trained technical men into a force between foremen and owners appear to hinder a workman's progress beyond a foremanship more than formerly.

New technical developments such as the automobile and multiplied uses of electricity have opened new doors to some working men, enabling them to become owners of garages, filling stations, or electrical shops. The sharp increase in size, complexity, and cost of the modern machine-equipped shop, however, makes the process of launching out for oneself as a small manufacturer somewhat more difficult than a generation ago.

In general, the greater accessibility of those on the lower business rungs to sources of credit through lodge, club, church, and social contacts would seem to make fresh opportunities through the starting of a small industrial shop, retail store, or business of their own easier for them than for the working class. No direct study was made of the chance for promotion among the business group, and the local sentiment is such that one may not talk to business men and their wives about their personal advancement as one may to the working class. Close contact with Middletown's small shopkeepers and clerks as well as with the more powerful members of the business group throughout nearly a year and a half, however, yielded a distinct impression that psychologically the business families of the city tend to live, in the main, not on a plain stretching unbroken to the horizon, but on ground sloping upward, however gently. Contact with the working class, supplemented by interviews with the sample of wives and some of their husbands regarding the latter's chances of advancement, brought an equally clear impression that psychologically the outlook of the working class is somewhat flatter. The new rush of the children of the business man to college and of the working man's children to high school and college is increasing the vertical mobility of the children by offering all manner of short-cuts to the young man or woman with an education, but once established in a particular job, the limitations fixing possible range of advancement seem to be narrower for an industrial worker.

Vocational accidents are yet another differential accompaniment of getting a living for the two groups. Such accidents are practically unknown among the business class. For an average of 7,900 working men and women in the thirty-six factories constituting the industrial population of the city during the first half of 1923,[34] however, 824 accidents serious enough to involve a loss of time from work were recorded during this six-month period. If this period can be taken as representative, roughly one in each five persons of the working class employed in factories in Middletown has an accident serious enough to make him stop getting a living for a while each year. Fifty-seven per cent. of these injured workers lost less than eight

[34] Payrolls of 7,743 and 9,655 on December 31, 1922, and June 30, 1923, respectively, were averaged, and since none of the accidents recorded concerned a member of the office staff, an estimated total of 799 office employees was deducted from the payroll average of 8,699, the above figure of 7,900 resulting. A few very small industrial plants for whom records were not available are not included in the thirty-six above, also such groups of workers as the building trades, a few railroad mechanics, and other workers not in factories. The records of accidents were taken directly from the cards in the files of the State Industrial Board which administers the State Workmen's Compensation Law.

days, 13 per cent. lost eight days to two weeks, 1 per cent. two to three weeks, and the remaining 29 per cent. three weeks or more. Three of the 824 injured during these six months were killed, one other was expected to die at the time the figures were tabulated, two lost one eye and three lost permanent partial use of an eye, three lost a hand and six partial use of a hand, eight lost a finger and sixteen partial use of a finger, and so on.

We can only infer a trend toward fewer accidents. In view of the fact that in the year ending September 30, 1920, there were only 922 amputations out of a total of 42,994 accidents reported throughout the entire state, numerous records like the following in the Middletown press in 1890 suggest a very different frequency: In one leading plant, employing about 200 hands, three men were injured in one day in three different accidents—one losing a hand, a second having a foot mashed, and a third losing a finger. The last-named is reported as "another to lose a finger in the machinery where no less than five have been nipped off in the past month or so." A superintendent in a leading plant employing about 200 men in 1890, when asked if working conditions then gave rise to a good many accidents, exclaimed:

"I should say they did! We kept a horse and buggy busy all the time taking men from the plant to the doctor."

"Not literally, of course?"

"No, not literally, but we used to have one almost every day."

The compulsory presence in each plant today of a first-aid kit undoubtedly reduces infections; [35] hernias are fewer, as there is less heavy lifting; plants are better built and aired, and such conditions, conducive to pneumonia and rheumatism, as those described by a glass worker in 1890, are far less common: "We worked dripping with sweat, burning up on the side facing the pots and freezing on the other side in winter in the draughty old plants." On the other hand the speed of the iron man has brought new health hazards all its own—nerve strain due to noise and speed, new types of localized ailments due to specialization of activity curtailing movement in many cases from the larger body segments to a few small muscles used over and

[35] One plant has a doctor in attendance, a second a graduate nurse but no doctor, two others have practical nurses, another a matron but no nurse, and a number use the local Visiting Nurses' Association, the company paying for the service in each case. All this is new since 1890.

over. Two under-officials in the packing room of a large glass plant agreed in saying that "there have been several nervous breakdowns since the installation of the belt conveyor bringing the jars to the women packers." And one added, "This system may be good for the plant, but it certainly isn't good for the girls."

Prior to 1897, when the first factory inspector was appointed in the state, the workman carried the full burden of accident under the common law principles of "assumed risk," "contributory negligence," and the "fellow servant" doctrine. In 1915 the trend towards group participation in such matters eventuated in a State Workmen's Compensation Law under which the industrial plant, and thus ultimately the general public, bear a share of the burden.[36]

This process of the socialization of accident hazard is a phase of a larger trend towards impersonality in industrial operations in Middletown. Under the existing type of corporate ownership the presidents of three of the seven largest Middletown industrial plants today reside in other states, and two of the three plants are controlled by directors few of whom have ever even been in Middletown. This wide separation between a plant and the real authority over it combines with the increasing extent and complexity of the units of operation and the introduction of technically trained personnel to make it, in general, farther from the "floor" of a Middletown shop to the "office" today than a generation ago. Thus one plant whose sixty men in 1890 were officered by a president, a secretary who was also the chief engineer, a superintendent, and no foremen, today has for a force less than three times as large, a president, a vice-president (both largely inactive), a treasurer and general manager, a secretary who is also chief engineer, a superintendent, assistant superintendent, and three foremen. A second plant whose 200 men in 1890 were officered by a president, a vice-president

[36] The jungle of conflicting elements in a "social problem" is reflected in this case by two local situations: (1) The situation described in Ch. V in which the adoption of casualty insurance has led in one large plant to a policy of "firing" all employees at sixty. (2) The fact that the company which is perhaps doing more than any other among the largest half-dozen in the city to care for its aging workers was reported by the State Industrial Board as "not in good standing"; this company, owned and operated by public-spirited citizens, had in 1925, according to the State Board, been carrying its own risk for three and a half years without the legal permission of the Board.

who was also general manager, a secretary and treasurer, and two foremen, is operated today, with six times the original staff of workers, by a president, a vice-president and general manager, a treasurer, an assistant secretary, an assistant treasurer (largely inactive), an auditor, two superintendents, and thirty foremen. A third plant, a machine shop not locally owned and new since 1890, has a staff of 800 directed by a president (living out of the city), a resident vice-president who is also general manager, a second vice-president (inactive), a secretary, a comptroller, a factory manager, a general superintendent, three division superintendents, and twenty-five foremen.

More than one manufacturer said that he was no longer able to know his working force and their problems as he used to. One gains an impression of closer contact between many managers and their workers thirty-five years ago; we read in the press of 1890 of a plant closing down and owners and 176 workmen attending the funeral of one of the workers. On another occasion the management, unable to dismiss the force for a day at the county fair, ordered into the plant one hundred pounds of taffy from the fair grounds. Yet another old-time employer when he sold his plant a few years ago stipulated in his contract with the purchasers that the latter were to take over the entire force and keep all employees long enough to learn their worth before discharging any of them. This same man is reported to have sent $500 to each of his foremen when he sold out, and he endowed a room at the local hospital for his old workers and their families.

A few Middletown industrial plants make an attempt to bridge the gap between shop and office by such devices as shop committees and short term training groups, including lectures on engineering and metallurgy by extension lecturers from the state university. One factory has a safety committee and another a nominal "council of foremen," with an appointed head. The character of these groups appears in the exclamation of one leading manufacturer, representative in this respect of the entire group, when asked about his "shop committee": "You don't mean collective bargaining or anything of that sort, I hope? *We're* running this plant and want no mistake about that. We won't tolerate any shop councils or anything of that sort." This plant is reported on reliable authority to have "thrown all sorts of obstructions in the way of the insurance people getting together with their foremen to talk over safety means in the plant." Personnel and welfare managers, appointed by four plants, occasionally exercise a personal oversight of the workers' problems; in one prominent plant, however, the kind of personnel adjustment work done is reflected by the emphatic statement of the personnel manager: "If a man is fired by a foreman, he stays fired. A thing a man does once in one department he'll do again in another."

These various devices, together with the carrying by at least three plants of a blanket life-insurance policy for all employees, the passage of the State Workmen's Compensation Law, and the appointment of state factory inspectors, represent tendencies to diminish somewhat the disparity between the accompaniments of getting a living for the working class and for the business group. But, while these new devices are attempting to solve the "social problems" involved in getting a living, the long arm of the job in this swiftly changing culture is touching the lives of workers as well as business class with new problems.

Chapter VIII

WHY DO THEY WORK SO HARD?

One emerges from the offices, stores, and factories of Middletown asking in some bewilderment why all the able-bodied men and many of the women devote their best energies for long hours day after day to this driving activity seemingly so foreign to many of the most powerful impulses of human beings. Is all this expenditure of energy necessary to secure food, clothing, shelter, and other things essential to existence? If not, precisely what over and beyond these subsistence necessaries is Middletown getting out of its work?

For very many of those who get the living for Middletown the amount of robust satisfaction they derive from the actual performance of their specific jobs seems, at best, to be slight. Among the business men the kudos accruing to the eminent in getting a living and to some of their minor associates yields a kind of incidental satisfaction; the successful manufacturer even tends today to supplant in local prestige and authority the judge, preacher, and "professor" of thirty-five to forty years ago. But for the working class both any satisfactions inherent in the actual daily doing of the job and the prestige and kudos of the able worker among his associates would appear to be declining.

The demands of the iron man for swiftness and endurance rather than training and skill have led to the gradual abandonment of the apprentice-master craftsman system; one of the chief characteristics of Middletown life in the nineties, this system is now virtually a thing of the past.[1] The master mechanic was the aristocrat among workmen of 1890—"one of

[1] Less than 1 per cent. of those listed by the 1920 Census as engaged in manufacturing and mechanical industries in Middletown were apprentices. Of 429 workers in Middletown wood, glass, and iron and steel industries in 1891, 51 per cent. were apprentices or had served apprenticeships. If the laborers be excluded from the group, 64 per cent. of the remaining 342 either were apprentices or had been apprenticed, and, taking the iron and steel

the noblest of God's creatures," as one of them put it. But even in the nineties machinery was beginning to undermine the monopolistic status of his skill; he was beginning to feel the ground shifting under his feet. The State Statistician recorded uneasy protests of men from all over the State.[2] Today all that is left of the four-year apprentice system among 9,000 workers in the manufacturing and mechanical industries is three or four score apprentices scattered through the building and molding trades.[3] "It's 'high speed steel' and specialization and Ford cars that's hit the machinist's union," according to a skilled Middletown worker. "You had to know how to use the old carbon steel to keep it from gettin' hot and spoilin' the edge. But this 'high speed steel' and this new 'stelite' don't absorb the heat and are harder than carbon steel. You can take a boy fresh from the farm and in three days he can manage a machine as well as I can, and I've been at it twenty-seven years."

With the passing of apprenticeship the line between skilled and unskilled worker has become so blurred as to be in some shops almost non-existent. The superintendent of a leading Middletown machine shop says, "Seventy-five per cent. of our force of 800 men can be taken from farm or high school and trained in a week's time." In the glass plant whose shift in processes is noted in Chapter VI, 84 per cent. of the tool-using personnel, exclusive of foremen, require one month or less of training, another 4 per cent. not more than six months, 6 per

plants and the glass houses separately, this percentage mounts to 79 per cent. and 70 per cent. respectively. These figures must be used cautiously, since the manner of selecting the sample of 429 workers is not stated. (*Fourth Biennial Report of* [*the State*] *Department of Statistics, 1891-2*, pp. 57, 130, 317.)

Cf. in Chapter V the discussion of the tendency of young workers to displace the older master craftsmen.

[2] "It is getting harder to find employment at molding every year." "Machinery and specialty men are used in large establishments altogether. I think in about twenty years a mechanic will be a scarce article in this country." "Men are put in as master mechanics who could not build a wheelbarrow. . . . *Any one can do one thing over and over, so he is just put on a machine at $1.00, or perhaps $1.25, a day.*" [Italics ours.] (*Fourth Biennial Report*, 1891-2, pp. 26-41.)

[3] The Personnel Department of a leading local automobile parts plant listed but four apprentices (all in the tool room) among the more than 2,000 employees on their "normal force." Another plant listed only two apprentices (also in the tool room) in a normal force of 1,000.

cent. a year, and the remaining 6 per cent. three years.[4] Foundry workers have not lost to the iron man as heavily as machinists, but even here the trend is marked. In Middletown's leading foundry in the early nineties, 47 per cent. of the workers (including foremen) had three to six years' training. This trained group today is half as great (24 per cent.) and 60 per cent. of all the castings produced are made by a group of newcomers who cast with the help of machines and require only a fortnight or so of training.

"Do you think the man who runs a complicated machine takes pride in his work and gets a feeling of proprietorship in his machine?" a responsible executive in charge of personnel in a large machine shop was asked.

"No, I don't," was his ready reply. "There's a man who's ground diameters on gears here for fifteen years and done nothing else. It's a fairly highly skilled job and takes more than six months to learn. But it's so endlessly monotonous! That man is dead, just dead! And there's a lot of others like him, and I don't know what to do for them."

"What," asked the questioner, "do you think most of the men in the plant are working for?—to own a car, or a home, or just to keep their heads above water?"

"They're just working. They don't know what for. They're just in a rut and keep on in it, doing the same monotonous work every day, and wondering when a slump will come and they will be laid off."

"How much of the time are your thoughts on your job?" an alert young Middletown bench molder was asked.

"As long as there happens to be any new problem about the casting I'm making, I'm thinking about it, but as soon as ever I get the hang of the thing there isn't 25 per cent. of me paying attention to the job."

The shift from a system in which length of service, craftsmanship, and authority in the shop and social prestige among one's peers tended to go together to one which, in the main, demands little of a worker's personality save rapid, habitual reactions and an ability to submerge himself in the performance of a few routinized easily learned movements seems to have wiped

[4] Nearly half the three-year group are carpenters and plumbers, i.e., not primarily factory workers but members of the strongly organized building trades.

out many of the satisfactions that formerly accompanied the job. Middletown's shops are full of men of whom it may be said that "there isn't 25 per cent. of them paying attention to the job." And as they leave the shop in the evening, "The work of a modern machine-tender leaves nothing tangible at the end of the day's work to which he can point with pride and say, 'I did that—it is the result of my own skill and my own effort.'"

The intangible income accruing to many of the business group derives in part from such new devices as membership in Rotary and other civic clubs, the Chamber of Commerce, Business and Professional Women's Club, and the various professional clubs.[5] But among the working class not only have no such new groups arisen to reward and bolster their work, but the once powerful trade unions have for the most part either disappeared or persist in attenuated form.

By the early nineties Middletown had become "one of the best organized cities in the United States."[6] By 1897, thirty "locals" totaling 3,766 members were affiliated with the A. F. of L. and the city vied with Detroit and other cities as a labor convention city. In 1899 the first chapter of a national women's organization, the Women's Union Label League, was launched in Middletown. At this time organized labor formed one of the most active coördinating centers in the lives of some thousands of Middletown working class families, touching their getting-a-living, educational, leisure-time, and even in a few cases religious activities. On the getting-a-living sector the unions brought tangible pressure for a weekly pay law, standardized wage scales, factory inspection, safety devices and other things regarded as improvements, and helped in sickness or death, while crowded mass meetings held in the opera house collected large sums for the striking workers in Homestead and elsewhere. A special Workingmen's Library and Reading Room,[7] with a paid librarian and a wide assortment of books, was much frequented. Undoubtedly the religious element in the labor movement of this day was missed by many, but a Middletown

[5] For discussion of these groups see Ch. XIX.
[6] From a letter from the Secretary of the Glass Bottle Blowers' Association of the U. S. and Canada, Sept. 27, 1924. A member of the executive board of this Association, who came to Middletown in 1893, says the city was "next to Rochester, N. Y., the best organized town in the country."
[7] Cf. Chapter XVII.

old-timer still refers enthusiastically to the Knights of Labor as a "grand organization" with a "fine ritual," and a member of both iron and glass unions during the nineties is emphatic regarding the greater importance of the ceremonial aspects of the unions in those days, particularly when new members were received, as compared with the bald meetings of today. As centers of leisure time the unions ranked among the important social factors in the lives of a large number of workers. Such items as these appear in the Middletown press all through the nineties:

A column account of the Ball and Concert given by Midland Lodge No. 20, Amalgamated Association of Iron and Steel Workers in Shirk's Hall, described it as "the largest event of its kind ever given in [Middletown] or the Gas Belt . . . 1,200 to 1,500 present."

An account of the installation of officers and banquet of the Painters' and Decorators' Union records the presence of 200 visitors, including wives and children. A "fine literary program was rendered." The Chief of Police was the guest of honor, and the ex-president and secretary of the Middletown Trades Council spoke. Nearly every member of the police force was present. The hall was decorated with American flags. There was singing, and the new invention, the gramophone, was featured. After the literary program came dancing.

"The Cigar Makers' 'Blue Label' nine played a very hotly contested game with union barbers' nine yesterday [Sunday] P.M."

"Yesterday P.M. [Sunday] the Bakers met at Hummel's Hall on invitation of Aug. Waick, our president, who set up a keg and lunch. We had a meeting, installed officers, then a good time."

Labor Day, a great day in the nineties, is today barely noticed.[8]

From the end of the nineties such laconic reports as "Strike defeated by use of machinery" mark increasingly the failing

[8] In 1891 the entire city participated in the first Labor Day celebration—commencing at 4 A.M. with an "artillery signal of forty-four rounds" and proceeding throughout a crowded day of bands, parade, greased pole, bicycle races in the street, pie-eating contest, reading of Declaration of Independence, two orations, greased pig, baseball, dancing all day, to a grand finale of fireworks at the fair grounds. But today the parade has been abandoned entirely. In 1923 an effort was made to draw a crowd to hear a speaker, free ice cream being used as an inducement, but in 1924 no ceremonies were even attempted.

status of organized labor in Middletown. According to the secretary of one national union, "the organized labor movement in [Middletown] does not compare with that of 1890 as one to one hundred." [9] The city's civic clubs boast of its being an "open shop town."

The social function of the union has disappeared in this day of movies and automobile, save for sparsely attended dances at Labor Hall. The strong molders' union, e.g., has to compel attendance at its meetings by making attendance at one or the other of the two monthly meetings compulsory under a penalty of a dollar fine. There is no longer a Workingmen's Library or any other educational activity. Multiple lodge memberships,[10] occasional factory "mutual welfare associations," the diffusion of the habit of carrying life insurance, socialized provision of workmen's compensation, and the beginning of the practice in at least three factories of carrying group life-insurance for all workers, are slowly taking over the insurance function performed by the trade unions. Of the 100 working class families for whom income distribution was secured, only eleven contributed anything to the support of labor unions; amounts contributed ranged from $18.00 to $60.00.

Likewise, public opinion is no longer with organized labor. In the earlier period a prominent Middletown lawyer and the superintendent of schools addressed an open meeting of the Knights of Labor, and the local press commended the "success of the meeting of this flourishing order." When Samuel Gom-

[9] From a letter from the Secretary of the Glass Bottle Blowers' Association of the United States and Canada, September 27, 1924.

Numerically at least this is an overstatement, though it may reflect the power of the organized group in the community in the two periods. In 1893 there were 981 union members in Middletown, as against 815 in 1924, when the city was two or three times as large. Some of the present total are aging workers who keep up old union affiliations for the sake of insurance benefits. The building trades, typographical workers, pattern makers, and molders are still well organized, though the first are feeling the competition of non-union workers from outlying small towns who invade the town daily in their Fords, while already, as pointed out above, in a leading foundry 60 per cent. of the castings are made by non-union men trained in a fortnight.

[10] Sickness, death, and old-age benefits are with many the sole reason for membership in working class lodges today. Mooseheart and Moosehaven, the two national "homes" for children and the aged maintained by the national order of Moose, are popular in Middletown and offer advantages which only a few of the older and richer unions, inaccessible to the great mass of local workers, can approach.

Cf. discussion of lodges in Ch. XIX.

pers came to town in ninety-seven he was dined in the mayor's home before addressing the great crowd at the opera house. The press carried daily items agitating for stricter local enforcement of the weekly pay law, or urging public support of union solicitations for funds for union purposes, or calling speeches at labor mass-meetings "very able and enjoyable addresses." The proceedings of the Glass Workers' Convention in Baltimore in 1890 were reported in full on the first page. Such a note as this was common: "During the last few months there have been organized in this city several trade organizations and labor unions . . . and much good has resulted therefrom." At a grand Farmers and Knights of Labor picnic in 1890, "a perfect jam, notwithstanding the rain," the speaker "ably denounced trusts, Standard Oil, etc.," according to the leading paper. The largest men's clothing firm presented a union with a silk parade-banner costing nearly $100.[11] Today the Middletown press has little that is good to say of organized labor.[12] The pulpit avoids such subjects, particularly in the churches of the business class, and when it speaks it is apt to do so in guarded, equivocal terms.[13] A prevalent attitude among

[11] It should not be inferred that the workers had things all their own way. Strikes and lock-outs were frequent, and the boycott was freely and effectively used against local business men who sold non-union goods. The diary of one elderly merchant complains that "a great many were compelled to show a left-handed sneaking approval." But the significant point is that labor was powerful and class-conscious, and the workers apparently gained added stature in many of their vital activities from their membership in this powerful union movement.

[12] One of the two daily papers spoke editorially on one recent occasion of unions as "fine things for those who work with their hands" but went on to decry any activity by the local union "composed of our own 'folks'" in trying "to drag into its own affairs the 'folks' that are international or national, and do not know our own local problems."

[13] Cf. the following press report, sent to the paper by the minister himself, which summarizes a sermon in the largest church in Middletown on national "Labor Sunday," 1924: "[The preacher] based his sermon on that portion of the Lord's Prayer which calls upon God 'to give us this day our daily bread.' He pointed out that when this prayer is repeated one does not ask God 'to give me my daily bread' but is broader and takes in all mankind in the words 'us' and 'our.'

"The speaker took up briefly the labor situation in the United States as regards the laborer and the employer, and declared that 'we do not have in [Middletown] the conditions that exist elsewhere,' implying that no serious labor problem is in existence here. Brotherhoods among laboring men have done much good for the laborer and have brought to him certain rights, he said, but in some cases, especially where public welfare is involved, they have gone too far."

the business class appears in the statement of one of the city's leaders, "Working men don't need unions nowadays. There are no great evils or problems now as there were fifty years ago. We are much more in danger of coddling the working men than abusing them. Working people are just as well off now as they can possibly be except for things which are in the nature of industry and cannot be helped."

This decrease in the psychological satisfactions formerly derived from the sense of craftsmanship and in group solidarity, added to the considerations adduced in the preceding chapters, serves to strengthen the impression gained from talk with families of the working class that, however it may be with their better-educated children, for most of the present generation of workers "there is no break through on their industrial sector." It is important for the consideration of other life-activities to bear in mind this fact, that the heavy majority of the numerically dominant working class group live in a world in which neither present nor future appears to hold as much prospect of dominance on the job or of the breaking through to further expansion of personal powers by the head of the family as among the business group.

Frustrated in this sector of their lives, many workers seek compensations elsewhere. The president of the Middletown Trades Council, an alert and energetic molder of thirty and until now the most active figure in the local labor movement, has left the working class to become one of the minor officeholders in the dominant political machine. Others who do not leave are finding outlets, if no longer in the saloon, in such compensatory devices as hooking up the radio or driving the "old bus." The great pressure toward education on the part of the working class is, of course, another phase of this desire to escape to better things.[14]

For both working and business class no other accompaniment of getting a living approaches in importance the money received for their work. It is more this future, instrumental aspect of work, rather than the intrinsic satisfactions involved, that keeps Middletown working so hard as more and more of the activities of living are coming to be strained through the bars

[14] Cf. in Ch. XIII the importance to working class parents of education for their children.

of the dollar sign.[15] Among the business group, such things as one's circle of friends, the kind of car one drives, playing golf, joining Rotary, the church to which one belongs, one's political principles, the social position of one's wife apparently tend to be scrutinized somewhat more than formerly in Middletown for their instrumental bearing upon the main business of getting a living, while, conversely, one's status in these various other activities tends to be much influenced by one's financial position. As vicinage has decreased in its influence upon the ordinary social contacts of this group,[16] there appears to be a constantly closer relation between the solitary factor of financial status and one's social status. A leading citizen presented this matter in a nutshell to a member of the research staff in discussing the almost universal local custom of "placing" newcomers in terms of where they live, how they live, the kind of car they drive, and similar externals: "It's perfectly natural. You see, they know money, and they don't know you."

This dominance of the dollar appears in the apparently growing tendency among younger working class men to swap a problematic future for immediate "big money." Foremen complain that Middletown boys entering the shops today are increasingly less interested in being moved from job to job until they have become all-round skilled workers, but want to stay on one machine and run up their production so that they may quickly reach a maximum wage scale.[17]

The rise of large-scale advertising, popular magazines, movies, radio, and other channels of increased cultural diffusion from without are rapidly changing habits of thought as to what

[15] Cf. Maynard Keynes on "the habitual appeal" of our age "to the money motive in nine-tenths of the activities of life, . . . the universal striving after individual economic security as the prime object of endeavor . . . the social approbation of money as the measure of constructive success, and . . . the social appeal to the hoarding instinct as the foundation of the necessary provision for the family and for the future." (*New Republic*, Nov. 11, 1925.)

[16] Cf. in Ch. XIX the places where both business and working class men and women see their friends.

[17] According to one veteran foundry foreman: "In the old days of the nineties a boy was shaped and trained by his foreman. When he started his apprenticeship for the molder's trade he was lucky to make $3 or $4 a week. At the end of the first year he was making, maybe, a dollar or $1.25 a day; at the end of the second year perhaps $1.50 or $2.00; the third year, $2.25; and then at the end of the fourth year he received his card and $2.75 a day. Meanwhile his foreman had shifted him about from job

things are essential to living and multiplying optional occasions for spending money.[18] Installment buying, which turns wishes into horses overnight, and the heavy increase in the number of children receiving higher education, with its occasions for breaking with home traditions, are facilitating this rise to new standards of living. In 1890 Middletown appears to have lived

to job until, when he became a molder and went on a piece-work basis, he knew his job from every angle and could make big money. But the trouble nowadays is that within a year a machine molder may be making as much as a man who has been there fifteen or twenty years. He has his eyes on the money—$40 to $50 a week—and resists the foreman's efforts to put him on bench molding where he would learn the fine points of the molder's trade."

[18] It is perhaps impossible to overestimate the rôle of motion pictures, advertising, and other forms of publicity in this rise in subjective standards. Week after week at the movies people in all walks of life enter, often with an intensity of emotion that is apparently one of the most potent means of reconditioning habits, into the intimacies of Fifth Avenue drawing rooms and English country houses, watching the habitual activities of a different cultural level. The growth of popular magazines and national advertising involves the utilization through the printed page of the most powerful stimuli to action. In place of the relatively mild, scattered, something-for-nothing, sample-free, I-tell-you-this-is-a-good-article copy seen in Middletown a generation ago, advertising is concentrating increasingly upon a type of copy aiming to make the reader emotionally uneasy, to bludgeon him with the fact that decent people don't live the way *he* does: *decent* people ride on balloon tires, have a second bathroom, and so on. This copy points an accusing finger at the stenographer as she reads her *Motion Picture Magazine* and makes her acutely conscious of her unpolished finger nails, or of the worn place in the living room rug, and sends the housewife peering anxiously into the mirror to see if *her* wrinkles look like those that made Mrs. X—— in the ad. "old at thirty-five" because she did not have a Leisure Hour electric washer.

Whole industries are pooling their strength to ram home a higher standard of living, e.g., the recent nation-wide essay contest among school children on home lighting conducted by all branches of the electrical industry. In addition to the national prizes of a $15,000 house and university scholarships, local prizes ranging all the way from a radio set and dressing table to electric curling irons and basket-ball season tickets were given to the thirty best Middletown essays. In this campaign 1,500 Middletown children submitted essays on how the lighting of their homes could be improved, and upwards of 1,500 families were made immediately aware of the inadequacies of their homes as regards library table lamps, porch lights, piano lamps, and convenient floor sockets. As one of the winning local essays said: "I and all my family have learned a great deal that we did not know before, and we intend improving the lighting in our own home."

The "style show" is a new and effective form of Middletown advertising that unquestionably influences the local standard of living. On two successive nights at one of these local shows a thousand people—ten-cent store clerks, tired-looking mothers with children, husbands and wives—watched rouged clerks promenade languorously along the tops of the show cases, displaying the latest hats, furs, dresses, shoes, parasols, bags and other accessories, while a jazz orchestra kept everybody "feeling good."

on a series of plateaus as regards standard of living; old citizens say there was more contentment with relative arrival; it was a common thing to hear a remark that so and so "is pretty good for people in our circumstances." Today the edges of the plateaus have been shaved off, and every one lives on a slope from any point of which desirable things belonging to people all the way to the top are in view.

This diffusion of new urgent occasions for spending money in every sector of living is exhibited by such new tools and services commonly used in Middletown today, but either unknown or little used in the nineties, as the following:

In the home—furnace, running hot and cold water, modern sanitation, electric appliances ranging from toasters to washing machines, telephone, refrigeration, green vegetables and fresh fruit all the year round, greater variety of clothing, silk hose and underwear, commercial pressing and cleaning of clothes,[19] commercial laundering or use of expensive electrical equipment in the home,[20] cosmetics, manicuring, and commercial hair-dressing.

In spending leisure time—movies (attendance far more frequent than at earlier occasional "shows"), automobile (gas, tires, depreciation, cost of trips), phonograph, radio, more elaborate children's playthings, more club dues for more members of the family, Y.M.C.A. and Y.W.C.A., more formal dances and banquets, including a highly competitive series of "smartly appointed affairs" by high school clubs;[21] cigarette smoking and expensive cigars.

In education—high school and college (involving longer dependence of children), many new incidental costs such as entrance to constant school athletic contests.[22]

[19] In the Middletown city directory for 1889 there were no dry cleaners and only one dye house. Today a city less than four times the size has twelve dry cleaners and four dye houses. The habit of pressing trousers is said not to have "come in" until about 1895.

[20] The hand-washers of 1890 sold for $7.50-$10.00, while the modern machines cost $60.00 to $200.00.

[21] A dance no longer costs $0.50, as in the nineties, but the members of clubs are assessed about $4.00 for their Christmas dances today. Music used to be a two- or three-piece affair, but now it is an imported orchestra costing from $150 to $300. A boy has to take a girl in a taxi if he does not have the use of the family car. One does not go home after a dance but spends a dollar or so on "eats" afterwards. Expensive favors are given at annual sorority banquets.

[22] See Table VI for distribution of expenditures of 100 working class families.

In the face of these rapidly multiplying accessories to living, the "social problem" of "the high cost of living" is apparently envisaged by most people in Middletown as soluble if they can only inch themselves up a notch higher in the amount of money received for their work. Under these circumstances, why shouldn't money be important to people in Middletown? "The Bible never spoke a truer word," says the local paper in an editorial headed "Your Bank Account Your Best Friend," "than when it said: 'But money answereth all things.' . . . If it doesn't answer all things, it at least answers more than 50 per cent. of them." And again, "Of our happy position in world affairs there need be no . . . further proof than the stability of our money system." One leading Middletown business man summed up this trend toward a monetary approach to the satisfactions of life in addressing a local civic club when he said, "Next to the doctor we think of the banker to help us and to guide us in our wants and worries today."

Money being, then, so crucial, how much money do Middletown people actually receive? The minimum cost of living for a "standard family of five" in Middletown in 1924 was $1,920.87.[23] A complete distribution of the earnings of Middletown is not available. Twelve to 15 per cent. of those getting the city's living reported a large enough income for 1923 to make the filing of a Federal income tax return necessary.[24] Of the 16,000-17,000 people gainfully employed in 1923 —including, however, somewhere in the neighborhood of a thousand married women, some of whom undoubtedly made joint returns with their husbands—210 reported net incomes (i.e., minus interest, contributions, etc.) of $5,000 or over, 999 more net incomes less than $5,000 but large enough to be taxable after subtracting allowed exemptions ($1,000 if single, $2,500 if married, and $400 per dependent), while 1,036 more filed returns but were not taxable after subtracting allowed deductions and exemptions. The other 85-88 per cent. of those earning the city's living presumably received either less than

[23] Based on the budget of the United States Bureau of Labor and computed on the basis of Middletown prices. See Table VII.

[24] These income tax data, fallible as they are, owing to non-reporting and other possible errors, are used here simply as the best rough estimate available. There are at the outside probably not over two- or three-score people in Middletown who made income tax returns who are not actually engaged in getting a living.

$1,000 if single or less than $2,000 if married, or failed to make income tax returns. A cross section of working class earnings is afforded by the following distribution of 100 of the working class families interviewed according to their earnings in the preceding twelve months: [25]

	Distribution of Families by Fathers' Earnings Only	Distribution of Families by Total Family Earnings
Total number of families	100	100
	—	—
Earning less than minimum standard of $1,920.87		
Families of 5 members or more	42	39
Families of 4 or 3 members (including families of 2 foremen)	35	35
Earning more than minimum standard of $1,920.87		
Families of 5 members or more (including one foreman)	10	13
Families of 4 or 3 members (including 6 foremen)	13	13

The incomes of these 100 families range from $344.50 to $3,460.00, with the median at $1,494.75 and the first and third quartiles respectively at $1,193.63 and $2,006.00.[26]

The relative earning power of males and females in Middletown is indicated by the fact that in a characteristic leading Middletown plant during the first six months of 1924 the weighted average hourly wage of all females (excluding office force and forewomen) was $0.31 and of all males (excluding

[25] See Table VI for distribution of income of these 100 families by members of family earning and for distribution of certain major items of expenditure throughout the year.

Six of the twelve months (Oct. 1, 1923, to Oct. 1, 1924) covered by these income figures were good times in Middletown and six months were relatively bad times locally, though the latter was not a period of national depression. This would tend to make the 1924 average income less than on a "big year" like 1923—though 50 per cent. good and bad times is more representative of the actual chance to get a living in Middletown today than either a completely good or bad year would have been.

See Appendix on Method regarding choice of families in connection with the fact of the presence of nine foremen's families in the sample.

The minimum standard for a family of less than five members would be less than $1,920.87, and consequently certain marginal families of three or four grouped above with those earning less than the minimum would on a more exact calculation be transferred to the group earning more than the minimum standard.

[26] The incomes of the husbands alone of these 100 families exhibit a spread from $344.50 to $3,200.00, with the median at $1,303.10 and the first and third quartiles at $1,047.50 and $1,856.75 respectively.

office force and foremen) $0.55. The bulk of this plant is on a ten-hour basis, fifty-five hours per week, making the average annual income for fifty-two weeks, provided work is steady, $886.60 for females and $1,573.00 for males. In three other major plants similar average wages for males were $0.55, $0.54 and $0.59. In general, unskilled female labor gets $0.18 to $0.28 an hour and a few skilled females $0.30 to $0.50.[27] Unskilled males receive $0.35 to $0.40 an hour and skilled males from $0.50 to $1.00 and occasionally slightly more.

As over against these wages of women in industry in Middletown in 1924, ranging from $10.00 to $18.00 a week in the main, the younger clerks in the leading department store received $10.00 a week, and more experienced clerks a flat rate from $8.00 to $17.00 a week plus a bonus, if earned—the whole amounting occasionally "when times are good" for a veteran clerk to $30.00 to $40.00 a week.

A detailed calculation of a cost of living index for Middletown in 1924 on the basis of the cost of living in 1891 reveals an increase of 117 per cent.[28] A comparison of the average yearly earnings of the 100 heads of families in 1924 with available figures for 439 glass, wood, and iron and steel workers in Middletown in 1891 reveals an average of $1,469.61 in the former case and $505.65 in the latter, or an increase of 191 per cent. today.[29] Or if we take the earnings of school teachers as an index, probably conservative, of the trend in earnings, as against this rise of 117 per cent. in the cost of living, it appears that the minimum salary paid to grade school teachers has risen 143 per cent. and the maximum 159 per cent., and the minimum salary paid to high school teachers 134 per cent. and the maximum 250 per cent. The median salary for grade school teachers in 1924 was $1,331.25, with the first and third quar-

[27] Willford I. King says wages of females the country over are "about three-fourths those of males." (*Op. cit.*, p. 144.)

[28] See Table VIII for the increase by major items and also for the method of computing this index.

[29] The 1891 earnings are taken from the *Fourth Biennial Report* for the state in which Middletown is located, dated 1891-2, pp. 57, 130, and 317. This Report gives the average income of 225 Middletown adult male glass workers as $519.49, of sixty-nine wood workers as $432.32, and of 145 iron and steel workers as $519.06—or an average for the entire 439 of $505.65. Too much weight obviously cannot be put upon these 1891 figures, as nothing is known either as to the method of their collection or as to their accuracy.

tiles at $983.66 and $1,368.00 respectively. The median salary for high school teachers was $1,575.00, with the first and third quartiles at $1,449.43 and $1,705.50 respectively. Substantial increases in the incomes of persons in certain other representative occupations are suggested by the fact that the salary of a bank teller has mounted from $50.00 or $65.00 a month in 1890 to $166.67 a month in 1924, that of an average male clerk in a leading men's clothing store from $12.00 a week in 1890 to $35.00 today; a doctor's fee for a normal delivery with the same amount of accompanying care in both periods has risen from $10.00 to $35.00, and for a house call from $1.00 to $3.00.

Thus this crucial activity of spending one's best energies year in and year out in doing things remote from the immediate concerns of living eventuates apparently in the ability to buy somewhat more than formerly, but both business men and working men seem to be running for dear life in this business of making the money they earn keep pace with the even more rapid growth of their subjective wants. A Rip Van Winkle who fell asleep in the Middletown of 1885 to awake today would marvel at the change as did the French economist Say when he revisited England at the close of the Napoleonic Wars; every one seemed to run intent upon his own business as though fearing to stop lest those behind trample him down. In the quiet county-seat of the middle eighties men lived relatively close to the earth and its products. In less than four decades, business class and working class, bosses and bossed, have been caught up by Industry, this new trait in the city's culture that is shaping the pattern of the whole of living.[30] According to its needs, large numbers of people anxious to get their living are periodically stopped by the recurrent phenomenon of "bad times" when the machines stop running, workers are "laid off" by the hundreds, salesmen sell less, bankers call in loans, "credit freezes," and many Middletown families may take their children from school, move into cheaper homes, cut down on food, and do without many of the countless things they desire.

The working class is mystified by the whole fateful business. Many of them say, for instance, that they went to the polls and

[30] R. H. Tawney speaks of the rise of industry "to a position of exclusive prominence among human interests" until the modern world is "like a hypochondriac . . . absorbed in the processes of his own digestion."

voted for Coolidge in November, 1924, after being assured daily by the local papers that "A vote for Coolidge is a vote for prosperity and your job"; puzzled as to why "times" did not improve after the overwhelming victory of Coolidge, a number of them asked the interviewers if the latter thought times would be better "after the first of the year"; the first of the year having come and gone, their question was changed to "Will business pick up in the spring?"

The attitude of the business men, as fairly reflected by the editorial pages of the press which today echo the sentiments heard at Rotary and the Chamber of Commerce, is more confi-dent but confusing. Within a year the leading paper offered the following prescriptions for local prosperity: "The first duty of a citizen is to produce"; and later, "The American citizen's first importance to his country is no longer that of citizen but that of consumer. Consumption is a new necessity." "The way to make business boom is to buy." At the same time that the citizen is told to "consume" he is told, "Better start saving late than never. If you haven't opened your weekly savings account with some local bank, trust company, or building and loan, today's the day." Still within the same year the people of Mid-dletown are told: "The only true prosperity is that for which can be assigned natural reasons such as good crops, a demand for building materials, . . . increased need for transporta-tion," and ". . . advancing prices are due to natural causes which are always responsible for prices. . . . As all wealth comes from the soil, so does all prosperity, which is only an-other way of saying so does all business." But again, "natural causes" are apparently not the chief essential: "There can be no greater single contribution to the welfare of the nation than the spirit of hopefulness. . . ." "[This] will be a banner year because the people believe it will be, which amounts to the de-termination that it shall be. . . ." Still another solution for securing "good times" appears: "The most prosperous town is that in which the citizens are bound most closely together. . . . Loyalty to the home town . . . is intensely practical. . . . The thing we must get into our heads about this out-of-town buying business is that it hurts the individual who does it and his friends who live here. Spending your money at home in the long run amounts practically to spending it upon yourself, and buying away from home means buying the comforts and luxuries for the other fellow." "A dollar that is spent out of town never returns." One looking on at this procedure may begin to wonder if the business men, too, are not somewhat bewildered.

Although neither business men nor working men like the recurring "hard times," members of both groups urge the maintenance of the present industrial system. The former laud the group leaders who urge "normalcy" and "more business in government and less government in business," while the following sentences from an address by a leading worker, the president of the Trades Council, during the 1924 political campaign, sets forth the same faith in "free competition" on the part of the working class: "The important issue is the economic issue. We can all unite on that. We want a return to active free competition, so that prices will be lower and a man can buy enough for himself and his family with the money he makes." Both groups, as they order a lay-off, cut wages to meet outside competition, or, on the other hand, vote for La Follette in the hope of his being able to "do something to help the working man," appear to be fumbling earnestly to make their appropriate moves in the situation according to the rules of the game as far as they see them; but both appear to be bound on the wheel of this modern game of corner-clipping production. The puzzled observer may wonder how far any of them realizes the relation of his particular move to the whole function of getting a living.[31] He might even be reminded of a picture appearing in a periodical circulated in Middletown during the course of the study: A mother leans over her two absorbed infants playing at cards on the floor and asks, "What are you playing, children?"

"We're playing 'Putcher,' Mamma. Bobby, putcher card down."

In the midst of such a partially understood but earnestly followed scheme of getting a living, the rest of living goes on in Middletown.

[31] Cf. Walton Hamilton's *The Case of Bituminous Coal* (The Institute of Economics Series, *Investigations in Industry and Labor*. New York; Macmillan, 1925), pp. 251-2.

II. MAKING A HOME

Chapter IX

THE HOUSES IN WHICH MIDDLETOWN LIVES

The forty-three people in each hundred who get Middletown's living divide their lives regularly between two places: their best waking energies, five and a half out of every seven days, are spent in the buildings set apart for industry and business; their other activities traditionally center about the homes in which they and their families sleep and eat. In these homes the twenty-three people in each hundred who are engaged in making the homes of the city, the four in each hundred who are the very old or feeble, the eleven who are very young, and, to a considerably less extent, the nineteen receiving required school training, carry on their respective activities. Next to the places where people get a living, these homes form the most apparent locus of the lives of the community.

Middletown's 38,000 people live in 9,200 homes:[1] 86 per cent. of these homes are in one-family houses, each standing on a separate patch of ground, the latter called, with increasing significance in view of its shrinking size, "a yard"; 10 per cent. are in two-family houses, a more common type since 1890 as building costs have risen; 1 per cent. in apartments;[2] and 3 per

[1] See Appendix on Method for the basis of this population estimate.
The official count of the households of the city by the Middletown post office in August, 1924, was 9,240. The house-by-house map of the City Plan Commission for May, 1923, showed 9,163 (including supplementary counts in connection with the present study for the number in apartments and over stores in the downtown section). The count of the Commercial Department of the Indiana Bell Telephone Company, begun in the spring of 1922 and completed in the spring of 1923, showed 9,159 households within the city limits.

[2] In the ample Middletown of 1890 it was "poor-folksy" not to have a house with a yard, and local sentiment still frowns upon apartment life for families with children. Apartments tend to be inhabited by childless couples, elderly women, and similar atypical family groups. There were no apartment buildings in 1890, and there are less than a score today, but both apartments and duplexes are increasing in number.

According to the Federal Census the persons in each dwelling (any building under a single roof in which one or more persons live, including hotels,

cent. over stores, chiefly in the "downtown" section. The life of a house in Middletown is thirty to fifty years, and, as in each new generation the less well-to-do tend to inherit the aging homes of the group slightly "better off" in the preceding generation, the city lives, in the main, in houses fifteen to forty years old.[3] Although working class families tend to be larger than those of the business class (the 124 working class families interviewed average 5.4 members each and the forty business class families 4.7), it is the business group, generally speaking, who live in the larger dwellings; one in each three of the 124 working class families interviewed and four out of five of the forty business class families live in single-family houses large enough to have two floors.[4]

In the eighties the usual size of the plot of ground on which a house was built was sixty-two and a half feet fronting on the street, with a depth of 125 feet; the standard building lot today has a frontage of forty feet. Whereas the city blocks of 1890 usually contained eight lots, the same blocks now contain ten, twelve, and even fourteen. A common practice today is to saw off the back of a lot and insert an additional house fronting on the side street. The implications of this shift for play room for children, leisure-time activities for the entire family, family privacy, and even for the former sense of substantial pride in the appearance of one's "place" are obvious. Houses are crowding closer to the front paving-line, and flowers and shrub-

duplexes, and apartment houses) in Middletown decreased from 4.70 to 4.22 between 1890 and 1920. The families per dwelling rose from 1.05 to 1.10. (The figure for 1910 was the same as that for 1890; the post-war housing shortage may account for the 1920 figure.) At the same time persons per family decreased from 4.46 to 3.83.

[3] Since the population and presumably the number of dwellings today are roughly six times the number in 1887, three and one-third times that of 1890, not quite double that of 1900, and one and a half times that of 1910, the rough proportion of dwellings inherited from each decade is apparent, although allowance must be made for fires and the razing and improvement of old buildings.

[4] Another third of the working class sample live in small, one-floored cottages, a sixth in new and more prosperous bungalows, and the final sixth in two-family houses and "flats." Two of the forty business class families interviewed live in new, pretentious two-floored homes; four in pretentious two- and three-floored houses of an earlier period built of stone or brick; seven in unpretentious new houses (bungalows and small two-floored houses, in general less than ten or twelve years old); twenty-two in unpretentious houses more than ten or twelve years old (comfortable-looking places, all but one of them two-floored); two in apartments; two in duplex houses; one in a hotel.

bery must give way as the lawn shrinks to allow a driveway to the garage. The housewife with leisure does not sit so much on the front porch in the afternoon after she "gets dressed up," sewing and "visiting" and comparing her yard with her neighbors', nor do the family and neighbors spend long summer evenings and Sunday afternoons on the porch or in the side yard since the advent of the automobile and the movies. These factors tend to make a decorative yard less urgent; the make of one's car is rivaling the looks of one's place as an evidence of one's "belonging." [5]

Houses are usually built of wood.[6] Those of the poorest working class families appear essentially the same externally as they did in the middle of the last century—bare little one-story oblong wooden boxes with a roof and with partitions inside making two to four small rooms. At the other extreme are the large homes of the more prosperous members of the business class; these exhibit considerably greater simplicity of line than a generation ago, when an "elegant new residence" might be described in the press as a "conglomeration of gables, nooks, verandas, and balconies with three stone chimneys towering above, giving the appearance of an ancient castle." [7]

[5] Back yards are ceasing to be ample affairs with grass and fruit trees and grape arbors where the housewife sits and peels potatoes for supper or cuts up fruit for canning and the family spends Sunday afternoons around the barrel-stave hammock. In many working class families smaller yards and closer neighbors have reduced the back yard to an overflow storage place in this day of no attics, store rooms, or barns.

The large back-yard garden has either disappeared altogether or is considerably more limited in size and content. Among working class families, smaller yards, less home canning, lack of winter storage space for food, time spent riding and tinkering on the car, movies, and similar factors have been responsible for the decline of back-yard gardening, while among the business class families the high cost of labor, increased preoccupation of the son of the family with extracurricular school activities and "Y" summer camp, the tendency of a few families to move in summer to a cottage at the Lakes ninety miles distant, are additional factors operating in the same direction. A local seed store proprietor who has lived in the city since the Civil War estimated that certainly not over 40 or 50 per cent. of the local families today have even small back-yard gardens, whereas he thinks "75-80 per cent. anyway" had them in 1890, and the 1890 gardens were considerably larger. He attributes the difference to the automobile.

[6] Over ninety-five houses in each one hundred are made of wood. In 1924 licenses were issued in Middletown for the construction of 205 wooden houses, two one-floor brick houses, three two-floor brick houses, one tile, and three stucco houses.

[7] The disappearance of the ornate houses of the seventies and eighties, decked out in elaborate scrolls and "gingerbread" work in favor of the

In the eighties with their ample yards, porches were not urgently needed. Towards 1900, as smaller yards were driving the family closer to the house, people began to hear of porches in the state capital "fitted up like a room," and the era of porch furniture began; small wood-working plants that were losing their local trade in wagons and agricultural implements exploited this new specialty eagerly. Already, however, business class homes are leading the way in a reversion to porchless designs with glassed-in sun parlor and sleeping porch, the latter showing how far Middletown has moved from its "fear of the night air," which in 1890 prompted sleeping with windows down a gingerly six inches at the top. Today none of the workers' houses has a sleeping porch and relatively few of the homes of the business group, but according to a local building expert, "The man who puts $8,000 and up into a house today demands a sleeping porch." The trend is apparently to divert the money formerly put into front porches to sleeping porches, glassed-in dens, and other more private and more often used parts of the house.

A Middletown building expert estimates that the majority of homes constructed within the last ten to fifteen years have at least 50 per cent. more glass surface than in 1890. More air can be let in from without today because more heat can be secured within. Several hundred homes are heated by a central heating plant operated under city franchise and hundreds of others by furnace, steam or oil, although most of the working class still live in the base-burner and unheated-bed-room era.[8] With

plainer homes of today in which the emphasis is upon all manner of new interior devices in bathroom and elsewhere, probably reflects, among other things, a tendency commonly observable in human culture: in periods when improvements in effectiveness of the utility of a tool are at a standstill, human ingenuity tends to spend itself in decorating the tool, but in periods of evolution in the effectiveness of the major use, ornamentation tends to assume a place of secondary importance. The Middletown house of the eighties was still simply a box divided into rooms, with relatively few changes in process in its adaptation as a place in which to live comfortably. The coming of bathrooms, a wide range of electrical equipment, central heating, and other inventions, is today focusing attention upon a wide variety of changes in the interior livableness, with corresponding decline in exterior ornamentation.

[8] In the survey of Zanesville, Ohio, a city of 30,442, in 1925 it was found that 48.0 per cent. of the 11,232 homes of the city were heated by furnace. *Zanesville and Thirty-six Other American Communities* (New York; *Literary Digest,* 1927), p. 65. It is not unlikely that conditions in Middletown resemble roughly those prevailing in Zanesville in this respect.

furnaces have come basements which, in the sense of cement-floored and walled rooms beneath the house, were practically non-existent in the Middletown of 1890. This inconspicuous item of basement, plus foundation, adds $700 to the cost of even a small house; a basement today costs about what a small house cost in the early nineties.[9]

There was no running water prior to 1885, and by 1890 not more than 20 per cent. of the total mileage of the city's streets was underlaid with water mains. It is estimated that in 1890 only about one family in six or eight had even the crudest running water—a hydrant in the yard or a faucet at the iron kitchen sink. A leading citizen thought it sufficiently important to enter in his diary in 1890 that a neighbor "has a hydrant for his house." The minutes of the Board of Education for 1888 contain an item: "Eph Smell . . . 1 wooden pump for High School . . . $10.00." For the most part, Middletown pumped its water to the back door or kitchen from a well or cistern. By 1890 there were not over two dozen complete bathrooms in the entire city. For approximately ninety-five families in each hundred, "taking a bath" meant lugging a heavy wooden or tin tub into a bedroom, or more usually the warm kitchen, and filling it half full of water from the pump, heated on the kitchen stove. Today all new houses, except the very cheapest, have bathrooms, and many old houses are installing this improvement rapidly.[10] Many homes, however, still lack not only bathroom, but in January, 1925, approximately one in four of all the city's dwellings lacked running water.[11] This considerable use of a water supply from back-yard wells accompanies the persistence in even more working class homes of the old-fashioned backyard "privy." According to the City Engineer, only two-thirds of the houses had sewer connections in 1924. It is not un-

[9] Another aspect of this change from the old-fashioned fruit cellar to the modern basement is the increasing tendency to substitute the corner grocery for the family storage of food in quantities in cellar and attic. Instead of buying potatoes and apples by the barrel, the tendency is to purchase potatoes by the peck and fruit by the dozen or pound.

[10] The extent to which this improvement is being introduced into older houses is reflected in the fact that one of the dozen local plumbing firms alone claims to have installed 50 per cent. more bathrooms in 1923, when "times were good," than the entire total of new houses built during the year.

[11] Sixty-one and eight-tenths per cent. of the 11,232 homes of Zanesville have bathrooms, and 61 per cent. plumbing. (*Op. cit.,* pp. 55 and 63.)

common to observe 1890 and 1924 habits jostling along side by side in a family with primitive back-yard water or sewage habits, yet using an automobile, electric washer, electric iron, and vacuum cleaner. This unevenness in the diffusion of material culture becomes even more significant in the light of the community's public health service with its outwardly stringent prohibition upon back-yard water supplies and back-yard toilets and sewage disposal.

Electric lighting is so much a matter of course today that it is hard to recover the days before its advent. The Middletown press in 1895 regarded natural gas as the last word in home-making—"the millennium of comfort and cleanliness is at hand"; but most of the discoveries and virtually all of the enormous popular diffusion of modern electricity in terms of the home—electric lighting of a brilliance and steadiness undreamed of then, labor-saving devices for cleaning the house, washing clothes, and cooking—still lay in the future. Over 95 per cent. of Middletown's houses were without electricity in 1890; by 1916, 60 per cent. were using electricity for lighting purposes, and in June, 1925, 99 per cent. of the homes were wired and presumably at least lighted by electricity.[12] In slightly more than two out of each three families cooking is done with gas at the present time, the others using gasoline, coal, and a very few electricity.

No other changes in Middletown's homes have been as marked as the adoption by the bulk of the community of these various conveniences, used only by a few of the very wealthy in the nineties. The interior plan of the house has remained fundamentally the same, although there has been some trend toward fewer and larger rooms, the "parlor" and the "spare

[12] The implications of improvement in lighting should be borne in mind in the complex of factors surrounding the reading of printed matter. The Middletown press in 1890 carried an announcement of a new "gas burner giving the light of 250 candle power without a particle of flicker or shadow. Just think of it." Club women still recall the discomforts of occasional evening meetings when the members who were on the program stood sweltering under the heat of the jumbo burners.

The increase in the reading serviceability of the modern electric light may be gauged from the following list of relative brightness of common illuminants in terms of candle power per square inch: Tungsten lamp 1,500; Gem lamp 750; Carbon lamp 400; Welsbach light 30; gas flame 7; kerosene light 7; incandescent frosted lamp 6; candle 3. Louis Bell, *The Art of Illumination* (New York; McGraw, Hill, 1012), p. 12.

bedroom" being the casualties. In some working class homes the parlor, living room, and kitchen have become living room, dining room, and kitchen, but in many the parlor survives, in which case the family lives in the dining room. Among the business class and in the case of the newer bungalows of the working class, the tendency is to throw together much of the lower floor by means of large double doorways.[13]

The main type of interior furnishing, as well as of plan of house, is fairly definitely fixed in the habits of all groups; a higher standard of living exhibits itself, not in any radical departure from this general type, but in minor variations and more costly elaborations.[14]

The poorer working man, coming home after his nine and a half hours on the job, walks up the frequently unpaved street, turns in at a bare yard littered with a rusty velocipede or worn-out automobile tires, opens a sagging door and enters the living room of his home. From this room the whole house is visible—the kitchen with table and floor swarming with flies and often strewn with bread crusts, orange skins, torn papers, and lumps of coal and wood; the bedrooms with soiled, heavy quilts falling off the beds. The worn green shades hanging down at a tipsy angle admit only a flecked half-light upon the ornate calendars or enlarged colored portraits of the children in heavy gilt frames tilted out at a precarious angle just below the ceiling. The whole interior is musty with stale odors of food, clothing, and tobacco. On the brown varnished shelf of the sideboard the wooden-backed family hair brush, with the baby bottle, a worn purse, and yesterday's newspaper, may be half stuffed out of sight behind a bright blue glass cake dish.

[13] Certain implications of these changes in rooms should be noted. The disappearance of the spare bedroom in these days of high building costs is probably not unrelated to the diminishing tendency pointed out in Ch. XXVI for aging parents to come home to live with their children. Likewise, the tendency to throw the former parlor and sitting room together as one large exposed living room, doing away with the privacy that the daughter of the house and her caller used to find in the earlier parlor, may be related to the fact that many a daughter is leaving home with her "date" to "get away from the family" and find privacy in the anonymity and darkness of the movies or automobile. This is in turn reflected in the prominence of the number of evenings children are away from home during the week as a source of disagreement between children and their parents, noted in Ch. XI.

[14] The descriptions which follow are composite pictures based upon detailed accounts of the homes visited by staff interviewers.

Rust spots the base-burner. A baby in wet, dirty clothes crawls about the bare floor among the odd pieces of furniture.[15]

The working man with more money leeway may go home through a tidy front yard; whether his home is of the two-floor variety, a bungalow, or a cottage, there are often geraniums in the front windows, neat with their tan, tasseled shades and coarse lace curtains. A name-plate of silvered glass adorns the door. The small living room is light, with a rather hard brightness, from the blue- and pink-flowered rug, bought on installment, to the artificial flowers, elaborately embroidered pillows and many-colored "center pieces." The furniture is probably straight-lined "mission" of dark or golden oak or, if the family is more prosperous, "overstuffed."[16] The sewing machine stands in the living room or dining room, and the ironing board with its neat piles of clothes stretches across one corner of the kitchen. "Knickknacks" of all sorts are about—easeled portraits on piano or phonograph, a paper knife brought by some traveled relative from Yellowstone Park, pictures that the small daughter has drawn in school, or if the family is of a religious bent, colored mottoes: "What will you be doing when Jesus comes?" or "Prepare to meet thy God." There may even be a standing lamp with a bright silk shade, another recent installment purchase and a mark of prestige. Some magazines may be lying about, but rarely any books.

The homes of some head bookkeepers, owners of small retail stores, school teachers, and other less wealthy members of the business group convey an atmosphere of continual forced choices between things for the house and things for the children—between a hardwood floor for the front hall and living room or a much-needed rug and the same amount of money put into music lessons or Y.M.C.A. summer camp. These houses

[15] Incongruities appear in some of these homes. There is not invariably a correspondence between dirt and low wages; in one of the poorest shacks the man was earning $45 a week. In another very dirty house almost totally without furniture, the housewife, dressed in a soiled and badly torn gingham dress, with wide runs in her black cotton stockings and run-down heels, was at work at an electric washing machine.

[16] A leading local furniture store centers its selling talk for its expensive over-stuffed living-room suites about the claim that the home is judged by its living room—the place where guests and strangers are received. In that way, according to the manager, "working class families are persuaded to buy very expensive living-room suites and let the rest of the house slide a bit."

may be twenty years old and unadorned, with small rooms and a miscellany of used furniture. There is less likely to be a radio than in the more prosperous working class home, but one may come upon a copy of Whistler's portrait of his mother or a water-color landscape and a set of Dickens or Irving in a worn binding; the rugs are often more threadbare than those in the living room of a foreman, but text-books of a missionary society or of a study section of the Woman's Club are lying on the mission library table.

To some more prosperous members of the business group their homes are a source of pride as they walk up a neatly paved, tree-bordered street to homes which are "the last word in the up-to-date small house." The house may be shingled or stuccoed, in a trim terraced yard. Everything from the bittersweet in the flower-holder by the front door to the modern mahogany smoking table by the over-stuffed davenport bespeaks correctness. The long living room opens by a double doorway into the dining room. Colors in rugs, chair coverings, curtains, and the elaborate silk shades of the standing lamps match. There are three or four pictures—colored photographs or Maxfield Parrish prints—hung precisely at the level of the eyes, a pair of candle-sticks on the sectional bookcase, and a few bowls and trays; the kitchen cabinet has every convenience. Here one sees the complete small house.[17] "It's so hard to know what to give our relatives for Christmas any more," said one woman; "they have their homes and their knickknacks and their pictures just as we have. It's hard to find anything new that they haven't got. We've stopped giving to our friends except just cards, but we have to give to the family."

A group of wealthy families live in "fine old places" in the

[17] An ambitious specimen of this general type is the six-room "ideal home" built and furnished for public view on "Meadow Lane" in Middletown's smartest subdivision in 1924—a $9,700 house, exclusive of the furnishings. This "Tudor-bethan" house of "stone, brick, and stucco" was "built to demonstrate up-to-the-minute building materials and household equipment." "An attractive feature of the living room is the black and gold damask wall panel hung from the decorated pole on special cords, against which is placed a long console table on each end of which are attractive imported pottery lamps. The Tuxedo davenport, the Cotwell, Windsor, and colonial chairs suggest the English idea. The sunfast gold and red gauze net with overhangs of taffeta in burnt orange, henna, brown, and soft green, makes a very restful and soft effect, hung to the floor under a hand-decorated cornice. . . .

"The dining room furniture is of Spanish Renaissance period design with

"East End" of town, some of them still in the houses where the husband or wife was born. These houses may be large, heavy brick or stone affairs with perhaps two stone lions guarding the driveway near the old hitching post and carriage block bearing the owner's name. Other families live in rambling, comfortable frame houses in this section, while still others are following the movement out to the newer college district. Here they build low homes of brick or field stone or of the white Dutch colonial type with every convenience in the way of plumbing and lighting and with spacious glassed-in porches.[18]

Whether the father of one of these families comes home from office or bank to the large parlors and library of the older type of house or to the ample long living room of the new, he is greeted by an atmosphere of quiet and space. The wide rooms, soft hangings, old mahogany, one-toned rugs or deep-colored Orientals, grand piano, fireplaces, cut flowers, open book-shelves with sets of Mark Twain and Eugene Field and standard modern novels, the walls hung with prints of the Bargello, St. Mark's, "Mme. Lebrun and Her Daughter" [19] may be combined with certain individual touches, a piece of tapestry on the wall, a picture not seen elsewhere, a blue Chinese bowl.

Here, then, in this array of dwellings, ranging from the mean and cluttered to the spacious and restful, Middletown's most "sacred" institution, the family, works out its destiny.

chair seats of red mohair to blend with the Italian red wool damask draperies. Of especial note is the wrought-iron lantern torchier.

"The breakfast room is furnished in bright mandarin red in sharp contrast with the soft gray tones of the walls and woodwork.

"A pleasant atmosphere is imparted to the master's bedroom by the rose dotted Swiss curtains hung under a cornice to match the floral design of the furniture. The twin beds, vanity dresser and dresser with separate mirror are of the colonial turned-post design finished in buff and brown enamel. The guest bedroom is furnished in American walnut of Tudor design with draperies and bed-set of soft green satinette.

"A harmonious effect is created in the third bedroom with the imported English chintz curtains, the rag rugs, and a Jenny Lind spool-turned day-bed, toilet table, and chest of drawers finished in Sheraton brown mahogany."

[18] None of these families lives stylishly—with liveried chauffeurs and similar paraphernalia. The general absence of ostentation in standards of living is apparently due at least in part to the relative simplicity of a small group of the wealthiest families.

[19] Pictures serve as a species of furniture in most Middletown homes, as do the glassed cases of books in some. Cf. the discussion of art in Middletown life in Ch. XVII.

Within the privacy of these shabby or ambitious houses, marriage, birth, child-rearing, death, and the personal immensities of family life go forward. Here, too, as at so many other points, it is not so much these functional urgencies of life that determine how favorable this physical necessity shall be in a given case, but the extraneous detail of how much money the father earns.

Roughly speaking, a family has to sacrifice well over its entire income for a single year to own a home, or the income of every fourth week to rent a home. When the families, big and little, are shuffled about and sorted according to this monetary device there emerge the "East End," "Riverside," the "South Side," "Neely Addition," and "Industry" with an array of 9,200 homes as follows: [20]

Percentage of total homes	Value	Monthly rent
100		
7	$7,000 and over	$55 and over
22	$4,500-$7,000	$40-$55
44	$2,500-$4,500	$25-$40
27	Less than $2,500	$10-$25

In this connection it should be borne in mind that, as shown in Chapter VIII, only between 12 and 15 per cent. of those engaged in earning Middletown's living reported to the Federal Government in 1923 incomes, before making allowed deductions, of more than $1,000 if single or $2,000 if married; and the median income of the sample of 100 working class heads of families, including nine foremen, was $1,303.10.

The process by which under the institutional rules of the game a house gets built and lived in presents a complex pattern. There is a deep-rooted sentiment in Middletown that home ownership is a mark of independence, of respectability, of belonging, a sentiment strengthened by the lag in house building during the war years and the resulting housing shortage which has made purchase necessary in many cases to keep

[20] Based upon a study by the Commercial Engineering Department of the State Bell Telephone Company in 1922-23, revised by two local real estate experts to approximate more closely local valuations and rentals.

a house from being sold over a family's head. The greater purchasing power of current incomes noted in a preceding chapter,
the decline in the profitableness of renting as an investment,
and the widespread ownership of automobiles which makes
propinquity of work and home less essential, have all facilitated this drift toward ownership. Another important factor
working in the same direction is the development of credit
facilities for financing home-buying. In 1890, as noted above,
a young man of good character with $350 in cash failed to
discover any way to finance the purchase of a $750 building
lot. Today the building and loan association is the working
man's way *par excellence* of achieving a home: "The great
function of building and loan associations is to provide homes
for the masses of laboring people, to give them the opportunity
to pay for these homes with their daily and weekly savings." [21]
A worker who desires to purchase or build a house can do
so through a building and loan for a payment of $0.25 weekly
for each $100.00, the $0.25 covering both interest and principal.[22] The first infant building and loan association opened in
Middletown in 1889, when twenty-eight members paid in $230.
A year later its assets were $4,471.47. So slow was the diffusion
of the building and loan idea that five years later its assets had
risen only to $36,068.98; in 1900 it was $142,621.34; in 1910,
$678,428.50; while in 1924 it was $2,733,667.92. This one
company had, at the close of 1924, 7,090 members. The head
of the largest building and loan association in the city estimates
that 75-80 per cent. of the annual grist of new houses constructed locally today are financed through the four building
and loan organizations. According to this man's estimate, 85
per cent. of all those using the building and loan associations
in Middletown are working men. Another increasingly com-

[21] *Report of the* [State] *Building and Loan Department* ([State] Year
Book, 1920), p. 179.
[22] According to Ely, "The American method of acquiring a home is to
buy the site, gradually pay for it, then to mortgage it through a building
and loan association or otherwise, to construct the home with the aid of the
mortgage and gradually to extinguish the mortgage. We have no statistics
to give us accurate information about the number who acquire homes in
this way, but it is a familiar observation that this may be described as the
American method. The present writer, who has made careful observations
for a good many years, would say that in a city of 30,000 or 40,000 inhabitants in the Mississippi Valley this might represent the method in nine-
tenths of the cases where home ownership is attained." (Foreword to *Mortgages on Homes, 1920*, Census Monograph, p. 13.)

mon method today of becoming the owner of a home is the contract-for-deed plan whereby one pays for his home by the month at a rate approximately 50 per cent. higher than the rent for such a house would be, meanwhile occupying the house and keeping up taxes and all other costs. If one misses two months' rent in succession the home automatically reverts, with all improvements that the buyer may have put into it, to the holder of the title.[23]

Over against these factors inviting home ownership are the never before equaled array of bidders against the home for Middletown's dollars. The manager of a local company that has sprung up to finance automobile purchases states that a working man earning $35 a week and buying a car frequently aims to use one week's wages each month in paying for the car. Higher education for children is another serious new competitor.

A further important deterrent to ownership is the multiplication of institutional hurdles which must be cleared in building a house. A stranger to Middletown's ways might well be puzzled by repeated references to the "housing shortage" which hinders many Middletown families from finding suitable homes. The head of the Social Service Bureau might tell him of many families having to double up, of married children bringing their families home to live because they have no home of their own, of forty families living in shacks in a poor quarter called "Shedtown." Yet at the Advertising Club's Friday luncheons he might sit beside a Middletown builder of houses who has great quantities of wood and nails and all other materials as well as idle carpenters and masons skilled in fabricating houses. Across the table sits another man who speaks of owning many desirable building lots, suitable locations for homes, which he is

[23] The spread of this last method, like the facilitation of many other credit purchases, is leading certain families into difficulties: they get their living under a set of rules that require the suspension of play when the game becomes unprofitable to the owners of industrial plants, whereas coincidentally buying a house under this contract-for-deed system operates under rules requiring uninterrupted payment. Untrained in such matters as the amount a family can afford and coaxed by constant pressure to buy, it is not surprising that ventures in home ownership not infrequently end in shipwreck. There is much bad feeling among the working class of Middletown over the operation of this system. An influential business man spoke bitterly of a man prominent in the religious life of the city who "has grown rich selling houses that way—making people pay up the first of each month and squeezing them hard until eventually they may pay up enough to allow them legally to take the house out of his hands to a building and loan."

eager to dispose of. Why then a housing shortage if all the necessary ingredients for making an abundance of shelters are at hand and workmen who know how to build them are waiting idle?

Standardized large-scale production, the new habit in industry that makes Middletown's large automobile parts shops possible, is coming very slowly in the complex of tool-using activities concerned with making houses; the building of homes is still largely in the single-unit handicraft stage. Likewise continuous production, which is increasingly utilized in Middletown's industries, e.g., the abolition of the two-month summer shut-down and single-shift day in the glass factories of 1890 in favor of twenty-four-hour production twelve months in the year today, is not yet in operation in the building trades, for house-building is still carried on actively for only a part of the year.[24] All of which bears out Kipling's words—

> "I tell you this tale, which is strictly true,
> Just by way of convincing you
> How very little, since things were made,
> Things have altered in the building trade."

But it is not seasonal, single-unit production that Middle-town regards as the tight neck of the bottle:

"It is a widely known fact," says a Middletown newspaper editorially, "that it is almost impossible to induce anybody to construct rental houses here because it is difficult to obtain a reasonable percentage of profit out of them owing to the tremendous expense of owning them. Rents are high, far too high for the city's good, and still real estate owners are unable profitably to build dwellings even though when industrial conditions in Middletown are right there is always a great shortage of dwellings.

"It is doubtful if 10 per cent. of the new homes erected in [Middletown] last year [1923] were built for rental purposes."

And again the same paper throws up its hands in the face of this "long-standing need in the city" for "more and better homes" at a price local families can pay: "One who can solve this problem would be [Middletown's] principal benefactor." It decries recent efforts "to meet the situation . . . by the construction of

[24] The active building season, the country over, is said to be limited to less than 200 days a year. Permits were issued for only four new dwellings in Middletown in the month of February, 1924, although 214 such permits were issued during the entire year.

houses built of cheap material. . . . It is notably true that some of the dwellings erected here of late hardly can be expected to be serviceable fifteen years. It is not economy to buy a pair of two-dollar shoes that will give out in two months. . . . What is to be done? The answer may lie in some new method of financing. . . . We must eventually devise a way whereby men with the capital to spare can find it a good investment to build houses and apartment buildings for rental purposes." [25]

Middletown builders, bankers, and carpenters, in discussing the housing shortage, talk of "high rents," "first mortgages," "second mortgages," "plumbers' unions," "contracts for deed," "clear titles," "sub-contracts," "new real estate subdivisions," "street improvement assessments," and "fair returns on your investment." It seems to be not so much a lack of raw materials or skill that creates the "social problem" of a housing shortage as this intricate network of institutional devices through which a citizen of Middletown must pick his way in undertaking the building of houses for others or in trying to secure a home for his own use.

Although in the six-room "ideal home" described above the work was, for advertising purposes, somewhat more widely spread than usual among different groups of workers, the following list of those who helped construct it is illuminating as regards present-day conditions in the Middletown building trades:

The site was part of a new subdivision exploited by local real-estate men.

A landscape architect was brought from Chicago to lay out the subdivision.

The house was built by the owners of the real estate.

The project was financed by local banks.

The plan was made by an architect.

Construction was under the general supervision of a man whose business is described in the local directory as "Designer and Builder, Real Estate, Investments, General Insurance."

The above man is busy at many things and had therefore another man on the job as general supervisor of work and materials.

The foundation was laid by a bricklayer sub-contractor's men.

[25] Building houses to rent was a regular and lucrative type of investment in the Middletown of 1890. A local "boom book" of 1892 boasted to prospective citizens and investors that "A house costing $600 can easily be rented for $10 a month."

The stone base topping the foundation was sub-contracted to another contractor and his men.

The hardwood floors were sanded and surfaced by still another contractor specializing in this work and owning an expensive machine used for this work.

The stucco and plaster work was done by another sub-contractor and his men.

An electrical contractor installed the wiring.

Plumbing was installed by a plumbing contractor.

A painting contractor and his men did the painting and decorating.

A tinware contractor and his men did the tin work.

A group of heating specialists installed the furnace.[26]

The upshot of all the pullings and haulings back and forth of these and other institutional factors is that somewhat more people own their homes today than a generation ago. The Federal Census shows 54 per cent. of the homes of the city as rented in 1920 as against 65 per cent. in 1900.[27] The percentage of renters had probably dropped even further by 1924. Only a third of the 123 working class families for whom data on this point were secured rented their homes, and only a third of the eighty-one who owned their homes had mortgages on them. "For Rent" signs had become so scarce in 1924 that they were the subject of humorous comment in the press:

"Vacant bungalows are so rare that a small house with a 'For Rent' sign would make an attractive display in a museum."

[26] Speaking of conditions throughout the country, John M. Gries, Chief of the Division of Building and Housing of the United States Department of Commerce, says: "Under the present organization of labor more than a dozen skilled trades may be employed on the erection of even a small dwelling. Not many decades ago practically all work was done by four skilled trades. Most houses were built of wood and the carpenter was the principal skilled workman. Today work is widely subdivided and the scheduling of work is most difficult for the small builder. . . . Probably 75 per cent. of all the single-family houses are built by contractors erecting fewer than ten houses annually." ("Housing in the United States," *Journal of Land and Public Utility Economics*, Vol. I, No. 1, January, 1925.)

[27] Seventy-three per cent. of 247 Middletown working men in the wood, iron, and glass plants of the city in 1891 lived in rented homes, according to the *Fourth Biennial Report* of the State Department of Statistics, 1891-92 (pp. 57, 130, 319). These figures may be somewhat heightened by boom conditions. According to the same source, 62 per cent. of 4,009 workers throughout the state, chiefly in the same three groups of industries, lived in rented homes in 1891, while the Federal Census for 1890 showed 60 per cent. of the homes in cities in the state with 8,000 to 100,000 inhabitants to be rented.

Another item, under the heading "Times Have Changed," says: "Not so many years ago it was the custom for the 'to-be brides and bridegrooms' to go rent-hunting just prior to their marriage. In those days many lawns were decorated with signs advertising the fact that the house could be rented. But in these days, the young lovers would be forced to spend several months looking for one of these signs, since they have almost disappeared. The signs now found are 'For Sale.'"

If this slow trend towards home ownership represents a permanent drift, it may make itself felt in such ways as the diffusion throughout a wider section of the population of the place solidarity so ardently fostered by Chamber of Commerce and civic clubs, and the slowing up to some extent of the disintegration of the neighborhood as a social unit, to be noted below.[28] In this connection, however, attention should be called to the fact that the tendency is for the place-roots of the working class to be somewhat more shallow than are those of the business class: less than half of the group of business class families interviewed and three out of each five of the working class families interviewed had moved during the nearly five years from January 1, 1920, to October 1, 1924; one in ten of the business group had moved more than once as against one in four of the workers; none of the former had moved more than twice, whereas every sixth family of the working class sample had moved more than twice. The working class group of today appear to exhibit more mobility than the families of their mothers.[29]

[28] Cf. Ch. VII for the decline of the custom of workers' living in close proximity to the plants where they work.

[29] See Table IX. Data of this sort, based in part upon memory, must naturally be used carefully. The fact that a move is a rather large event in the family annals insures a reasonably good chance of the facts being remembered accurately.

Chapter X

MARRIAGE

In each of Middletown's homes lives a family, consisting usually of father, mother, and their unmarried children, with occasionally some other dependents. These family groups are becoming smaller. According to the Federal Census, which defines a family as one person living alone or any number of persons, whether related or not, who live together in one household, Middletown families shrank from an average of 4.6 persons in 1890 to 4.2 in 1900, to 3.9 in 1910, and 3.8 in 1920. Both the decrease in the number of children and the decline in the custom of having other dependents in the home are factors in this change.[1] The forty business class families interviewed by the staff in 1924 averaged 4.7 persons and the 124 working class families 5.4, but only those families were interviewed which had one or more children of school age.[2]

Within the walls of each house this small family group carries on the activities concerned with sex, child-rearing, food, clothing, sleep, and to some extent play and religion. These activities center about the institution of marriage.

The country over, a smaller percentage of the population is unmarried today than a generation ago, and while earlier fig-

[1] Fewer "old-maid" sisters live with married relatives now when women commonly work outside the home for pay, move about town freely at night unescorted, and live in small flats of their own. Smaller houses without "spare" rooms are diminishing the custom, according to the head of the Social Service Bureau, of having elderly parents live in the homes of married children. This modern tendency towards the non-support of parents was openly recognized in 1920 when, among the recommendations made by the State Board of Charities and Correction to the Legislature, was one stating that "We believe that there should be legislation to prevent the abandonment of parents by children who are able to support them."

The presence of outsiders as boarders at the family table is somewhat less sanctioned today, though roomers are still not uncommon; business class families who take in roomers are likely to do so to help pay children's expenses in college, and this explanation lessens any social stigma that may attach to this procedure.

[2] See Table X for distribution of these families by size.

ures by which the trend within Middletown can be observed are not available, in 1920 the city had a smaller proportion of both males and females single than had either state or nation:

	Percentage single males constitute of all males aged 15 and over		Percentage single females constitute of all females aged 15 and over	
	1890	1920	1890	1920
United States, all classes .	41.7	35.1	31.8	27.3
" " urban	Not available	35.5	Not available	29.0
State, all classes	38.9	30.9	29.6	23.9
" urban	Not available	31.1	Not available	24.5
Middletown	Not available	28.5	Not available	20.8

While only 22.8 per cent. of the population of the urban United States aged fifteen to twenty-four in 1920 were married, 31.4 per cent. of this group in Middletown were married. It would appear that more people are marrying young in Middletown today.[3] Explanation of this apparent drift toward more and earlier marriages may lie in part in such changes, noted elsewhere, as the cessation of apprenticeship, which gives a boy of eighteen a man's wages at a machine, the increased opportunities for wives to supplement the family income by working, the relatively greater ease and respectability of dissolving a marriage today, the diffusion of knowledge of means of contraception, and the growing tendency to engage in leisure-time pursuits by couples rather than in crowds, the unattached man or woman being more "out of it" in the highly organized paired social life of today than a generation ago when informal "dropping in" was the rule.

[3] While the single males in the United States declined between 1890 and 1920 from 41.7 to 35.1 per cent. of all males fifteen and over, and the percentage of single males aged thirty-five to forty-four showed an actual slight increase from 15.3 to 16.1, the single males aged fifteen to nineteen dropped from 99.5 to 97.7 of all males of this age and the twenty to twenty-four group from 80.7 to 70.7. In other words, the increase in marriage is occurring in the younger age groups. Likewise, throughout the state in which Middletown is situated the percentage of single men in the group aged thirty-five to forty-four increased from 11.5 per cent. to 13.6 between 1890 and 1920, while those aged fifteen to nineteen fell from 99.6 to 97.3 per cent. and those aged twenty to twenty-four from 80.6 to 66.5 per cent. It is noteworthy that Middletown had, in 1920, 64.9 per cent. of its population fifteen years old and over married, while in all cities in the United States of 2,500 and over there were but 58.3 per cent. of this age group married. Neither race nor age or sex distribution of the population accounts for the difference between the age of marriage in Middletown and in the urban United States. Probably various cultural factors are responsible.

Marriage consists in a brief ceremonial exchange of verbal pledges by a man and woman before a duly sanctioned representative of the group. This ceremony, very largely religious in the nineties, is becoming increasingly secularized. In 1890, 85 per cent. of the local marriages were performed by a religious representative and 13 per cent. by a secular agent, while in 1923 those performed by the religious leaders had fallen to 63 per cent., and the secular group had risen to 34 per cent. of the total.[4] A prominent local minister accounted for the prevalence of divorces in Middletown in 1924 by the fact that "there are too many marriages in secular offices away from the sanctity of the churches." The marriage ceremony relaxes the prohibition upon the mutual approaches of the two persons to each other's persons and as regard the sexual approach makes "the wrongest thing in the world the rightest thing in the world." [5] The pair usually leave the homes of their parents at once and begin to make a home of their own; the woman drops the name of her father for the name of her husband.[6]

A heavy taboo, supported by law and by both religious and popular sanctions, rests upon sexual relationships between persons who are not married. There appears to be some tentative relaxing of this taboo among the younger generation, but in general it is as strong today as in the county-seat of forty years ago. There is some evidence that in the smaller community of the eighties in which everybody knew everybody else, the group prohibition was outwardly more scrupulously observed than today. A man who was a young buck about town in the eighties says, "The fellows nowadays don't seem to mind being seen on the street with a fast woman, but you bet we did then!" That all was not serene underneath, particularly after the influx of population accompanying the gas boom, appears from various items in the local press in 1890:

"In looking over the Board of Health statistics . . . I noticed in the birth records . . . that some of our prominent citizens are

---

[4] Two per cent. of the 1890 officiating persons and 3 per cent. of the 1923 group could not be identified. Figures here are for the entire county.

[5] Advances prior to marriage are traditionally made entirely by the man, but, as suggested in Ch. XI, there is an increasing aggressiveness on the part of the girls in the activity preliminary to mating.

[6] There is occasional talk in the community of a woman's "keeping her own name," but no woman in Middletown follows this practice, and it is sharply frowned upon by the group.

given the credit or being the father of offspring of which women of very loose character are the mothers."

An editor "hopes that something can be done about the large number of 'street walkers' which are to be seen every evening."

"This morning an officer requested the *Times* to state that he had of late seen a number of married men in company with disreputable characters and that after this date every one so detected will be arrested and exposed."

By 1900 the rough-and-tumble industrial influx is reflected in the press in such headlines as "The Bowery Outdone in [Middletown]": "A traveling man who has visited both the universally notorious Bowery in New York City and the locally notorious High Street theater asserts that for downright lewdness and immorality the former is outdone by the latter."

Editorials posed the question, "What does [Middletown] need?" and answered: "Mayors and police officers who will guard boys and girls from dens of evil like High Street theater."

The city, according to a local historian, bore an "ill repute during its early career as a manufacturing city."

A former proprietor of one of the largest saloons in the city estimates the number of houses of prostitution about 1890 as twenty-five, and an old iron puddler estimates twenty, both agreeing on four to eight girls per house. In 1915 a state act was passed providing for injunction and abatement of houses of prostitution, thereby driving the institution underground.[7] Today conditions fluctuate in Middletown. Within ten years a group of public officials is reported to have conducted Middletown as a "wide-open town" in which city officials, later sent to the Federal penitentiary, were alleged to have a financial interest in "the red-light district." At the present time there are reported to be only two or three fly-by-night, furtively conducted houses of prostitution, catering exclusively to the working class, but a comparison with 1890 on this point is fruitless,

[7] The description in the *Cleveland Hospital and Health Survey* of this new type of prostitution to some extent among girls nominally employed at other occupations probably applies in a general way to Middletown; it concludes, "How far the activities of such amateur prostitutes make up for the reduction in the activities of the professionals no one knows. The doctors testify, however, that a large number of their men patients claim to have been infected by such amateurs." Part V, *Venereal Disease* (Cleveland; Cleveland Hospital Council, 1920), pp. 420-1.

because, as the judge of the juvenile court points out, "the automobile has become a house of prostitution on wheels."

The choice of a mate in marriage is nominally hedged about by certain restrictions—legal, religious, and customary. Legal stipulations, substantially the same as a generation ago, prohibit marriage between a white and a negro, by an insane person, an imbecile or epileptic, by a person having a transmissible disease, or, within certain limits, by a male who has within five years been a public charge, by a person whose former marriage has not been dissolved, by a person under the influence of liquor or narcotics, and by a man under eighteen and a woman under sixteen years. Other requirements implicitly recognized by law appear from the allowable grounds for dissolution of a marriage: sexual exclusiveness, living together in the same home, financial support of the wife by the husband, sufficient mutual consideration to exclude "cruel" treatment, sufficient "sobriety" and "morality" to avoid charges of "habitual drunkenness" and "criminal conviction." Religious requirements, today as in the nineties, vary somewhat from one religious group to another, but concern two main points: the nominal prohibition by Catholics of marriage "outside the Church" and a corresponding though weaker sentiment among Protestants against marriage to a Catholic; and, second, a varying but somewhat lessening emphasis upon the permanence of marriage whereby a few religious leaders refuse to remarry a divorced person. Some ministers would refuse to marry persons living in "open [sexual] sin," though the marriage ceremony is commonly regarded as the accepted means of regularizing such individuals.

Further informal demands, made by the fluid sentiments of the group, have apparently altered little since the nineties, although they have been given somewhat greater legal recognition.[8] Foremost among these is the demand for romantic love as the only valid basis for marriage. Theoretically, it is the mysterious attraction of two young people for each other and that alone that brings about a marriage, and actually most of Middletown stumbles upon its partners in marriage guided

---

[8] Loss of affection after marriage was not legally recognized as sufficient reason for dissolving a marriage until recent years, but in 1924 divorces were granted to couples who came into court frankly saying, "We have no affection for each other and do not want to live together," and "She says she does not love me and does not want to live with me."

chiefly by "romance." [9] Middletown adults appear to regard romance in marriage as something which, like their religion, must be believed in to hold society together. Children are assured by their elders that "love" is an unanalyzable mystery that "just happens"—"You'll know when the right one comes along," they are told with a knowing smile. And so young Middletown grows up singing and hearing its fathers sing lustily in their civic clubs such songs as:

> "It had to be you,
> It had to be you.
> I wandered around and finally found
> The Somebody who
> Could make me true, could make me blue,
> And even glad just to be sad
> Thinking of you.
> Some others I've seen might never be mean,
> Might never be cross, or try to be boss,
> But they wouldn't do,
> *For nobody else gave me a thrill.*
> *With all your faults I love you still.*
> It had to be you, wonderful you,
> It had to be you."

And yet, although theoretically this "thrill" is all-sufficient to insure permanent happiness, actually talks with mothers revealed constantly that, particularly among the business group, they were concerned with certain other factors; the exclusive emphasis upon romantic love makes way as adolescence recedes for a pragmatic calculus. Mothers of the business group give much consideration to encouraging in their children friendships with the "right" people of the other sex, membership in the "right" clubs, deftly warding off the attentions of boys whom

[9] Cf. the discussion in Ch. VI of the equally casual method by which Middletown youths stumble upon the kind of work by which they forever after earn their living. The casualness of procedure in both these cases is probably traceable in part to the same inherited conceptions regarding the individual's "freedom" and "rationality."

The close identification of love with the religious life of the group has tended to import into courtship some of the same inscrutability that envelops the religious life of Middletown. By the same token the religious taboo upon "carnal love" has carried over into the situation so that, although sexual exclusiveness in marriage is demanded both by law and custom, virtually no direct consideration is given prior to marriage to the physical and sexual compatibility of the two contracting parties. Cf. the discussion of sex instruction to the young in Chapter XI.

they regard it as undesirable for their daughters to "see too much of," and in other ways interfering with and directing the course of true love.

Among the chief qualifications sought by these mothers, beyond the mutual attraction of the two young people for each other, are, in a potential husband, the ability to provide a good living, and, in a wife of the business class, the ability, not only to "make a home" for her husband and children, but to set them in a secure social position. In a world dominated by credit this social function of the wife becomes, among the business group, more subtle and important; the emphasis upon it shades down as we descend in the social scale until among the rank and file of the working class the traditional ability to be a good cook and housekeeper ranks first.

"Woman," as Dorothy Dix [10] says, "makes the family's social status. . . . The old idea used to be that the way for a woman to help her husband was by being thrifty and industrious, by . . . peeling the potatoes a little thinner, and . . . making over her old hats and frocks. . . . But the woman who makes of herself nothing but a domestic drudge . . . is not a help to her husband. She is a hindrance . . . and . . . a man's wife is the show window where he exhibits the measure of his achievement. . . . The biggest deals are put across over luncheon tables; . . . we meet at dinner the people who can push our fortunes. . . . The woman who cultivates a circle of worth-while people, who belongs to clubs, who makes herself interesting and agreeable . . . is a help to her husband. . . ."

Not unrelated to this social skill desired in a wife is the importance of good looks and dress for a woman. In one of

[10] References will be made to Dorothy Dix from time to time in the following discussion of what the group demands of marriage. Day after day two columns of syndicated advice to "Desolate," "A Much-disturbed Husband," "Young Wife," etc., appear in the leading Middletown paper from this elderly lady. This is perhaps the most potent single agency of diffusion from without shaping the habits of thought of Middletown in regard to marriage and possibly represents Middletown's views on marriage more completely than any other one available source. Of the 109 wives of working men interviewed giving information on this point, 51 said that they read Dorothy Dix regularly and 17 occasionally, while of 29 wives of the business class answering on this point 16 read this column regularly and 10 occasionally. Her advice is discussed by mothers and daughters as they sew together at Ladies' Aid meetings and many of them say that her column is the first and sometimes the only thing which they read every day in the paper. Her remarks were quoted with approval in a Sunday morning sermon by the man commonly regarded as the "most intellectual" minister in town.

Marion Harland's *Talks,* so popular in Middletown in the nineties, one reads, "Who would banish from our midst the matronly figures so suggestive of home, comfort, and motherly love?" Today one cannot pick up a magazine in Middletown without seeing in advertisements of everything from gluten bread to reducing tablets instructions for banishing the matronly figure and restoring "youthful beauty." "Beauty parlors" were unknown in the county-seat of the nineties; there are seven in Middletown today.

"Good looks are a girl's trump card," says Dorothy Dix, though she is quick to add that much can be done without natural beauty if you "dress well and thereby appear 50 per cent. better-looking than you are . . . make yourself charming," and "cultivate bridge and dancing, the ability to play jazz and a few outdoor sports."

Emphasis upon the function of the man in marriage as "a good provider" and of the woman as home-maker, child-rearer, and, among the bulk of the business group, social pace-setter, is far-reaching as affecting the attitude of the sexes toward each other. In general, "brains" tend to be regarded as of small importance in a wife; as one of the city's most "two-fisted" young business men announced to the high school seniors at a Rotary high school "chapel," "The thing girls get from high school is the ability to know how to choose a 'real one' from a 'near one.' When a girl gets around eighteen or so I begin to expect her to get married."

Middletown husbands, when talking frankly among themselves, are likely to speak of women as creatures purer and morally better than men but as relatively impractical, emotional, unstable, given to prejudice, easily hurt, and largely incapable of facing facts or doing hard thinking. "You simply cannot criticize or talk in general terms to a woman," emphatically agreed a group of the city's most thoughtful men. "There's something about the female mind that always short-circuits a general statement into a personal criticism." A school official, approached regarding the possibility of getting a woman on the school board, replied that "with only three people on the board there isn't much place for a woman." In a group of prominent Middletown men a suggested new form of social grouping came up in conversation and was promptly downed because "the women couldn't abide by it. Woman is the most unselfish creature on earth within her family, but with outsiders she is quick to imagine snubs to her family, bristle up, and become unsocial."

Middletown wives appear in part to accept the impression of them that many of their husbands have. "Men are God's trees; women are his flowers," and "True womanliness is the greatest charm of woman," the recent mottoes of two of the local federated women's clubs, suggest little change from the prevailing attitude reflected in a commencement essay in 1891, "Woman Is Most Perfect When Most Womanly." [11] At a local political dinner the talk about one of the tables turned to women's smoking and a woman politician said with an air of finality: "Women have to be morally better than men. It is they who pull men up or cause their downfall." Women, on the other hand, are frequently heard to express the opinion, accompanied by a knowing smile, that "Men are nothing but big little boys who have never grown up and must be treated as such."

In general, a high degree of companionship is not regarded as essential for marriage.[12] There appears to be between Middletown husbands and wives of all classes when gathered together in informal leisure-time groups relatively little spontaneous community of interest. The men and women frequently either gravitate apart into separate groups to talk men's talk and women's talk, or the men do most of the talking and the women largely listen. Even since women have been allowed to vote with men some tendency persists for women of all classes to depend in such practical matters upon the opinions of their

[11] The following is part of a tribute to woman read with general approval at the close of a meeting of another of the local women's federated clubs in 1924: "There is a being, the image and reflection of whom is ever present in the mirror of my soul. Her words are like enchanted echoes in a beautiful dell and her laughter like the sweetness of the bursting magnolia and her beauty like the smiling violet and the laughing morning glory. The sound of her footstep is like that of a messenger bearing gifts from a queen, and her touch like the gentle zephyrs fanning the tired brow of a weary traveler, and her presence like an altar of holiness and benediction. That spirit has taught me to revere heaven's divinest gift to the world—womanhood."

[12] It may not be wholly fantastic to surmise that there may be some significance for the understanding of the basis of local marital association in the hierarchy of terms by which local women speak of their husbands. There is a definite ascent of man in his conjugal relations as one goes up in the social scale, from "my old man" through "the man," "he" (most frequent), "the mister," "John," "My husband," to "Mr. Jones." The first four are the common terms among the working class families and the last two among business class families.

husbands "coming in from the outside," as Dorothy Dix puts it, "with the breath of the fighting world about them."

Companionship between husband and wife in sheer play varies greatly in different families; among the business group who belong to the country club, e.g., husbands frequently play eighteen holes of "real golf" with a male "foursome" Sunday morning and possibly a concessionary "one round with the wife in the afternoon just to make her feel good." One wife who makes "a definite effort to do things" with her husband says that she has achieved this at the cost of cutting herself off from much of the routine social life of the community:

"My husband, my children, and what community work I have time for after them are my job. I have gradually withdrawn from the social activities of the wives of my husband's business associates because most of these women seem absorbed in activities that do not include their husbands. That is just the sort of thing that leads to the break-up of families and I don't see why I and my family should be exempt from the things that befall other people."

One of the commonest joint pursuits of husbands and wives is playing cards with friends. A few read aloud together, but this is relatively rare, as literature and art have tended to disappear today as male interests. More usual is the situation described by one prominent woman: "My husband never reads anything but newspapers or the *American Magazine*. He is very busy all day and when he gets home at night he just settles down with the paper and his cigar and the radio and just rests." The automobile appears to be an important agency in bringing husbands and wives together in their leisure, counteracting in part the centrifugal tendency in the family observable in certain other aspects of Middletown's life.[13]

Among the working class, leisure activities and other relations between married couples seem to swing about a somewhat shorter tether than do those of business folk. Not infrequently husband and wife meet each other at the end of a day's work too tired or inert to play or go anywhere together; many of them have few if any close friends.[14] In families where there is some financial leeway there are plans for an addition to the

[13] Cf. Chs. XVII and XVIII for fuller discussion of these leisure-time pursuits.

[14] Cf. Ch. XIX on the decline of intimate friendships.

house or perhaps the possibility of normal school for the children, which are spontaneous centers of interest and conversation between husband and wife; in the sixty out of 122 families reporting who had automobiles, these and the trips they make possible form a chief center of interest; in some cases there is talk of lodge affairs or the movies. But if, as in many families, the necessities of shelter and food overshadow other plans, such conversation as there is may be of a bickering sort, or may lapse into apathetic silence. In a number of cases, after the interviewer had succeeded in breaking through an apparently impenetrable wall of reserve or of embarrassed fear, the housewife would say at the close of the talk, "I wish you could come often. I never have any one to talk to," or "My husband never goes any place and never does anything but work. You can talk to him, but he never says anything. In the evenings he comes home and sits down and says nothing. I like to talk and be sociable, but I can hardly ever get anything out of him." [15]

This frequent lack of community of interests, together with the ideas each sex entertains regarding the other, appears in many families in a lack of frankness between husband and wife, far-reaching in its emotional outcome. "One thing I always tell my young men when they marry," said the only one of the six leading ministers who gives any instruction to people he marries, "is that they must get over any habit of thinking that they must be frank and tell everything they know to their wives." Dorothy Dix urges:

" 'Let well enough alone' is a fine matrimonial slogan and as long as husband and wife are good actors it is the part of wisdom for their mates not to pry too deeply into the motives that inspire their conduct. . . . What we don't know doesn't hurt us in domestic life, and the wise do not try to find out too much." And again, "Nothing does more to preserve the illusions that a man and woman have about each other than the things they don't know."

Traditionally this institution of marriage is indissoluble. "What God hath joined together let no man put asunder," commands the religious marriage ritual. But the trend toward

[15] This is the kind of maladjustment which may later figure in the divorce courts as "cold, grouchy, never says anything" or may lead to more violent reactions which figure as "cruel treatment."

secularization noted in the performance of the marriage ceremony appears even more clearly in the increased lifting of the
taboo upon the dissolution of marriage. With an increase, between 1890 and 1920, of 87 per cent. in the population of the
county in which Middletown is located, the number of recorded
divorces for the four years 1921-24 has increased 622 per
cent. over the number of divorces in the county in the four years
1889-92.[16] There were nine divorces for each 100 marriage
licenses issued in 1889 and eighteen in 1895.[17] After fluctuat-
ing about the latter figure for fifteen years, the total in 1909
first passed twenty-five divorces for each 100 marriage licenses,
and six years later thirty. In 1918 it was fifty-four for each
100 marriage licenses; 1919, thirty-nine; 1920, thirty-three;
1921, fifty-five; 1922, forty; 1923, thirty-seven; 1924, forty-
two.[18]

The frequency of divorces and the speed with which they
are rushed through have become commonplaces in Middletown.
"Anybody with $25 can get a divorce" is a commonly heard
remark. Or as one recently divorced man phrased it, "Any one
with $10 can get a divorce in ten minutes if it isn't contested.
All you got to do is to show non-support or cruelty and it's a
cinch." Following are typical Sunday morning headlines after
what the newspapers are wont to call the "usual Saturday morning divorce grind":

"Court hears eighteen divorce suits; ten decrees granted and
five dismissed, others pending." "Only one divorce case [out of
twelve] is dismissed; decrees granted in all other suits." "Another

[16] See Table XI.

[17] The State Biennial Reports do not give marriages prior to 1895; see N.
accompanying Table XI for source of these figures.

[18] These figures do not take into consideration the number of divorce suits
filed. In 1924, e.g., 477 married couples sought divorces in the county, while
644 couples married—seventy-four couples seeking to dissolve marriage for
each 100 couples seeking to marry. In March, 1925, fifty divorce actions were
filed and only forty-four marriage licenses issued. Middletown's behavior in
this regard is part of a national trend; according to the *Journal of Social
Hygiene,* "Thirty-one years ago the ratio of divorce to marriage was 1 to
14.8. Ten years later, in 1905, the ratio was one divorce for every 6.7 marriages. For approximately the last fifty years, the period for which fairly
accurate statistics are obtainable, the number of divorces has steadily tended
to approach the number of marriages." (Vol. XIII, No. 1, January, 1927,
p. 42.)

Middletown is not a Reno. There is no appreciable influx of people from
outside the county seeking divorces.

grist of divorces ground out by Judge ——." "Judge —— hears six complaints for divorce; five decrees issued." "Twenty-one divorce suits scheduled for hearing next Saturday." "Woes of nine couples heard; five husbands and three wives granted divorces; one suit dropped."

Such casual comment is in marked contrast to the attitude of a generation ago when the State Statistician in presenting the annual divorce figures for the state in the *Third Biennial Report* for 1889-90 says, "The following is the repulsive exhibit." Apparently this growing flexibility in attitude towards the marriage institution reacts back upon itself; one factor in the increasing frequency of divorce is probably the growing habituation to it.[19]

Some conception of the maladjustments underlying the decision of so many Middletown couples to terminate their marriage may be gained from the records of the courts,[20] combined with the comments of the women interviewed on their efforts at marital adjustment. "Cruel treatment" is the "cause" assigned for the largest percentage of divorces for the county in which Middletown is located; a comparison of the stated causes for which divorces were granted in the two four-year periods 1889-92 and 1919-22 shows that marriages dissolved because of alleged cruelty have increased from 30 per cent. to 52 per cent. of the total.[21] A charge of "cruelty" may cover al-

[19] Of the sixty-one wives of working men who expressed opinions concerning reasons for the prevalence of divorce, seventeen thought that marriage is "too easy" and five that divorce is "too easy." Fourteen gave "women's working" as the chief factor in divorces and fourteen "frivolity." The other eleven women gave as the principal reason: "unemployment," five; "sin," three; "extravagance," two; "wanting things they can't afford," one.

[20] See Table XII.

It is not intended here to take the conventional forms under which divorce cases are pled as anything more than very roughly suggesting the real issue between the husband and wife. Often the real reason is pretty thoroughly disguised. Thus one man who was divorced for "non-support" stated privately that "She and I split up over the G—— d—— Klan. I couldn't stand them around any longer."

[21] This trend towards relatively more divorces for "cruelty" is apparently part of a long-term national trend. According to *Marriage and Divorce 1887-1906*, "A comparison of the earliest five-year period, that from 1867 to 1871, with the latest, that from 1902 to 1906, shows that adultery has decreased in relative importance as a cause while cruelty has increased." (Department of Commerce Bulletin, No. 96, Second Edition, Revised and Enlarged, 1914, pp. 30-1.)

most any variety of marital maladjustment, and the increase in
divorces on this charge probably indicates chiefly a growing
flexibility which allows divorces on other than specific charges
such as "adultery" and "abandonment." It is impossible to say
to what extent this charge is connected with the sex relation,
but it seems probable that, in some cases at least, the connection is close. Many of the women interviewed voiced opinions
similar to the statement of one woman: "Women never used to
talk to each other about such things. Every woman used to
think that other women had to put up with what she did because
that is the way men are. Now they are beginning to wonder."
Time and time again the wife of a working man spoke with
obvious emotion of the fact that the responsibility for the prevention of pregnancy was placed entirely upon her by a non-
coöperative husband.

Traditionally, voluntary control of parenthood is strongly
tabooed in this culture, as is all discussion of sexual adjustment involved in mating, but this prohibition is beginning to be
somewhat lifted, a fact perhaps not unrelated to the increasing
secularization of marriage noted above. The widely divergent
habits of different persons in regard to control of parenthood
reveal strikingly the gap found in so many cases to exist between the habits of different groups of people living together
in the same community. All of the twenty-seven women of the
business class who gave information on this point used or believed in the use of some method of birth control and took it
for granted. Only one woman spoke of being uncertain as to
whether she had been wise in limiting her family as she had.
Of the seventy-seven wives of workers from whom information was secured on this subject, only thirty-four said that
they used any means of birth control; of these twelve were
"careful," two used primitive practices, only twenty used artificial means that might be considered moderately scientific,
and only half of these last employed means of the sort utilized by the business class. Of the forty-three not using any
means of birth control, fifteen vaguely approved of contraception but did not use it because they thought it unnecessary in
their individual cases; fifteen definitely disapproved; four were
ignorant of all contraceptives except such as their husbands
were unwilling to use; and nine were eager for some means of
control but totally ignorant of any.

The following answers are characteristic of the women who definitely disapprove of use of contraceptives:

"God punishes people who deliberately try not to have children."

"Large families are hard, but it's wrong to do anything about it. I'll take what's sent me."

"Abortion is murder and birth control is just as bad."

"I'd like a small family but it's not good for a person's health to do anything to prevent having children."

"God told Eve to be fruitful and multiply and if he had wanted her to regulate the number of children he would have told her so. I have had eleven children and there ain't nobody has better health than I have. He often says how much better off I am than if I had taken things as some do."

Judging from the fact that a number of those who had tried to use some kind of contraceptives had been unsuccessful in preventing conception and from certain vague statements which had to be listed as "Not answered," the following comments of some of these women seem to suggest an underlying bewilderment considerably more widespread and more pervasive of the rest of their lives than the figures taken by themselves would indicate:

One wife "hopes to heaven" she'll have no more children. She said that people talked to her about contraceptives sometimes, and she told "him" what they said, but he said it was none of their business. She had never dared ask him what he thought about birth control, but thought he disapproved of it. She would "die" if she had any more children but is doing nothing to prevent it.

"Men will not stand for use of things to prevent having children," said another. "This is one cause of divorce."

Another woman complained that the contraceptives which she obtained had not prevented her having a sixth child, and when asked why she did not try something else said, "I'm ashamed; I would never have asked for these if some one had not told me about them."

A mother of eight children had found that certain home-made contraceptives worked for three years and then she became pregnant again. "I could not believe it. I went to the doctor just all of a tremble. My sister-in-law says these things helped her but I guess nothing can help me. And I've been sick ever since the baby came."

A woman who has had five children in six years knew of no way to prevent pregnancy and said that her husband would not find out for her. "He doesn't care how many children I have."

A wife of twenty-two replied to the question about number of children: "We haven't any. Gracious, no! We mustn't have any till we get steady work. No, we don't use anything to prevent children. I just keep away from my husband. He don't care—only at times. He's discouraged because he's out of work. I went to work but had to quit because I was so nervous." [22]

Another woman, the wife of a molder, a tired-looking, neatly dressed woman of forty-seven with seven children, had recently had to go to work because her married daughter came back home with two children and the burden of a family of eleven was too much for the husband. "My daughter and her husband—he's a machinist—didn't know anything about birth control, and they had a second baby and then she insisted that they keep apart until his work was regular enough to support a larger family. He wouldn't and she left him and came home to us here while she was still nursing her second baby. I certainly believe in birth control! But I don't know anything about it. I never even heard of it until a little while ago. I sure wish I had known of it when I was young, for then he wouldn't be slaving away to support this big family and my daughter wouldn't be in all the trouble she's in."

The behavior of the community in this matter of the voluntary limitation of parenthood—in this period of rapidly changing standards of living, irregular employment, the increasing isolation and mobility of the individual family, growing emphasis upon child-training and upon education and other long-term family plans such as insurance and enforced home ownership on a time payment basis—presents the appearance of a pyramid. At the top, among most of the business group, the use of relatively efficacious contraceptive methods appears practically universal, while sloping down from this peak is a mixed array of knowledge and ignorance, until the base of ignorance is reached. Here fear and worry over pregnancy frequently

[22] This woman was not included in the 124 sample families as she had no children, but was one of those interviewed on unemployment.

walk hand in hand with discouragement as to the future of the husband's job and the dreaded lay-off.[23]

Although the court records show that the proportion of divorces for a listed cause of "non-support" has remained stationary since 1890, according to a Middletown lawyer handling many divorce cases, "seventy-five per cent. of the people seeking divorces are women, and two-thirds of the women ask divorce for non-support." Talks with the women interviewed would seem to indicate that economic considerations figure possibly more drastically than formerly as factors in divorce. This does not necessarily mean that the husband has failed to provide food and shelter for his wife, but it does indicate that in some way their economic adjustment has broken down and that the wife would rather get her own living than continue trying to seek an adjustment with her husband. The frequent lack of frankness between husband and wife noted above appears often in the handling of money matters; adjustments like the following, if not actually the rule, are common:

A business man's wife said that she "used to believe there was no reason why a woman should not work if she wanted to and that it isn't necessary for a man to have a woman dependent upon him to make a fine character of him, as husband and wife could be held together simply by their mutual affection. I believe now, though, that a man simply gets into the habit when his wife is not economically dependent of believing that she doesn't need money, and so he goes and spends it on some other woman."

A wife of one of the less prominent business men, a self-respecting woman with children in college, said in a tone of finality: "I never let my husband tell me anything about business matters."

The economic status of many of these women is reflected in the following appeal by one of them to the members of the missionary society of the largest church in town: "Ask your husbands, who spend much on cigars and tobacco, for pennies for your mite boxes."

The wife of a foreman, a capable-looking woman with children in high school, said to the interviewer, "Marriage ought to be a

[23] It might be suggestive to inquire into the possible relation between a highly competitive medical profession in a pecuniary society (cf. Ch. XXV) and the lack of diffusion of knowledge of methods of contraception.

partnership, but we started out wrong by not sharing money matters. My husband doesn't believe in telling such matters. I don't know either how much he is earning or how much we save. I just know we are saving and that's all. It was because of this that I went to work. I liked having my own money and my husband hated my having it. Men are to blame for women going out to work. They haven't treated their wives fairly." [24]

With the spread of the habit of married women's working, women are less willing to continue an unsatisfactory marital arrangement. Said the lawyer quoted above, "If a woman has ever worked at all she is much more likely to seek a divorce. It's the timid ones that have never worked who grin and bear marriage. Unemployment always increases the number of women seeking a divorce." [25] In one sturdy, self-respecting family the daughter of twenty-three and the son of twenty-one have both been married and divorced and are living at home again:

"My daughter's husband," said the mother, "is at the house six days a week and they have an awfully good time together, but they just couldn't get on about money matters. Now she has her own money and he has his and there's no trouble. That was the trouble with my son and his wife, too. They split up over money." This boy and girl had gone to public school until they were fourteen and fifteen respectively, but in this matter of handling money the thing that had shaped them had apparently been their parents' example. The mother replied to the question as to her husband's income, "Would you believe it, I don't know any more than you

[24] There are a few families at the other extreme in which the husband turns over his pay check to his wife and she has entire charge of the household economy, but these are rare. In between these two extremes are all manner of provisional, more or less bickering agreements.

[25] How attractive an alternative staying unmarried and earning one's own living appears to some Middletown women is reflected in the comment of the wife of a worker who has had no unemployment since 1917, a woman of forty-one with four children, owning a car and a two-story home free of encumbrances: "A woman doing housework is never through. She can't ever dress up, and she hasn't any interest except in her home. A friend of mine with a good job in [the leading department store] married not long ago at thirty-four and I said to her, 'Why in the world do you want to get married at your age when you've got a good position?'"

The growing independence of women was succinctly described by a dismayed citizen recently divorced by his "woman": "Everybody's getting a divorce. Why should a woman stay married if she don't like a man and can get a job? Why, there's so many working it's getting so a man can't get a woman to fry him a piece of meat or bake a pie!"

Cf. N. 10 above, regarding the relation of women's working and divorce.

do—and married to him all these years! Mrs. —— was saying the other day she wouldn't stand it. But she doesn't know my man! He's that close-mouthed! He gives me $10 a week for myself and I get the children's board money. He pays some bills. I only know we aren't saving anything. I looked in the bank book and there hasn't been an entry in two years."

The older generation in this family had weathered a marriage that was not economically satisfactory, but the present generation is less content with such maladjustment today when spending money is more in demand, jobs for women common, and divorce easy.

Like "cruelty" and "non-support," "adultery" and "abandonment" as alleged grounds for divorce indicate not isolated phenomena but a wide complex of factors leading to the failure of husband and wife to achieve an adjustment. " 'Before the deserter there was a broken man,' said a district secretary. . . . By this characterization she meant not necessarily a physical or mental wreck, but a man bankrupt for the time being in health, hopes, prospects, or in all three." [26] It is significant that divorces for these specific causes are relatively less frequent today while divorces for "cruelty" and unclassified causes have increased.[27]

The increasing number of divorces today for any of these alleged causes may indicate that a larger proportion of husbands and wives than formerly are failing to achieve a marital adjustment, or that people are tending to demand more of a tolerable marriage, and, failing to achieve an adjustment, seek the divorce courts. Many people in Middletown would probably agree with Dorothy Dix that:

"The reason there are more divorces is that people are demanding more of life than they used to. . . . In former times . . . they expected to settle down to a life of hard work . . . and to putting up with each other. Probably men are just as good husbands now as they ever were, but grandmother had to stand grandpa, for he was her meal ticket and her card of admission to good society. A divorced woman was a disgraced woman. . . . But now we view the matter differently. We see that no good purpose is achieved by keeping two people together who have come to hate each other."

[26] Joanna C. Colcord, *Broken Homes* (New York; Russell Sage Foundation, 1919), p. 17.
[27] See Table XII.

These greater demands on life, emphasizing inadequacies in marriage, rest back upon many other changes in the life of the city.

The way in which these antecedents of divorce are imbedded in the whole complex of Middletown's culture touching the adjustments between a man and his wife is suggested by comparing what Middletown regards as minimum essentials of marriage with conditions actually existing in many Middletown homes, particularly those of the working class, among whom, according to lawyers handling divorce cases, divorce is more frequent. The husband must "support" his family, but, as pointed out above,[28] recurrent "hard times" make support of their families periodically impossible for many workers; the wife must make a home for her husband and care for her children, but she is increasingly spending her days in gainful employment outside the home; husband and wife must cleave to each other in the sex relation, but fear of pregnancy frequently makes this relation a dread for one or both of them; affection between the two is regarded as the basis of marriage, but sometimes in the day-after-day struggle this seems to be a memory rather than a present help. Not one of the sixty-eight working class wives mentioned her husband in answering the question as to the things that give her "courage to go on when thoroughly discouraged." [29] More than one wife seems to think of her husband less as an individual than as a focus of problems and fears—anxiety about loss of job, disappointment over failure in promotion, fear of conception—the center of a whole complex of things to be avoided. To many husbands their wives have become associated with weariness, too many children, and other people's washings. It is out of such situations that the incidents which appear in the divorce courts as "cruel treatment," "abandonment," and "adultery" frequently arise. The effect of these various types of emotional maladjustment upon getting a living, child-rearing, and other activities can hardly be over-estimated.

[28] Cf. Chapter VII.
[29] Cf. Ch. XX.
The nearest approach of any woman to thought of her husband was in the answer of one who said, "I think that when evening comes we'll all be together again." Very few business class women answered this question. No woman of either business or working class mentioned spending time with her husband as among the things she would like to do with an extra hour in the day. Cf. Ch. XIX.

And yet it is easy by emphasizing only the more obvious and accessible facts to form a distorted picture of these Middletown homes. For a large number of them disappointment and anxiety may lurk steadily in the background, but they are not forever in the foreground. However drab or shadowed by fear these homes may be, there are always the plans for today and to-morrow, the pleasures of this half-hour, the "small duties and automatic responses to the custom of the daily round of living [which] imperceptibly but surely mitigate the tragedies and disappointments of existence."

Indeed it might seem that in many homes this response to custom is the most marked characteristic of marriage. There are some homes in Middletown among both working and business class families which one cannot enter without being aware of a constant undercurrent of sheer delight, of fresh, spontaneous interest between husband and wife. But such homes stand out by reason of their relative rarity. In others where this quality is less apparent, marriage is doubtless the deepest reality in the lives of the pair. For many couples, however, for whom thought of the divorce court may never figure even as a remote possibility, marriage seems to amble along at a friendly jog-trot marked by sober accommodation of each partner to his share in the joint undertaking of children, paying off the mortgage, and generally "getting on."

Chapter XI

CHILD-REARING

Child-bearing and child-rearing are regarded by Middletown as essential functions of the family. Although the traditional religious sanction upon "fruitfulness" has been somewhat relaxed since the nineties, and families of six to fourteen children, upon which the grandparents of the present generation prided themselves, are considered as somehow not as "nice" as families of two, three, or four children,[1] child-bearing is nevertheless to Middletown a moral obligation. Indeed, in this urban life of alluring alternate choices, in which children are mouths instead of productive hands, there is perhaps a more self-conscious weighting of the question with moral emphasis; the prevailing sentiment is expressed in the editorial dictum by the leading paper in 1925 that "married persons who deliberately refuse to take the responsibility of children are reasonable targets for popular opprobrium." But with increasing regulation of the size of the family, emphasis has shifted somewhat from child-bearing to child-rearing. The remark of the wife of a prosperous merchant, "You just can't have so many children now if you want to do for them. We never thought of going to college. Our children never thought of anything else," represents an attitude almost universal today among business class families and apparently spreading rapidly to the working class.

[1] The phenomenon of fewer children per married couple today is closely linked with many trends in the behavior of the community: higher standards of living; the relation of the wear and tear of pregnancy upon women physically to the prevalent emphasis upon a woman's "staying young"; the desire to release time and energy for club work and social pace-setting; the growing tendency for both husband and wife to engage in getting a living; and so on.

Figures on infant mortality in Middletown are not available for 1890 and even those for recent years are not reliable. Obviously any shift here affects directly the size of family. Middletown has probably shared in the decline in infant mortality experienced by the rest of the country. Cf. Ch. XXV for discussion of the care of health in the city.

The birth of the child tends to give him his place in the group;[2] many of the most important activities of his life, as noted in Chapter IV, are determined by the fact of his being born into a family of workers' or of business class habits and outlook. The rearing of the child in the home goes far toward shaping his more critical lifelong habits; the "significance of the family as a transfer point of civilization cannot be overestimated."[3] This transfer takes place not only through any training which his parents consciously set out to give him, but still more through the entire life of the home. From birth until the age of five or six a child is reared almost entirely in the individual home by his parents, under whatever conditions or according to whatever plan or lack of plan their habits and inclinations may favor. He may live in a home where getting a living is the dominant concern of both parents or where the mother, at least, devotes much of her time to her children; in a home of affection or of constant bickering; of any variety of religious or political affiliation or use of leisure; he may be "made to mind" by spanking or bribing, or he may rule the house; he may be encouraged to learn or told "not to ask so many questions"; he may be taught to tell the truth or laughed at as "cute" when he concocts little evasions—unless he is "cruelly treated" no one interferes. From five or six to twelve or thirteen the home still remains the dominant formal agency responsible for the child, but supplemented by compulsory schooling and by optional religious training and the increasing influence of playfellows. After the age of twelve or thirteen the place of the home tends to recede before a combination of other formative influences, until in the late teens the child is regarded as a kind of junior adult, increasingly independent of parental authority.

Child-rearing is traditionally conceived by Middletown chiefly in terms of making children conform to the approved ways of the group; a "good" home secures the maximum of conformity; a "bad" home fails to achieve it. But today the swiftly moving environment and multiplied occasions for con-

---

[2] Rivers speaks of the highly important function of marriage "as the means by which every individual born into a society is assigned a definite place in that society. . . . Each child, by virtue of being born as a child of a marriage, takes its place in the social structure." (*Social Organization*, pp. 37-8.)

[3] Goldenweiser, *op. cit.*, p. 239.

tacts outside the home are making it more difficult to secure adherence to established group sanctions, and Middletown parents are wont to speak of many of their "problems" as new to this generation, situations for which the formulae of their parents are inadequate. Even from the earliest years of the child's life the former dominance of the home is challenged; the small child spends less time in the home than in the ample days of the nineties. Shrinkage in the size of the yard affords less play space.[4] "Mother, where *can* I play?" wailed a small boy of six, as he was protestingly hauled into a tiny front yard from the enchanting sport of throwing ice at passing autos. "We had a large family," said one mother, "and when things got jangled my father used to take one of us and say, 'Let us go out under the stars and meditate.' I'd like to do that with my children, but we'd have to go up to the roof to see the stars!" The community has recently begun to institute public playgrounds, thereby hastening the passing of the time when a mother could "keep an eye on" the children in the home yard. The taking over of the kindergarten by the public schools in 1924 offers to children of four and five an alternative to the home. "Why, even my youngster in kindergarten is telling us where to get off," exclaimed one bewildered father. "He won't eat white bread because he says they tell him at kindergarten that brown is more healthful!"

Nor can parental authority reassert itself as completely as formerly by the passing on of skills from father to son. Less often does a son learn his trade at his father's work bench, perhaps being apprenticed under him,[5] nor do so many daughters learn cooking or sewing at their mothers' side; more than a few of the mothers interviewed said unhappily that their daughters, fresh from domestic science in school, ridicule the mothers' inherited rule-of-thumb practices as "old-fashioned." [6]

The growing tendency for working class mothers to work

[4] Cf. Ch. IX.

[5] Cf. in Ch. VI the tendency for sons of Middletown workers to desert their fathers' lines of work.

[6] In addition to courses in cooking and dress-making, a new course in child care and nutrition has recently been introduced into one school. Cf. Ch. XIV on vocational and domestic science courses.

In the nineties the mother-daughter method of handing down household knowledge was emphasized perhaps even more than today by the conservative influence of the physical shell of the home. As each fresh generation of married couples could afford only older houses and tended to have their

outside the home has accelerated the assumption by the group of even some of the more intangible functions of parents. Following the war a "Dean of Women" was appointed in the high school to stand, according to the local press, *in loco parentis:*

"It was found impossible for mothers who worked during the day and were busy with household duties during the evening to give proper time to the boy and girl in school. . . . It was deemed necessary to have women in the schools who were sufficiently interested in boys and girls . . . to devote their entire time to working with and for them. . . . It is the dean's business to help solve their problems along every line—social, religious, and educational." [7]

And, with entry into high school, the agencies drawing the child away from home multiply. Athletics, dramatics, committee meetings after school hours demand his support; Y.M.C.A., Y.W.C.A., Boy Scouts, Girl Reserves, the movies, auto-riding —all extra-neighborhood concerns unknown to his parents in their youth—are centers of interest; club meetings, parties or dances, often held in public buildings,[8] compete for his every evening. A "date" at home is "slow" compared with motoring, a new film, or a dance in a near-by town. It is not surprising that both boys and girls in the three upper years of the high school marked the number of times they go out on school nights and the hour they get in at night more frequently than any other sources of friction with their parents,[9] and that approximately

habits more or less fixed in this form by the time they were able to build, home-making processes tended to become solidified in the mold of an earlier generation. The recent rapid diffusion of credit devices for building new houses and of new labor-saving devices for use in the home operates with the domestic science work in the school and the spread of woman's magazines to interrupt the sequence of mother-daughter inheritance.

[7] The actual work performed by the Dean of Women falls as yet considerably short of this statement, but the significant point here is that the new office is actually so conceived.

[8] Cf. Ch. XVIII and XIX for a discussion of the multiplication of these formal leisure-time activities.

[9] Of 348 boys and 382 girls, high school sophomores, juniors, and seniors, checking a list of twelve possible sources of disagreement with their parents, 45 per cent. of the boys and 43 per cent. of the girls checked "the hour you get in at night" and 45 per cent. of the boys and 48 per cent. of the girls "the number of times you go out on school nights." Most persons checked more than one item, making the totals more than 100 per cent. See Table XIII and Ch. XVIII. The boys answering these questionnaires were almost evenly divided between working class and business class parents (cf. Ch. VI for distribution of these boys).

half of the boys and girls answering the question say that they are at home less than four evenings out of the week.[10] "I've never been criticized by my children until these last couple of years since they have been in high school," said one business class mother, "but now both my daughter and older son keep saying, 'But, Mother, you're so old-fashioned.' " "My daughter of fourteen thinks I am 'cruel' if I don't let her stay at a dance until after eleven," said another young mother. "I tell her that when I was her age I had to be in at nine, and she says, 'Yes, Mother, but that was fifty years ago.' "

With the diminishing place of the home in the life of the child comes the problem of "early sophistication," as business class parents put it, or "children of twelve or fourteen nowadays act just like grown-ups," in the words of workers' wives. A few of the wealthier parents have reluctantly sent their children away to school, largely in order that they may avoid the sophisticated, early-maturing social life which appears to be almost inescapable. As one listens to the perplexity of mothers today, the announcement in the local press in 1900 that "Beginning March first, curfew bell will be rung at 9 P.M. instead of 8 P.M." seems very remote.

"What can we do," protested one mother, "when even church societies keep such late hours? My boy of fifteen is always sup-

[10] Three hundred and ninety-six boys and 458 girls, high school sophomores, juniors, and seniors answered the question, "How many evenings in the past seven were you home all evening from supper to bed time?" as follows:

	Per cent. of Boys	Per cent. of Girls
Home no evenings	19	5
Home one evening	11	9
Home two evenings	12	12
Home three evenings	13	18
Home four evenings	15	18
Home five evenings	14	19
Home six evenings	9	11
Home seven evenings	7	8
Total	100	100

In other words, 55 per cent. of the boys and 44 per cent. of the girls had been at home less than four evenings out of the last seven. The somewhat greater mobility in the case of the boys is possibly significant.

In order to guard against securing data for an exceptional week, the questionnaire asked the students also to state the number of evenings they were usually at home during a typical week if it differed from the week just passed. This showed practically no difference.

Cf. in Ch. XVIII the figures on movie attendance.

posed to be home by eleven, but a short time ago the Young People's Society of the church gave a dance, with the secretary of the Mothers' Council in charge, and dancing was from nine to twelve! And so few mothers will do anything about it. My son was eleven when he went to his first dance and we told him to be home by ten-thirty. I knew the mother of the girl he was taking and called her up to tell her my directions. 'Indeed, I'm not telling my daughter anything of the kind,' she said; 'I don't want to interfere with her good time!' " [11]

"We haven't solved the problem," said another conscientious mother. "Last year we seriously considered sending our daughter away to school to get away from this social life. We try to make home as much a center as possible and keep refreshments on hand so that the children can entertain their friends here, but it isn't of much use any more. There is always some party or dance going on in a hotel or some other public place. We don't like the children to go out on school nights, but it's hard always to refuse. Last night it was a Hallowe'en party at the church and tonight a dramatic club dance at the high school. Even as it is, we're a good deal worried about her; she's beginning to feel different from the others because she is more restricted and not allowed to go out as much as they do."

Almost every mother tells of compromise somewhere. "I never would have believed I would have let my daughter join so many clubs," said one thoughtful mother of a high school girl. "I have always criticized people who did it. But when it comes right down to it, I want to minimize the boy interest, and filling her life full of other things seems to be the only way to do that. She belongs to three high school clubs besides the Matinee Musicale and a Y.W.C.A. club."

Another woman, criticized by her neighbors for letting her children "run wild," insists that the only difference between her and other mothers is that she knows where her children are and the other mothers don't: "I wish you could know the number of girls who come over here and then go to —— [a much-criticized public dance resort fifteen miles away]. They say, 'Well, it's perfectly all right if you keep with your own crowd, but I can't explain it to mother, so I just don't tell her.' "

One working class mother said that she no longer lets her children go to church on Sunday evening "because that's just an excuse to get out-of-doors and away from home."

[11] Cf. in Ch. XII the discussion of the emphasis upon clothing at dances as a part of this early sophistication.

Late nights away from home bring further points of strain over grades at school and use of the car. The former is ranked by both boys and girls as third among the sources of disagreement with their parents.[12]

"That crowd of girls was as fine as any in school two years ago," lamented one high school teacher. "Now they all belong to two or three clubs and come to me morning after morning, heavy-eyed, with work not done, and tell of being up until twelve or one the night before. Their parties used to begin earlier and end earlier, but now it isn't a party if it breaks up before midnight."

Use of the automobile ranks fifth among the boys and fourth among the girls as a source of disagreement.[13] The extensive use of this new tool by the young has enormously extended their mobility and the range of alternatives before them; joining a crowd motoring over to dance in a town twenty miles away may be a matter of a moment's decision, with no one's permission asked. Furthermore, among the high school set, ownership of a car by one's family has become an important criterion of social fitness: a boy almost never takes a girl to a dance except in a car; there are persistent rumors of the buying of a car by local families to help their children's social standing in high school.

The more sophisticated social life of today has brought with it another "problem" much discussed by Middletown parents, the apparently increasing relaxation of some of the traditional prohibitions upon the approaches of boys and girls to each other's persons. Here again new inventions of the last thirty-five years have played a part; in 1890 a "well-brought-up" boy and girl were commonly forbidden to sit together in the dark; but motion pictures and the automobile have lifted this taboo, and, once lifted, it is easy for the practice to become widely extended. Buggy-riding in 1890 allowed only a narrow range of mobility; three to eight were generally accepted hours for riding, and being out after eight-thirty without a chaperon was largely forbidden. In an auto, however, a party may go to a city halfway across the state in an afternoon or evening, and unchaperoned automobile parties as late as midnight, while sub-

---

[12] See Table XIII. Forty per cent. of the boys and 31 per cent. of the girls marked it as a source of friction.

[13] Thirty-six per cent. of the boys and 30 per cent. of the girls marked it as a source of friction.

ject to criticism, are not exceptional. The wide circulation among high school students of magazines of the *True Story* variety and the constant witnessing of "sex films" tend to render familiar postures and episodes taken much less for granted in a period lacking these channels of vivid diffusion.[14]

The relaxing of parental control combines with the decrease in group parties to further the greater exclusiveness of an individual couple. In the nineties, according to those who were in high school then, "We all went to parties together and came home together. If any couple did pair off, they were considered rather a joke." Today the press accounts of high school club dances are careful to emphasize the escort of each girl attending. The number of separate dances at a dance is smaller and there is much more tendency for each individual to dance with fewer partners, in some cases to dance the entire evening with one person. "When you spend four or five dollars to drag a girl to a dance," as one boy put it, "you don't want her to spend the evening dancing with every one else."

In such a grown-up atmosphere it is hardly surprising that the approaches of the sexes seem to be becoming franker. Forty-eight per cent. of 241 junior and senior boys and 51 per cent. of 315 junior and senior girls marked "true" the extreme statement, "Nine out of every ten boys and girls of high school age have 'petting parties.' " [15] In the questionnaire given

---

[14] Cf. Chs. XVII and XVIII for discussion of these leisure-time activities. The impact of these new streams of diffusion is apparent in the habits of such a girl as the following, a healthy seventeen-year-old high school girl, popular in school and the daughter of a high type of worker, who happened to be known personally to members of the research staff. She attends the movies twice a week (she had been home only one evening in the last seven) and reads regularly every week or month *Snappy Stories, Short Stories, Cosmopolitan, True Story, Liberty, People's Popular Monthly, Woman's Weekly, Gentlewoman,* and *Collier's.* She and her parents are at loggerheads most often, she says, about the way she dresses, and after that, about her use of the family Ford and about her boy and girl friends. Along with these evidences of divergence from the ways of her parents, she still maintains the family religious tradition, being an indefatigable church worker and Sunday School teacher.

The possible relation of popular songs to the courtship habits of 1890 and today should not be overlooked. In the ballads of a generation ago—"After the Ball," "Airy, Fairy Lillian," "On a Bicycle Made for Two"—lovers might sit together in the moonlight, "hands touching lightly," or perhaps in the bolder songs there was a solitary kiss or "squeeze," but rarely was there the "I'll-hold-you-enfold-you" quality of the songs to which young Middletown dances dreamily today.

[15] Thirty-five per cent. of the boys marked this "false," 15 per cent. "uncertain," and 2 per cent. did not answer. Thirty-eight per cent. of the girls

to sophomores, juniors, and seniors, 44 per cent. of the 405 boys answering the questionnaire (88 per cent. of the 201 answering this question) and 34 per cent. of the 464 girls answering the questionnaire (78 per cent. of the 205 answering this question) signified that they had taken part in a "petting party" by checking one or another of the reasons listed for doing so—though, of course, data of this sort are peculiarly open to error.[16] There is a small group of girls in the high school who are known not to allow "petting." These girls are often "respected and popular" but have less "dates"; the larger group, "many of them from the 'best families,'" with whom "petting parties" are not taboo, are said to be much more frequently in demand for movies, dances, or automobile parties. Stimulated in part, probably, by the constant public watching of love-making on the screen, and in part, perhaps, by the sense of safety in numbers, the earlier especially heavy ban upon love-making in public is being relaxed by the young. Such reasons as the following given by high school students for taking part

marked it "false," 10 per cent. "uncertain," and 1 per cent. did not answer.

There is, of course, no way to check the number of flippant answers to such a question. Students were urged to leave questions blank rather than not answer them seriously. Informal conversation with some of the students afterwards confirmed in general the feeling of those who gave the questionnaires to the classes that most of the students were trying to answer them seriously. One boy said, "I think I answered one question wrong. I put 'true' after 'nine out of every ten boys and girls of one high school age have petting parties,' and I really don't believe it's more than three out of four." All such data must be regarded as simply suggesting tendencies.

Cf. Appendix on Method.

[16] Group compulsion is apparently a potent factor, particularly among the girls. Forty-seven per cent. of 241 junior and senior boys and 65 per cent. of 315 girls marked "true" the statement, "Most girls allow 'petting' not because they enjoy it but because they are afraid they will be unpopular if they refuse." Thirty-four per cent. of the boys and 24 per cent. of the girls marked it "false," 17 and 9 per cent. respectively "uncertain," and 2 per cent. of each did not answer. Eighty-six per cent. of the 177 boys who had "taken part in 'petting parties'" said that they did it for the sake of "having a good time," and 8 per cent. because they were "afraid of being unpopular," while 48 per cent. of the 159 girls said that they did it for the "good time" and 36 per cent. for fear of unpopularity.

Such answers as these were given by the boys: "To be one of the gang," and "Because the others did"; while among the girls' answers, these were characteristic: "They laugh at you if you don't, and I see no harm as long as you don't go too far," "Forced to," "Afraid of being sneered at if I don't—but because of good time once," "Hard to keep from it when there was insistence," "Because the boy I go with wants to."

The remaining 6 per cent. of the boys and 16 per cent. of the girls distributed their answers under "Curiosity," "I cared for him (her)," etc.

in "petting parties" suggest the definite group connotation of the term to them: "I did not know it was going to be that kind of a party," "I did not know what it was going to be like; I did not stay long," "I do not believe in them, but have gone to one or two," "Just to see what they were like."

Mothers of both working and business class, whether they lament the greater frankness between the sexes or welcome it as a healthy sign, agree that it exists and mention the dress and greater aggressiveness of girls today as factors in the change. Such comments as the following from the mothers of both groups are characteristic:

"Girls aren't so modest nowadays; they dress differently." "It's the girls' clothing; we can't keep our boys decent when girls dress that way." "Girls have more nerve nowadays—look at their clothes!" "Girls are far more aggressive today. They call the boys up to try to make dates with them as they never would have when I was a girl." "Last summer six girls organized a party and invited six boys and they never got home until three in the morning. Girls are always calling my boys up trying to make dates with them." "Girls are bolder than they used to be. It used to be that if a girl called up and asked a boy to take her somewhere she meant something bad by it, but now they all do it." "My son has been asked to a dance by three different girls and there is no living with him." [17] "When I was a girl, a girl who painted was a bad girl—but now look at the daughters of our best families!"

The declining dominance of the home and early sophistication of the young bring still another difficulty to Middletown parents in the increased awkwardness of the status of the child, particularly the boy, as he nears adulthood. Socially, children of both groups are entering earlier into paired associations with members of the other sex under a formalized social system that makes many of the demands for independence of action upon them that it does upon self-supporting adults. Sexually, their awareness of their maturity is augmented by the maturity of their social rituals and by multiplied channels of diffusion, such

[17] The rise of the telephone plays an important part in this shift since it affords a semi-private, partly depersonalized means of approach to a person of the other sex. Then, too, the diffusion downward into the conglomerate high school population of today of a social life conducted, as in the case of their elders, in large part by the females, has greatly facilitated the aggressiveness of the girls: only five of the fourteen Christmas dances were male-conducted, the girls seeking out the boys for the other nine.

as the movies and popular magazines. But meanwhile, economically they are obliged by the state to be largely dependent upon their parents until sixteen, the age at which they may leave school, and actually the rapidly spreading popular custom of prolonged schooling tends to make them dependent from two to six years more. The economic tensions inherent in this situation are intensified by the fact that, as in the case of their parents, more of their lives than in any previous generation must surmount intermediate pecuniary hurdles before they can be lived. Expenses for lunches purchased in the high school cafeteria, for movies and athletic games, as well as for the elaborate social life—with its demands for club dues, fees for formal dances and banquets, taxis, and variety and expense of dress—mean that children of all classes carry money earlier and carry more of it than did their parents when they were young. Thirty-seven per cent. of the 348 high school boys and 29 per cent. of the 382 girls answering the question checked "spending money" as a source of disagreement between them and their parents.[18]

"One local youth sighs for the return of the good old days when one could sit on the davenport at home with one's 'best girl' and be perfectly contented," says a local paper. "You can't have a date nowadays," he says, "without making a big hole in a five-dollar bill."

And again, "The coal dealer and the gas men may fear the coming of summer—but florists aren't much worried over the fact that spring flowers will soon be seen growing in every front yard. 'It don't mean anything,' says one local florist. 'The days have gone by when a young man may pick a bouquet of flowers from his own yard and take them to his best girl. Nowadays she demands a dozen roses or a corsage bouquet in a box bearing the name of the best florist.'"

"There are still some youths," says another note, "who believe that a girl is overjoyed when they take her into a soda fountain after they have been to a moving picture show. That is what one calls 'the height of being old-fashioned.' Nowadays one has to have a six-cylindered seven-passenger sedan to take her joy-riding in, and one must patronize the most expensive shows and take her to the most exclusive restaurants to cap the evening off."

[18] See Table XIII.

If such statements represent journalese hyperbole, they nevertheless reflect a powerful trend affecting the young of every economic level. A wide variety in the kinds of adjustment different families are attempting to effect in regard to these new demands for money appears in the answers of 386 boys and 454 girls, high school sophomores, juniors, and seniors, to a question on the source of their spending money: 3 per cent. of the boys and 11 per cent. of the girls receive all their spending money in the form of an allowance; 15 per cent. of the boys and 53 per cent. of the girls are dependent for all their spending money upon asking their parents for it or upon gifts; 37 per cent. of the boys and 9 per cent. of the girls earn all their spending money.[19] It is perhaps significant that, while over three-fourths of these Middletown boys are thus learning habits of independence as regards money matters by earning and managing at least a part of their money, over half of the girls are busily acquiring the habits of money dependence that characterize Middletown wives by being entirely dependent upon their parents for their spending money without even a regular allowance.[20] At no point is parental influence more sharply challenged than by these junior-adults, so mature in their demands and wholly or partially dependent upon their parents economically but not easily submitting to their authority.

A natural reaction to these various encroachments upon parental dominance and shifts in the status of children is the vigorous reassertion of established standards. And in Middletown the traditional view that the dependence of the child carries with it the right and duty of the parents to enforce "discipline" and "obedience" still prevails.

"Study the lives of our great men. Their mothers were true home-makers who neither spared their prayers nor the rod and spoiled the child," says a paper read before one of the federated women's clubs in 1924. "It is the men and women who have been taught obedience from the cradle, who have been taught self-

[19] In addition 9 per cent. of the boys and 5 per cent. of the girls both earn money and have an allowance; 31 per cent. of the boys and 17 per cent. of the girls supplement earnings in some other way than by an allowance; 5 per cent. of the boys and 5 per cent. of the girls supplement allowances in some other way than by earnings.

[20] As one senior girl put it, "Some of us don't want an allowance; you can get more without one."

control and to submit to authority and to do things because they are right who are successful and happy in this world."

A prominent banker and a prominent physician agreed in a dinner-table discussion that there must be once in every child's life a brisk passage at arms that "will teach them where authority lies in the family. You have to teach them to respect parental authority. Once you've done this you can go ahead and get on the best possible relations with them." "My little grandchild has been visiting me," said a teacher in a Sunday School class in a leading church, "and he's a very bad little boy; he's so full of pep and energy that he doesn't do what I want him to do at all." "I am going to bring my little girl up just as strict as I can," said one perplexed working class mother; "then if she does go bad I won't feel that I haven't done my duty."

And yet not only are parents finding it increasingly difficult to secure adherence to established group sanctions, but the sanctions themselves are changing; many parents are becoming puzzled and unsure as to what they would hold their children to if they could. As one anxious business class mother said:

"You see other people being more lenient and you think perhaps that it is the best way, but you are afraid to do anything very different from what your mother did for fear you may leave out something essential or do something wrong. I would give anything to know what is wisest, but I don't know what to do."

As a possible index of the conscious emphases of Middletown mothers in training their children as well as of the points at which this generation is departing from the ways of its parents, the mothers interviewed were asked to score a list of fifteen habits according to their emphases upon them in training their children, and each was asked to give additional ratings of the same list as her own home training led her to believe her mother would have rated it thirty years ago when she was a child.[21] To the mothers of the last generation, according to both groups,

[21] See Table XIV. The list consisted of fifteen habits observed to be stressed in greater or less degree in Middletown in the training of children: frankness; desire to make a name in the world; concentration; social-mindedness (defined as "a sense of personal responsibility for those less fortunate"); strict obedience; appreciation of art, music, and poetry; economy in money matters; loyalty to the church; knowledge of sex hygiene; tolerance (defined as "respect for opinions opposed to one's own"); curiosity; patriotism; good manners; independence (defined as "ability to think and act for oneself"); getting very good grades in school. The lists were given

"strict obedience" and "loyalty to the church" were first in importance as things to be emphasized.[22] The working class mothers of the present generation still regard them as preëminent, but with closer competitors for first place. In the ratings of the group of business class mothers of the present generation, however, "strict obedience" is equaled and "loyalty to the church" is surpassed by both "independence" and "frankness." "Strict obedience does not accomplish anything at all," said one business class mother, marking it an emphatic zero. And another commented: "I am afraid that the things I really *have* emphasized are obedience, loyalty to the church, and getting good grades in school; those are the things easiest to dwell on and the things one naturally emphasizes through force of habit. But what I really believe in is the slower but surer sort of training that stresses concentration, independence, and tolerance."

A more democratic system of relationships with frank exchange of ideas is growing up in many homes: "My mother was a splendid mother in many ways, but I could not be that kind of mother now. I have to be a pal and listen to my children's ideas," said one of these mothers who marked obedience zero for herself and "A" for her mother. One worker's wife commented, "Obedience may be all right for younger children, but, now, take my boy in high school, if we tried to jerk him up like we used to be he'd just leave home." And another, "We are trying to make our boy feel that he is entitled to his own opinion; we treat him as one of us and listen to his ideas." The value that the children apparently place upon this policy is indicated by the fact that "respecting children's opinions" is rated by 369 high school boys and 415 high school girls second only to "spending time with children" as a quality desirable in a father; this trait was rated fourth and fifth in importance respectively by these boys and girls as a quality desirable in a mother.[23] The different ratings for father and mother in this

to the women and they first marked the habits which they themselves regard as most important, rating the three most important "A," the five next most important "B," any of third-rate importance "C," and any which they regarded as unnecessary or undesirable zero. They then set down in another column what they thought, in the light of their own home training, their mothers' ratings would have been. This procedure is, of course, precarious. It represents verbalizations only, but every effort was made to check up on a woman's memories of her own training and to secure careful consideration.

[22] See Chs. XX and XXII regarding shifting religious emphases.

[23] See Table XV.

regard may reflect the fact that, although the father is less immediately concerned with the daily details of child-rearing, it is
he who puts the family's "foot down" periodically.

A shift in emphasis upon knowledge of sex hygiene appearing in the ratings of these mothers may be in part a response
to fear of the consequences of the present freedom between the
sexes. Mother after mother said in substance, "The only thing
that can be done is to teach boys and girls about such things."
Only four of the sixty-seven working class wives who rated
these items for both themselves and their mothers said that they
do not believe in teaching sex hygiene to their children, while
thirty-two of their mothers are reported not to have believed in
it; of the thirty-seven wives of the business group reporting
for themselves and for their mothers, there is not one who does
not believe in such instruction, although twenty-two of their
mothers were reported by them not to have believed in it. Some
of the wives of the working class believe that instruction should
be given by the schools but lack the courage to undertake it
themselves.[24] The teaching of the mothers who do attempt to
talk with their children apparently varies all the way from the
old wives' tales of some of the workers' wives to the practice
of one business class mother who took her small son of four
to see the birth of some puppies and gave full answers to all his
questions.

This trend in the direction of more teaching of sex hygiene
in the home appears more significant in the light of the fact that
there is no other accepted means of giving the young sex
knowledge or initiating them into the sanctions and prohibitions of the group regarding sex. No formal instruction in sex
hygiene is given anywhere in the Middletown schools. "I am an
old-fashioned mother," said the dean of women at the high
school, "and I don't believe in suggesting those things to children. The more we put such things into their heads the more
they will think of them." The six leading Protestant ministers

---

[24] This parental failure of nerve is illustrated in the following comment by
a working class woman in her middle thirties, the mother of a sixteen-year-
old girl: "I believe children ought to be taught such things. I'm not much
for talking about them. I've never talked to my daughter at all, though I
suppose she knows more than I think she does. She's the only one I've got
and I just can't bear to think of things like that in connection with her.
I guess I wouldn't even talk to her if she was going to be married—I just
couldn't!"

of the city and the Y.M.C.A. and Y.W.C.A. secretaries say that no formal teaching in this matter is afforded by any of their organizations.[25] That actual sex instruction by parents is still far from universal appears from the fact that, of 264 boys and 344 girls, high school sophomores, juniors, and seniors, answering the question, "From what source did you receive most of whatever information you have on sex matters?" 32 per cent. of the boys and 68 per cent. of the girls named their parents as their chief source of information.[26]

Despite the difficulty of holding children to established sanctions and the shifting of the sanctions themselves, not all of the currents in the community are set in the direction of widening the gap between parents and children. It is the mother who has the chief responsibility in child-rearing, and many Middletown mothers, particularly among the business class, are devoting a part of their increasing leisure to their children.[27] Such comments as the following represent many of the business class wives interviewed:

"I accommodate my entire life to my little girl. She takes three music lessons a week and I practice with her forty minutes a day. I help her with her school work and go to dancing school with her."

"My mother never stepped inside the school building as far as I can remember, but now there are never ten days that go by without my either visiting the children's school or getting in touch with their teacher. I have given up church work and club work since the children came. I always like to be here when they come

[25] The only exceptions to this generalization are discussions of the desirability of "keeping clean" in Y.M.C.A. Bible classes, informal camp-fire talks touching on sex matters among other things at the Y.M.C.A. summer camp, a Y.W.C.A. club of girls who sometimes ask their leader questions touching sex, and a sporadic effort by the minister of the smallest of the above six leading churches to talk to some of the boys in his church if their fathers are willing.

[26] Forty-two per cent. of the boys and 22 per cent. of the girls stated that they had received most of their information from boy or girl friends. Eleven per cent. of the boys and 0.6 per cent. of the girls named "Y.M.C.A." or "Y.W.C.A." workers. One per cent. of the boys and 2 per cent. of the girls answered "Sunday School teacher" and 4 per cent. of the boys and 4 per cent. of the girls "School teacher." Other answers were scattered. Movies and the so-called "sex magazines" were not mentioned specifically by any of the children, but according to many parents, as noted elsewhere, these play an important part.

[27] Cf. Ch. XII on the decreased time spent by women of both groups on housework.

home from school so that I can keep in touch with their games and their friends. Any extra time goes into reading books on nutrition and character building."

"Every one asks us how we've been able to bring our children up so well. I certainly have a harder job than my mother did; everything today tends to weaken the parents' influence. But we do it by spending time with our children. I've always been a pal with my daughter, and my husband spends a lot of time with the boy. We all go to basketball games together and to the State Fair in the summer."

"We used to belong to the Country Club but resigned from that when the children came, and bought a car instead. That is something we can all enjoy together." [28]

At the opposite pole from the most-leisured mothers of the business group are a considerable group of the working class wives for whom the pressure of outside work or of housework never done prevents the giving of much time and thought to the day-by-day lives of their children:

"I would like to play with the children more than I do, but I'm too tired to do it even when I have the time," was one comment. "I just can't get up any energy. My man is so tired when he comes

[28] The amount of time actually devoted to their children, even by mothers of the business class, should not be over-estimated. Clubs, bridge, golf, and other leisure-time outlets make heavy inroads upon women's time. One woman spoke of having played eighteen holes of golf on three afternoons during the preceding week with a mother of three children. Sand piles and other devices are provided at the Country Club where the children of members may be parked. The small daughter of one member said with evident bitterness, "I hate the Country Club because Mother is out there all the time." More than one mother who spoke of devoting most of her time to her children considered herself exceptional.

A very rough check of the time spent by mothers of the two groups with their children is afforded by the following summaries of their estimates. Of the forty business class mothers, none reported no time at all spent by her on a usual day with her children, two spend less than an hour a day, nineteen spend more than an hour a day but less than sixteen hours a week, nineteen spend sixteen or more hours a week. Of the eighty-five working class mothers answering this question, seven said they spend no time, thirteen less than one hour a day, twenty-six at least one hour a day and less than sixteen hours a week; thirty-nine spend sixteen hours or more a week. Sunday time was added to time spent on week days and the total divided by seven to secure these daily averages. Meal times were not included in these totals. Answers of women of both groups on the amount of time their mothers spent with them would suggest a trend in the direction of more time spent with children today, but these data are too rough to carry much weight.

home from work that he just lies down and rests and never plays with the children."

Another woman replied blankly to question after question about her eight children and their future, "I ain't ever give it a thought."

Certain others, less hard-pressed, give to the rearing of their children scarcely less time and consideration than do the most conscientious of the business class mothers:

One foreman's wife, wrapped up in her only child of eight, spends over an hour practicing with him each day at the piano, goes with him twice a week to his music lessons, works in the garden with him, and visits school once a month.

"I just can't afford to grow old," said another wife. "I have a boy of fifteen in high school and another of thirteen. I put on roller skates with the boys and pass a football with them. In the evenings we play cards and on Sundays we go to ball games. My mother back East thinks it's scandalous, but I tell her I don't think anything very bad can happen to boys when they're there with their father and mother."

The rôle of the father in child-rearing is regarded by Middletown as less important than that of the mother. "It is much more important for children to have a good mother than it is for them to have a good father," says Dorothy Dix, "because the mother not only establishes their social position, but because her influence is the prepotent one." A minority group of the men shift virtually the whole job to their wives, and, in general, the business men seem to find little, if any, more time for their children than do the working men.[29] "My husband has to spend time in civic work that my father used to give to us children," said one business class mother. But fathers of the business group seem somewhat more aware of the concern over parenthood being diffused to the city through civic clubs and other channels, and one meets less often among them than among the working class the attitude, "My husband never pays any attention to the children." There is a busy, wistful uneasi-

[29] Of the forty business class fathers only three were reported by their wives as spending no time with their children and eleven less than an hour a day, including Sunday, and of the ninety-two working men for whom such data were secured, the totals were respectively nine and twenty. Again it should be noted, however, that such answers are highly fallible.

ness about not being a better parent among many of the city's leaders:

"I'm a rotten dad," lamented one of these fathers. "If our children amount to anything it's their mother who'll get all the credit. I'm so busy I don't see much of them and I don't know how to chum up with them when I do."

Another remarked some time after a questionnaire had been given the high school children asking among other things for the time spent weekly by them in having a good time with their parents:

"You know, I don't know that I spend any time having a good time with my children, and it hit me all in a heap when they came home and repeated that question. And the worst of it is, I don't know how to. I take my children to school in the car each morning; *there* is some time we could spend together, but I just spend it thinking about my own affairs and never make an effort to do anything with them."

The evenings and long Sunday afternoons when the whole family is together in the elbow-to-elbow contact of the family automobile are giving many fathers of both classes a chance to "be a dad." Tinkering around the car and going to the weekly basket-ball games together offer other points of contact; one father has fitted up an old barn as a club room for his boys; another has made a stage for his little daughter where she carries on endless pantomimes with her dolls; a third has fitted up one room as a study for his daughter of eight so that she can arrange her own books and paints on low desk and shelves in a place that is her own. It is significant that the 369 high school boys and 415 girls chose "spending time with his children" among a list of ten possible desirable qualities in a father far more often than any of the other nine, and rated the corresponding item as second only to "being a good cook and housewife" among the qualities desirable in a mother.[30]

The attitude that child-rearing is something not to be taken for granted but to be studied appears in parents of both groups. One cannot talk with Middletown mothers without being continually impressed by the eagerness of many to lay hold of every available resource for help in training their children: one busi

[30] See Table XV.

ness class mother took a course in the Montessori method in a near-by city before the birth of her daughter; another reads regularly the pamphlets of the Massachusetts Society for Mental Hygiene and such books as *A System of Character Training for Children;* a few get informal help from the head of the Home Economics Department of the schools and from occasional state demonstrations on child care;[31] a handful get hold of government bulletins. Some mothers found help in a "mothers' training class" conducted for a time by a minister's wife, and a score of others are enthusiastic over a Mothers' Sunday School Class in another church; a few look to the Mothers' Council, but many of the supporters of these groups say that they get little concrete help from them.[32] Forty mothers, many of them from the working class, paid over forty dollars for an installment set of ten volumes on child-training entitled *Foundation Stones of Success;* with the purchase went membership in a mothers' club where child-training programs were to have been studied, had not the club died after a meeting or two.[33] Some working class mothers receive advice in the physical care of their children from the Visiting Nurses' Association and some through the schools, although the latter, like the medical profession, appear to be chiefly concerned with remedial rather than preventive work.[34] Most important of all new sources of information are the widely-read women's mag-

[31] The State Division of Infant and Child Hygiene coöperating with the Children's Bureau of the Department of Labor conducted a conference and demonstration in a local church in 1923. Cf. Ch. XIV on the work of the Domestic Science Department in the schools and on the evening classes attended by mothers; most of this work is confined to cooking and sewing. A course on nutrition has recently been introduced into one of the schools.

[32] Cf. the account of the Mothers' Council in Ch. XIX.

[33] Cf. Ch. XVII on the number of books purchased by Middletown parents to help them in the rearing of their children.

Nowhere is the isolation of many Middletown housewives more apparent than in such a case as this. They are eager to help their children and feel that a way out lies through books, but do not know where to go for advice in selecting them. One foreman's wife pointed to a set of books bought of an agent and said bitterly, "The agent came when I'd been putting up fruit all day and was tired and worried about the children. I finally said 'Yes' and now they're not what I need. And the worst of it is that all that money's gone into those books that might have gone into really good ones."

[34] Cf. Ch. XXV on the care of health in the community. It is indicative of the uneven diffusion of the best the community knows in such matters as child health that this instruction on pre-natal care, care of infants, etc., is not infrequently resented by the poorest mothers.

azines.[35] "There was only one weak magazine thirty-five years ago from which we got help in child-training," according to one mother, "and it was nothing like the fine women's magazines we have today." Such "baby books" as Holt's *Care and Feeding of Infants* are also supplanting the family "recipe book" of 1890.

And yet a prevalent mood among Middletown parents is bewilderment, a feeling that their difficulties outrun their best efforts to cope with them:

"Life was simpler for my mother," said a thoughtful mother. "In those days one did not realize that there was so much to be known about the care of children. I realize that I ought to be half a dozen experts, but I am afraid of making mistakes and usually do not know where to go for advice."

One working class wife, deeply concerned over her oldest boy of eighteen, said, "We thought we'd have an easier time when we moved in from the farm three years ago, but now my husband is laid off and can't get work anywhere. The boy is still working, but we never know anything about him any more; he don't pay any attention to Mom and Pop. He might be better if we had stayed on the farm, but I suppose he'd be in town all the time anyway. I helped him to buy his Ford by not taking board money for three months; we didn't want him to get it, but he was paying out so much on renting cars to go out with parties, it seemed cheaper to let him buy it. Now he wants a Studebaker so he can go seventy-five miles an hour. I told him I'd take in washing if he'd go to our church college, but he wanted to go to work. We wanted him to be a missionary; to run all over the country, that's his mission now! We never were like this when we were young, and we don't know what to do about him."

The following discussion among eighteen high school boys and girls at a young people's meeting in a leading church on the general topic, "What's Wrong With the Home?" reveals the parents' perplexity as seen by the children:

Boy. "Parents don't know anything about their children and what they're doing."

Girl. "They don't want to know."

Girl. "We won't let them know."

Boy. "Ours is a speedy world and they're old."

Boy. "Parents ought to get together. Usually one is easy and one is hard. They don't stand together."

[35] Cf. Ch. XVII for distribution of these magazines.

Boy. "Parents ought to have a third party to whom they could go for advice." [Chorus of "Yes."]

Boy. "This is the first year I've wanted to dance. Dad wanted me to go to only two this Christmas. [Triumphantly.] I'm going to five and passing up four!"

One shrewd business man summarized the situation: "These kids aren't pulling the wool over their parents' eyes as much as you may think. The parents are wise to a lot that goes on, but they just don't know what to do, and try to turn their backs on it."

Chapter XII

FOOD, CLOTHING, AND HOUSEWORK

The single household in Middletown today, as in 1890, is the unit for both the preparation and the eating of food. "Meal-time" serves a double function—nutrition and social intercourse. The day-by-day social life of the individual family as a group centers around meal-times and, to a considerably less extent, the family automobile.[1] The dining-room table is the family's General Headquarters; here activities on every sector come to focus: "Next Sunday's picnic," "The house needs painting—it looks shabby beside the Smiths'," "But, Marian, I don't like your going out three school nights in succession," "Jim, don't you want to go to prayer meeting with me to-night?" "Pretty punk report card, Ted; you'd better do better next month"—and so on, day after day. Meal-time as family reunion time was taken for granted a generation ago; under the decentralizing pull of a more highly diversified and organized leisure—in which Hi-Y, basket-ball games, high school clubs, pedro and bridge clubs, civic clubs, and Men's League dinners each drain off their appropriate members from the family group—there is arising a conscious effort to "save meal-times, at least, for the family." As one mother expressed it at a meeting of the Mothers' Council, "Even if we have only a little time at home together, we want to make the most of that little. In our family we always try to have Sunday breakfast and dinner together at least." "I ate only seven meals at home all last week and three of those were on Sunday," said one father.

[1] While the automobile is the less important of the two in this connection, it is making noticeable inroads upon the traditional prestige of the family's meal-times at certain points; it has done much to render obsolete the leisurely Sunday noon dinner of a generation ago at which extra leaves had to be put in the table for the company of relatives and friends who sat down to the great Sunday roast, and during half the year when "getting out in the car" is pleasant, it often curtails the evening meal to an informal "bite."

"It's getting so a fellow has to make a date with his family to see them."

Morning, noon, and evening meal eaten within the home are still the rule. A brisk skirmish is on, however, for possession of the noon meal: on the one hand, the diffusion of the ownership of automobiles is facilitating the husband's coming home to lunch despite the growing size of the city and the increasing remoteness of homes from place of work, while on the other, high school cafeteria, a factory cafeteria in at least one large plant, and business men's luncheon clubs are drawing many away from home.

Despite the fact that children in Middletown, as noted above, place "being a good cook and housekeeper" foremost among the qualities desirable in a mother, cooking occupies a less important place today than formerly among both groups, but especially among the business class. The preparation of food in the nineties was one of a woman's chief glories. Luncheon and dinner clubs meeting in the homes were favorite social events; referring to one of these clubs at which each hostess in turn strove to surpass the other members with a new dish, the local press exclaimed, "The ladies will resume the giving of those fine suppers which so pleased their husbands last winter." Not only have social clubs built around cooking greatly declined, but the trend is away from the earlier attention to elaborate food. According to a local butcher:

"The modern housewife has lost the art of cooking. She buys cuts of meat that are easily and quickly cooked, whereas in the nineties her mother bought big chunks of meat and cut them up and used them in various ways. Folks today want to eat in a hurry and get out in the car."

Something of what lies behind this change is reflected in the emphatic comment of a foreman's wife:

"It keeps me hustling just to keep up with my husband and boys. I go to high school games and root with my boy. Sundays I go to baseball games with my husband. I don't like Sunday sports, but he does, and it's our one chance to enjoy things together. Yesterday my husband said, 'Wouldn't some pumpkin pie go good?' But I said, 'Goodness, no, I have this custard because I can't take the time to make fancy things and still keep up with my family.' "

The abbreviated time spent today in preparation of food is reflected in such things as the increased use of baker's bread and the corresponding decrease in home baking. Eighty-one (more than two-thirds) of the 119 working class wives who answered the question and thirty-four of the thirty-nine business class wives spend no time today baking bread and rolls as against only three of the mothers of the former and sixteen of the mothers of the latter. According to the head of the leading Middletown bakery, not over 25 per cent. of the bread eaten in the city was commercially baked in 1890, while today 55 to 70 per cent. of the total annual consumption, varying with the season, is baked by commercial bakers.[2]

"The people who lived in the better homes and kept help in 1890 never thought of buying bread unless to fill in when the home-baked bread ran short," according to an old-time baker. "They regarded baker's bread as poor-folksy; it was the working class who bought baker's bread. It was in the factory districts that groceries first began about 1890 to keep baker's bread for the accommodation of their customers.[3] Few rolls were baked commercially then and no cakes except for weddings or big parties. People ate as many of them as they do today, but they weren't used to having them baked in a bakery."[4]

[2] An interesting fact pointed out by this man is the cycle occurring year after year whereby, taking the October output of local commercial bakers as 55 to 60 per cent. of the total consumption of bread of all kinds, this percentage increases to about 70 per cent. during the busy fall months from October through January; from January to March the figure falls back, only to increase again to 70 per cent. from April through July, followed by a decrease back to 55 or 60 per cent. from July to October.

[3] The direction of diffusion of the habit of using bakers' bread, from workers to business class, should be noted as a trend in the opposite direction from most of those observed in Middletown.

[4] The baking industry has been one of the slowest to yield to the centralizing, mechanizing trend of the industrial revolution; it ranked only eighteenth among the country's industries in 1909, but it was eleventh in 1914 and eighth in 1919. Even as late as 1890, only one or two Middletown bakers sold bread at wholesale to grocers. Most people preferred to buy directly from a local baker. Today the industry in Middletown exhibits all stages in the transition from isolated hand industry to centralized production; there are at least fourteen local bakers, producing all the way from a couple of hundred to 7,000 loaves daily apiece.

In 1923 the national baking industry witnessed the patenting of an oven capable of producing 5,000 loaves an hour—thirty times the output of the standard oven up to 1923. Already a big city bakery sixty miles from Middletown using the quantity production methods of modern industry is beginning to figure largely in the local bread supply.

Increased use of commercially canned goods has meant not only less time spent in home canning but a marked spread in the variety and healthfulness of the diet of medium- and low-income families throughout the bulk of the year when fresh garden products are expensive.[5] As one housewife expressed it, "You just spent your summer canning in 1890, but canned goods you buy today are so good that it isn't worth your while to do so much." Well-to-do families do least canning, since they can afford to buy the better grades of canned goods, but tomatoes and fruits, particularly jellies, are still "put up" in quantities by housewives, notably those of the medium- and smaller-income groups. A prejudice lingers among these latter against feeding one's family out of cans. One of the federated women's clubs recently gave over a program to a debate of the question, "Shall a Conscientious Housewife Use Canned Foods?"

New mechanical inventions such as the development of refrigeration and cold storage and of increasingly rapid transportation have also furthered sweeping changes in the kind of food eaten.[6] In 1890 the city had two distinct diets: "winter diet" and "summer diet." The "winter diet," as described by one housewife, was:

"Steak, roasts, macaroni, Irish potatoes, sweet potatoes, turnips, cole slaw, fried apples, and stewed tomatoes, with Indian pudding, rice, cake, or pie for dessert. This was the winter repertoire of the average family that was not wealthy, and we swapped about from one combination to another, using pickles and chow-chow to make the familiar starchy food relishing. We never thought of having fresh fruit or green vegetables and could not have got them if we had."[7] "In 1890 we had meat three times a

[5] According to the United States *Census of Manufactures, 1921,* the capital invested in the commercial canning of fruits and vegetables in the United States, expressed in thousands of dollars, increased from 15,316 in 1889 to 223,692 in 1919, and the value of the product, again in thousands of dollars, from 29,862 to 402,243.

[6] Mendel points out that not until 1893 did the then recently perfected ammonia process enable a sufficient quantity of food materials to be placed in cold storage to affect appreciably market conditions in the United States. *Changes in the Food Supply* (New Haven; Yale University Press, 1916), p. 12.

[7] The limited variety of foods available helped to keep housewives at work for long hours in their kitchens: "The lack of variety of vegetable foods in winter and of fresh meat in summer was without doubt the reason for the great abundance of preserves and pickles which every housewife deemed necessary and for the great number of kinds of pastry, cake, and similar dishes. In other words, there was a craving for variety and it was

day," says a Middletown grocery proprietor. "Breakfast, pork chops or steak with fried potatoes, buckwheat cakes, and hot bread; lunch, a hot roast and potatoes; supper, same roast cold."

Following "winter diet" came "spring sickness." "Nearly everybody used to be sick because of lack of green stuff to eat." [8] In the spring the papers carried daily advertisements of Sarsaparilla "to cure boils, sluggishness, thick blood, and other ailments resulting from heavy winter food." A thumbed volume, Chase's *Receipt Book,* still treasured as a current practical tool by one prominent Middletown woman "because it is the book my mother brought up her large family on thirty years ago," says of this "spring sickness":

"The matter is that the blood is thoroughly vitiated, and improving it must be a matter of time. Spring diet should do the work of medicine, largely. First in importance are salads of all sorts."

In those days the first beans or tomatoes began to be shipped in from the South in May. Then came the local gardens—and, according to a local seed man, "They used to *be* gardens in those days!" Then with the late October frosts green things disappeared and the families settled down once more to "winter diet."

Today, more and more Middletown housewives are buying fresh vegetables and fruits throughout the winter. Such supplies are shipped in regularly from large cities two or three hundred miles distant. These dietary changes have been facilitated also by the development of the modern women's magazines. Under the old rule-of-thumb, mother-to-daughter method of passing down the traditional domestic economy, when the same family recipe and doctor book—with "Gravy" in its index followed by "Gray hair, how to treat"—was commonly cherished by both mother and daughter, the home tended to resist

satisfied by using in many different ways the comparatively small number of food materials which were most commonly obtainable" C. F. Langworthy, *Food Customs and Diet in American Homes* (United States Department of Agriculture, Office of Experiment Stations, Circular 110, 1911).

Apropos of the ubiquitous cole slaw of 1890, a news note in October, 1890, records the arrival of "a carload of excellent cabbage" to be "sold in large or small quantities." Readers of Sherwood Anderson's autobiographical *A Story Teller's Story* will recall the ruse adopted by the impecunious mother of the writer to secure the family's winter supply of cabbages.

[8] George W. Sloan, *Fifty Years in Pharmacy* (Indianapolis; Bobbs-Merrill), 1903.

the intrusion of new habits. There were practically no house-keeping magazines bringing knowledge of new skills and different methods. In March, 1923, however, the *Ladies' Home Journal* was taken in 355 of the 9,200 homes in Middletown, and 1,152 other persons bought it on the news-stands; 366 subscribed to *Good Housekeeping,* and 202 copies were purchased on news-stands; 364 subscribed to *McCall's,* and 1,280 bought it on news-stands, 1,709 bought the *Woman's Home Companion,* 1,432 the *Delineator,* and 872 the *Pictorial Review,* the last three including both subscriptions and news-stand purchases. Through these periodicals, as well as through the daily press, billboards, and other channels, modern advertising pounds away at the habits of the Middletown housewife. Whole industries mobilize to impress a new dietary habit upon her.[9] Supplementing them are the new home economics courses in the schools; the old three- and four-day "cooking schools" that came to town every couple of years have vanished, and in their stead 1,500 girls and women from the 9,200 homes of the city are being taught new habits in regard to diet and other aspects of home-making in day and evening classes each school term.

But changes in the idea as to what constitutes a suitable meal come slowly. So close is the tradition of the farm and the heavy-muscled tool-using of 1890 that, like the New Englander who defends doughnuts for breakfast on the ground that lighter foods "digest away on you," Middletown tends to retain the earlier habit of meat and potatoes and hot bread for its breakfast table. The habit of thought still persists widely that a meal is not "right" without meat. A leading butcher still advertises "a tender cutlet for breakfast, a nice roast for dinner, and a good beefsteak for supper," and "most" of his customers follow him to the extent of eating meat "about twice a day when they can afford it."

[9] "The flour millers are trying with an 'eat more bread' propaganda to raise per capita consumption from 200 to 220 pounds per year. . . . Milkmen are trying to bring up per capita milk consumption to one quart per capita per day—it is scarcely two-thirds that now. The butter makers want to bring up per capita butter consumption to Australia's high level of ten pounds above our own. Cheese makers want to raise our cheese consumption by twenty-two pounds per capita, to equal Switzerland's. . . . The tremendous impetus given to fruits, vegetables, nuts, apples, raisins, and prunes by growers' coöperative marketing associations is developing a dietary change." Christine Frederick, "New Wealth" (*Annals of the American Academy of Political and Social Science,* Vol. CXV, No. 294, September, 1924), p. 78.

Just as the eating of food has also a social utility in family
and club, so clothing plays more than the mere workaday rôle
of protecting the body against inclement elements. Shop windows, newspaper advertisements, the throngs on the streets,
sermons, the eagerly read, canny advice of Dorothy Dix in the
daily paper, the clatter of talk in high school corridors, and
indignant remarks of Middletown mothers all make this trans-
parently patent.

In this era of furnace-heated houses and enclosed automobiles the utility of clothing in protecting the body is declining.
"It is important that the skin be kept as near a uniform temperature as possible," wrote Marion Harland. "Nothing will do
this so well as clothing made of pure wool worn next to the
skin" and "covering the body from the extremities to the neck."
The diary of a leading merchant records for the twenty-second
day of May, 1891, "Changed my flannels for cotton this morning." Today flannel underwear is almost as obsolete as the long
black equestrian tights, high-necked, long-sleeved nightgowns
for women, and the heavily-lined trousers of the working men
of a generation ago.

Were the sole use of clothing that of protection of the body,
the urgent local discussion of the "morality" of women's clothing would probably never have arisen. Today, men's clothing
still covers the body decorously from chin to soles of the feet.
Among women and girls, however, skirts have shortened from
the ground to the knee and the lower limbs have been emphasized by sheer silk stockings; more of the arms and neck are
habitually exposed; while the increasing abandonment of petticoats and corsets reveals more of the natural contours of the
body. The Middletown woman of 1890 lived her waking life
from adolescence to death in a tight brace worn under the
clothing from the hips to "within a hand's breadth of the arm-
pit."

"Women will wear corsets," says Marion Harland; "they always have and they always will. We may as well consider this a
settled fact. There are good and sufficient reasons why they are
a necessity to the women who dress as civilized women have for
the past two hundred years." [10]

[10] *Talks Upon Practical Subjects* (New York; The Warner Brothers
Company, 1895), p. 119.

Over the indispensable corset was worn an abundance of long petticoats and other garments that leave many a person of the older generation gasping today before the outfit of a high school girl, often consisting only of brassière, knickers, knee-length dress, low shoes, and silk stockings. Every step in this change has been greeted as a violation of morals and good taste. The editor of the leading Middletown paper in 1890 excoriated "the loose woolen shirts being worn by girls for coolness and comfort in the summer" under the headline, "A Disgusting Fashion":

"We had just as soon see a girl on the street in a night gown as in one of these female blouse rags. No girl possessing any self-respect or any regard for the opinion of respectable people will wear one."

What would this editor and his generation, when no "good man" went to shows featuring girls dressed in tights, have said of a current *revue* staged in the high school auditorium by a group of recent high school graduates in which these locally prominent girls danced with backs bare to the waist and bare thighs? All of which reveals the fact that the moral function of clothing, while it has persisted without variation among the males, has undergone marked modification among the females. As one high school boy confidently remarked, "The most important contribution of our generation is the one-piece bathing suit for women."

All through life clothing apparently occupies a position of less importance among the males than among the females; when the latter reach the age of twelve or thirteen, clothes suddenly leap into a position of dominating importance which lasts until marriage and then slowly tapers off thereafter as clothing is forced to compete with automobiles, home ownership, children, and college educations in the family budget. The greater attention given by girls and women to dress, with increasing brightness of color and variety of costume, suggests the use of clothing most important in the thought of the community. While the moral concern for covering one's nakedness and the physical necessity for securing at least a modicum of warmth and protection are still largely taken for granted, what most people in Middletown mean when they say they "need clothes" is neither of these but this third use of wearing apparel: to gain social recognition. Witness the frank statements of the advertising manager of a leading department store and of the head of a women's ready-to-wear store before the local Ad. Club of their thoroughly pragmatic advertising policy of "appealing first to style, second to price, and last to quality." Clothing, in other words, is addressed to impressing other persons and thus it still maintains one of its most primitive uses.[11]

Even among the men, including working class men, it is apparently less common today than in the nineties to renounce any effort at appearing well-dressed by speaking scornfully of "dudes." "They're no longer content with plain, substantial, low-priced goods," according to the leading men's clothier, "but demand 'nifty' suits that look like those every one else buys and like they see in the movies." Speaking of children's clothing, the same merchant said, "Poor mothers come in today with the attitude that 'My boy is just as good as any other boy,' and they often spend for a suit alone enough to buy a plainer suit, shoes, stockings, and cap." As to the diffusion of more "dressy" clothing habits among women of the working class, according to a business man, "I used to be able to tell something about the background of a girl applying for a job as stenographer by her clothes, but today I often have to wait till she speaks, shows a gold tooth, or otherwise gives me a second clew."

This spread of the habit of being, as one housewife remarked with an ominous shake of her head, "dressed up all the time," is probably not unrelated to such factors as increased clothing advertising;[12] the movies where the woman with a machinist husband sees a different kind of man—"a more dressy man, one more starchy," to quote the evidence in a divorce case; the decrease in individualism and increase in type-consciousness in the local business world; the rise of men's civic clubs with their noon luncheons, and of the women's club movement, and the

[11] John Dewey speaks of the origin of clothing "in situations of unusual awe or prestigious display." *Experience and Nature* (Chicago; Open Court, 1925), p. 385.

[12] See Table XXII. In October, 1923, the leading Middletown paper carried 55,277 lines of women's wear advertising as over against 863 lines in the same month in 1890; 66,474 lines of department store "copy" as against 22,586; 25,289 lines of men's clothing "copy" as against 7,523; and 18,426 lines of shoe advertising as against 6,366. On some mornings the clothing advertisements total two-thirds of the hundred or more columns of advertising in the local daily.

increase in extra-home activities of a civic and social nature; the increased rôle of the wife as the social pace-setter for the family; the diffusion of the habit of thought, in male parlance, that "You've got to spend money to earn money," or in the words of Dorothy Dix, "The world judges us largely by appearance. If we wish to be successful we have got to look successful." The upshot of it all is that, to quote an editorial in the local press: "People weren't as particular in former days about what they wore Monday to Saturday," but "The Sunday suit of clothes is one of the institutions that is vanishing in our generation. . . . Even the overall brigade is apt to wear the same suit week-day evenings as on Sunday."

The early sophistication of the young includes the custom of wearing expensive clothing; as in other social rituals, entrance to high school appears to be the dividing line. The cotton stockings and high black shoes of 1890 are no longer tolerated.[13] The wife of a working man with a total family income of $1,638, said as a matter of course, "No girl can wear cotton stockings to high school. Even in winter my children wear silk stockings with lisle or imitations underneath." Similarly, the invariable dark flannel waists and wool skirts of the nineties, with a silk waist for "dress up," have given way before an insistence upon a varied repertory in everything from sweaters to matching hose. As one business class mother said, "The dresses girls wear to school now used to be considered party dresses. My daughter would consider herself terribly abused if she had to wear the same dress to school two successive days." Another business class mother said that she started her thirteen-year-old daughter to high school wearing fine gingham dresses and lisle hose, thinking her very suitably dressed. "Mother, I am just an object of mercy!" wailed the child after a few days at school, and her mother, like many others, provided silk dresses and silk stockings rather than have her marked as peculiar.

The taboos of this intermediate generation on plain dress are

[13] A Middletown woman of the working class who thirty-five years ago bought three pairs of cotton stockings for herself or for a daughter and made them last a whole year may today sacrifice durability and pay thirty-five cents, eighty cents, a dollar or more, for meretricious silk hose. Only 1,570,000 pounds of artificial silk was produced in the United States in 1913. In 1923 the country produced 33,500,000 pounds, though not all of this was for domestic consumption.

the effective taboos of their elders: the males stay away. Girls fight with clothes in competition for a mate as truly as the Indians of the Northwest coast fight with the potlatch for social prestige. Since one of the chief criteria for eligibility for membership in the exclusive girls' clubs is the ability to attract boys, a plainly dressed girl feels the double force of this taboo by failing to receive a "bid" which she might otherwise get. "We have to have boys for the Christmas dances, so we take in the girls who can bring the boys," explained a senior member of the most exclusive high school girls' club. A girl who had just been asked in her junior year to join this club said, "I've known these girls always, but I've never been asked to join before; it's just clothes and money that make the difference. Mother has let me spend more money on clothes this last year." A fifteen-year-old son, wise in the ways of his world, protested to his mother because his sister of fourteen in the eighth grade wore lisle stockings to school: "Well, if you don't let her wear silk ones next term when she goes to high school," was his final retort, "none of the boys will like her or have anything to do with her." "I never thought I would dress *my* daughter this way, but it's a concession I had to make for her happiness," is a remark heard over and over again; in many cases "happiness" is frankly accepted as meaning popularity. Summer and winter the contest keeps up. "I want to get my daughter of fifteen off to a summer camp," insisted one harried mother. "I dread summer particularly because so many youngsters spend all their time worrying about the proper way to dress."

Daughters of families of the working class do what they can to keep up with the procession, and if they fall too far behind, frequently leave school. A working class mother of five children, with a family income of $1,363, complains that her oldest daughters of eleven and twelve are "so stuck up I can't sew for them any more," and despite poor health, she has tried to get any kind of work outside her home, "so as to hire their sewing done." In discussing the question of their children's leaving school, many workers' wives said:

"She stopped because she was too proud to go to school unless she could have clothes like the others"; or "Most of my time goes into sewing for my daughter. She's sixteen and I do want her to 'keep on until she graduates from high school and she wants to too, but she won't go unless she has what she considers proper clothes." [14]

Numerous dances and elaborate commencement activities, over and above the everyday social contacts at high school, exert further pressure for expense in dress. A working man and his wife, both of whom work in order to finance the education of their children, spent $150 in 1923 and again in 1924 to outfit their two sons for their respective high school commencements. "It takes two suits and many accessories to be properly dressed for the different occasions," the father explained. Just because they are somewhat *déclassé* socially, the compulsion to dress as the crowd does frequently rests as a doubly compelling necessity upon children of this class.

Only a minority of the junior and senior boys attending the big dances wear Tuxedos, but the obligatory nature of special evening dress for the girls is much more marked. "My daughter will have four evening dresses this winter," said one mother of the business group, "two left over which were new last year, and two new ones." For commencement a girl "needs three or four new dresses." In addition to these, evening dresses are required for the ten to fifteen Christmas fraternity, sorority, and other dances to which a popular girl goes. One girl in moderate circumstances had "three evening dresses and a couple of afternoon dresses," while "the most popular girl in the class has half-a-dozen evening dresses." [15]

All of which helps to explain the observed trend in the preparation of clothing in Middletown. The providing of clothing for individual members of the family is traditionally an activity of the home, but since the nineties it has tended to be less a hand-skill activity of the wife in the home and more a part

[14] Cf. discussion of reasons for leaving school in Ch. XIII.

[15] The importance of the right dress in this battle of clothes was naïvely described by a recent Middletown high school graduate who had appeared at one of the Christmas dances in a $90 "rust-colored gown." "I was so pleased that nobody spoke of how nice it looked. You know if your mother makes a dress at home people always come up to you and say, 'How ni-ice you look! Isn't it pretty!' and you know they're condescending. I was sorry though that my dress wasn't described in the paper, for you always like that. But it isn't the best dresses that are usually described."

Such descriptions as the following appear in the Middletown paper month after month after every high school dance: "Many fashionable dresses were

of the husband's money-earning. One of the housewives interviewed lived in 1890 on a farm just south of town, where wool
was clipped from sheep and practically all the family clothing
spun by the women of the family. The common practice a generation ago, however, was to buy "goods" and make the garments at home. As late as 1910 there was practically no advertising of women's dresses in the local newspapers, and goods
by the yard was prominently featured. Today the demand for
piece goods is, according to the head of the piece goods department of Middletown's largest department store, "only a
fraction of that in 1890." This store conducted a sale in 1924
at which two bolts of the featured material were sold, "but in
1890 we'd have sold ten bolts the first day."

Twenty-five of the 112 wives of working men for whom data
were secured and twenty of the thirty-nine wives of the business class, as over against estimates of four out of sixty-six
and eleven out of thirty-five respectively for their mothers,
spend not over two hours a week on all sewing and mending for
themselves, their husbands, and their children; while seventy-
two (nearly two-thirds) of the working class women and
twenty-nine (nearly three-fourths) of the business class women
today spend not over six hours. Of this sewing that continues
to be done in the home, the greater part is concentrated today
on the clothing of the female members of the family and of the
small children. Ninety-two (four out of five) of the 112 working class wives do no sewing other than current mending for
their husbands, as against thirty-eight (one in three) of their
mothers for whom estimates were made by them; while thirty-
five (nine in ten) of this group of thirty-nine business class
wives do no sewing for their husbands today, as against twenty-
nine (three in four) of their mothers. According to the leading
men's and boys' outfitters, virtually none of the boys' winter

worn by the guests, among them being a green crêpe built over gold satin,
which was worn by Miss ——. The blouse was made basque effect and
trimmed with gold lace. A gold headband, gold slippers, and gold hose were
worn with this dress. Miss —— was lovely in a smart frock of flowered
pussy-willow taffeta made over white crêpe de chine. It was fashioned tunic
effect and caught with turquoise blue ribbons. She wore white hose and
black satin slippers and a blue headband. Another attractive gown was that
worn by Miss ——, which was of green brocaded chiffon velvet, fashioned
with flower trimmings. She wore black satin slippers and matching hose.
Miss —— was stunning in a gown of dark blue chenille, fashioned with dark
brown fur trimmings; she wore gold hose and gold slippers."

suits worn in Middletown today is home-made, whereas he estimated that about half of them were made in the home in 1890.

It is characteristic of the customary lags and friction of institutional life in a period of change like the present that women,
thus forced into the market to buy and urged on by a heavily
increased advertising appeal, must as yet perform this task
dependent almost entirely upon the counsel of the selling agent,
whose primary concern is to capture the market. In the absence
of the knowledge requisite for buying on the basis of quality,
the Middletown housewife must in the main, as noted above,
depend upon "looks" and "price." The following frank statement by a local women's ready-to-wear dealer in the course of a
discussion at the Ad. Club reveals the helplessness of the buyer
who has to depend upon such criteria:

The question was raised as to women's liking higher-priced garments and buying by price, provided a garment has style. One
dealer said, "You men can't blame us: of two garments of the
same quality the women will buy the more expensive; they don't
know, but *think* it must be better. You wouldn't have us lose that
sale to her, would you? She mightn't buy at the lower price because her friend may have paid $70 for her coat and she is sus-
picious of a $40 coat. Of course, we have to put the price where
she'll buy under those circumstances."

This reflects the problem of the women with moderate money
leeway; the very poorest group are hammered at directly on a
low-price basis by one large store specializing in this trade.[16]

To the student of cultural change the battle royal going on at
present about this institution of clothing in Middletown raises
interesting questions. Men depend less than women upon this
device of dress to establish them among the group; will the
present feverish preoccupation of the younger women with this
activity of personal ornamentation diminish as their educa-

[16] It is noteworthy that the home economics work in the schools still con-
centrates largely upon sewing rather than upon buying. Cf. the conclusion of
the elaborate clothing survey conducted by the Junior and Senior Home
Economics Curriculum Revision Committee of the Denver Public Schools
in 1924: "This shows very clearly that the major problem in this field
is one of clothing selection, as the garments which are bought ready-made
in almost every instance greatly outnumber those made in the home." The
Denver situation probably does not differ greatly from that in Middletown.

Cf. Henry Harap's *Education for Consumption* (New York; Macmillan,
1924) for a discussion of the infinite variety of substitute materials that
the housewife buys blindly under such blanket titles as "all wool" and
"mahogany."

tion becomes more realistic, their lives more active, and their interests less confined to current household drudgeries ignored by men? As the woman becomes less the passive recipient of the man's advances in mating, will her present use of dress appeal diminish? The physical house and yard in which Middletown lives have yielded somewhat to the automobile as the outward emblem of conspicuous membership in this competitive society; will the automobile, the new religion of higher education, and other new means to fuller living cut down the money spent on clothing—as some shoe and men's clothing dealers complain the automobile is already doing in Middletown? What emotional outlet in these days of quickly assembled home sewing is compensating Middletown's women for the pride in needle craftsmanship, in the neat seam, overcast and bound, dear to the housewife of a generation ago? What does the trend towards greater simplicity of line foreshadow? There are fewer ruffles and tucks, plainer flounces, straight instead of gored skirts, no fitted bodices and sleeves—and in their place are straight-line one-piece dresses "as easy to make as a doll's dress." Will this tend to reinstate sewing in the home as advertising of new wants tightens the competition for the family income? Already the newer freedom and aggressiveness of women is insisting upon more comfort, looser clothing, and greater freedom of limb;[17] in a neighboring small city there was a lawsuit during 1925 over the right of a school-girl to wear knickers instead of a skirt to school. "Where," the editor of the 1890 Middletown *News* might ask, "will this freedom end?"

At no point can one approach the home life of Middletown without becoming aware of the shift taking place in the traditional activities of male and female. This is especially marked in the complex of activities known as "housework," which have always been almost exclusively performed by the wife, with more or less help from her daughters. In the growing number of working class families in which the wife helps to earn the family living, the husband is beginning to share directly in

[17] The substantial Middletown housewife of today does her morning housework in a light, loose-fitting, short-sleeved wash dress; in 1890, even in summer, she wore a shirtwaist with a high collar and long sleeves, and a wool skirt over a flannel petticoat, with a broken whalebone in her second-best corset gouging her somewhere down underneath all these clothes.

housework. Even in families of the business class the manual activities of the wife in making a home are being more and more replaced by goods and services produced or performed by other agencies in return for a money price, thus throwing ever greater emphasis upon the money-getting activities of the husband. This is simply another instance of the shuffling about of "men's ways" and "women's ways" observable among all peoples, for "it is partly a matter of accident as to how culture is adjusted to the two parts of the group." [18]

As noted earlier, the rhythm of the day's activities varies according to whether a family is of the working or business class, most of the former starting the day at six or earlier and the latter somewhat later. Of the group of 112 working class wives who estimated the amount of time they spend on housework "on an average week-day when there is 'nothing extra,' " [19] eight spend less than four hours a day, seventy-seven (more than two-thirds) four or more but less than seven hours, and twenty-seven (less than a fourth) seven or more. Of the forty wives of the business class making a similar estimate, nine spend less than four hours a day, twenty-two (more than half) four or more but less than seven hours, and nine seven or more. Only one of the working class wives interviewed does no housework in the afternoon, eight do less than one hour, and forty-eight less than two hours, while twelve (nearly a third) of the wives of the business class interviewed do no housework in the afternoon, fifteen less than one hour, and thirty less than two hours.

All housewives desire to reduce their hours spent in housework; starting from this general attitude, at least five factors may be observed as tending singly or in varying combinations to determine a given wife's daily routine: the amount of the family income, servants or available women kinsfolk in the home, the number of her children, age of her children, and her morale, i.e., whether or not she is discouraged with her outlook. The upshot of the interaction of these factors is three main types of housewives: the few to whom housework is a

[18] Wissler, *op. cit.*, p. 95.

[19] These estimates are, of course, only roughly accurate. An effort was made to secure the greatest possible care, by beginning at the rising hour and setting down what the housewife did in different chunks of time throughout the day. Housework was defined as routine cleaning, preparing of food, etc., exclusive of "extra" cleaning or baking or laundry work.

minor concern engaging but an incidental part of each day; the large group who by careful management fit everything somehow into the morning and an afternoon hour or two and contrive to keep many afternoons and evenings relatively free for children, social life, and civic activities; and finally the group, as large as or larger than the second, for whom each day is a nip-and-tuck race to accomplish the absolute essentials between morning and bedtime, with occasional afternoons or evenings free only by planning in advance.

Dorothy Dix catches the traditional situation in her remark, "Marriage brings a woman a life sentence of hard labor in her home. Her work is the most monotonous in the world and she has no escape from it." Many working class housewives, struggling to commute this sentence for their daughters if not for themselves, voiced in some form the wish of one mother, "I've always wanted my girls to do something other than housework; I don't want *them* to be house drudges like me!" And both groups are being borne along on the wave of material changes toward a somewhat lighter sentence to household servitude. Of the ninety-one working class wives who gave data on the amount of time their mothers spent on housework as compared with themselves, sixty-six (nearly three-fourths) said that their mothers spent more time, ten approximately the same, and fifteen less time. Of the thirty-seven wives of the business group interviewed who gave similar data, seventeen said that their mothers spent more time, eight about the same, and twelve less time.[20]

The fact that the difference between the women of this business group and their mothers is less marked than that between the working class women and their mothers is traceable in part to the decrease in the amount of paid help in the homes of the business class. It is apparently about half as frequent for Middletown housewives to hire full-time servant girls to do their housework today as in 1890. The thirty-nine wives of the business group answering on this point reported almost precisely half as many full-time servants as their mothers in 1890, and this ratio is supported by Federal Census figures;[21] thir-

[20] The lower percentage of farmers' wives among the mothers of the business class should be borne in mind. Cf. Ch. V, N. 8.

[21] According to the Federal Census, the number of families in the State to each "servant" increased from 13.5 in 1890 to 30.5 in 1920. (Middletown

teen of the thirty-nine have full-time servants, only two of them more than one. But if the women of the business class have fewer servants than their mothers, they are still markedly more served than the working class. One hundred and twelve out of 118 working class women had no paid help at all during the year preceding the interview, while only four of the thirty-nine women of the business group interviewed had had no help; one of the former group and twenty-five of the latter group had the equivalent of one or more days a week. Both groups of house-wives have been affected by the reduction in the number of "old maid" sisters and daughters performing the same duties as do-mestic servants but without receiving a fixed compensation. Prominent among the factors involved in this diminution of full-time servants are the increased opportunities for women to get a living in other kinds of work; [22] the greater cost of a "hired girl," ten to fifteen dollars a week as against three dollars in 1890; and increased attention to child-rearing, making mothers more careful about the kind of servants they employ. "Every one has the same problem today," said one thoughtful mother. "It is easy to get good girls by the hour but very diffi-cult to get any one good to stay all the time. Then, too, the best type of girl, with whom I feel safe to leave the children, wants to eat with the family." The result is a fortification of the tend-ency to spend time on the children and transfer other things to

figures for 1890 are not available, but in 1920 there were 25.5 families for each "servant.") Likewise the number of persons to each "servant" was, for the United States, 43.0 in 1890 and 83.2 in 1920; for the State 63.6 in 1890 and 121.2 in 1920; for Middletown in 1920 it was 97.9.

[22] Cf. the discussion of women's work in Ch. V. Throughout the United States, 31.1 per cent. of all women ten years of age and older gainfully employed in 1890 were "servants" as over against 11.8 in 1920.

A study of servant girl labor in eight cities in the state is reported in the *Fifth Biennial Report of the [State] Department of Statistics: 1893-4*, pp. 173 ff. Two cities of about the size of Middletown and situated in the same general section of the state probably reflect Middletown conditions fairly accurately. Two servant girls were selected for each 1,000 of popu-lation in each city. It is not related how this sample of seventy-six servants from these two cities was chosen. Seventy of the seventy-six, however, were white; of the seventy for whom ages were secured, twenty-nine were twenty or younger, forty-seven were twenty-five or younger, fifty were thirty or younger, sixty-one were thirty-five or younger, and only nine over thirty-five, including but two over fifty. In other words, these are precisely the age groups from which women in industry in Middletown today are most heavily recruited. When times are bad in Middletown a want ad. in the paper brings a long list of applicants for work as servants, but when times are good replies are scarce.

service agencies outside the home. A common substitute for a full-time servant today is the woman who "comes in" one or two days a week. A single day's labor of this sort today costs approximately what the housewife's mother paid for a week's work.

Smaller houses, easier to "keep up," labor-saving devices,[23] canned goods, baker's bread, less heavy meals, and ready-made clothing are among the places where the lack of servants is being compensated for and time saved today. Working class housewives repeatedly speak, also, of the use of running water, the shift from wood to coal fires, and the use of linoleum on floors as time-savers. Wives of the business class stress certain non-material changes as well. "I am not as particular as my mother," said many of these housewives, or, "I sometimes leave my supper dishes until morning, which my mother would never have thought of doing. She used to do a much more elaborate fall and spring cleaning, which lasted a week or two.[24] I consider time for reading and clubs and my children more important than such careful housework and I just don't do it." These women, on the other hand, mention numerous factors making their work harder than their mothers'. "The constant soot and cinders in this soft-coal city and the hard, alkaline water make up for all you save by new conveniences." A number feel that while the actual physical labor of housework is less and one is less particular about many details, rising standards in other respects use up the saved time. "People are more particular about diet today. They care more about having things nicely served and dressing for dinner. So many things our mothers didn't know about we feel that we ought to do for our children." [25]

Most important among these various factors affecting

[23] At least two of the business class wives interviewed said, in comparing the time they spend at housework with their mothers, "My labor-saving devices just about offset my lack of a maid."

[24] The diary of a leading merchant records in October, 1887: "Wed. 19th, Bessie commenced cleaning house yesterday morning. . . . Fri. 21st. Still cleaning house. . . . Sat. 22nd. Still cleaning house; done with upper part of house (except our room), parlors and lower hall. . . . Mon. 24th. House cleaning still going on; our room in the mess today. . . . Tues. 25th. House cleaning finished last night." The disappearance of carpets and heavy draperies combine with the custom of "keeping the house up" all the time to make such an orgy unnecessary today.

[25] Cf. Benjamin R. Andrews, *Economics of the Household* (New York; Macmillan, 1923), pp. 54-5, 408.

women's work is the increased use of labor-saving devices. Just as the advent of the Owens machine in one of Middletown's largest plants has unseated a glass-blowing process that had come down largely unchanged from the days of the early Egyptians, so in the homes of Middletown certain primitive hand skills have been shifted overnight to modern machines. The oil lamp, the gas flare, the broom, the pump, the water bucket, the washboard, the flatiron, the cook stove, all only slightly modified forms of some of man's most primitive tools, dominated Middletown housework in the nineties. In 1924, as noted above, all but 1 per cent. of Middletown's houses were wired for electricity.[26] Between March, 1920, and February, 1924, there was an average increase of 25 per cent. in the K.W.H. of current used by each local family. How this additional current is being used may be inferred from the following record of sales of electrical appliances by five local electrical shops, a prominent drug store, and the local electric power company for the only items it sells, irons and toasters, over the six-month period from May first to October thirty-first, 1923: curlers sold, 1,173; irons, 1,114; vacuum cleaners, 709; toasters, 463; washing machines, 371; heaters, 114; heating pads, 18; electric refrigerators, 11; electric ranges, 3; electric ironers, 1.[27] The manager of the local electric power company estimates that nearly 90 per cent. of Middletown homes have electric irons.

It is in part by compelling advertising couched in terms of certain of women's greatest values that use of these material tools is being so widely diffused:

"Isn't Bobby more important than his clothes?" demands an advertisement of the "Power Laundries" in a Middletown paper.

[26] The nineties witnessed the development of electric fans, heaters, and toasters, and the next decade the electric clothes washer and vacuum cleaner. As late as 1914, however, only a million dollars' worth of washing machines and about a million and a third dollars' worth of vacuum cleaners were sold in the United States. By 1916 washing machine sales were $7,000,000; 1919, $50,000,000; 1920, $85,000,000; and 1923, $82,000,000. In 1923, $50,000,000 of vacuum cleaners were sold.

[27] These figures do not pretend to be complete for the entire city, as they omit, e.g., the leading department store, which includes among its electrical lines the popular Hoover vacuum sweeper; they probably indicate roughly, however, the distribution of the various appliances. The count omits electrically driven sewing machines, percolators, possibly one or two dish washers, and electric cookers, of which last there are reported to be approximately 200 of one locally manufactured brand alone in use in the city.

The advertisement of an electrical company reads, "This is the test of a successful mother—she puts first things first. She does not give to sweeping the time that belongs to her children. . . . Men are judged successful according to their power to delegate work. Similarly the wise woman delegates to electricity all that electricity can do. She cannot delegate the one task most important. Human lives are in her keeping; their future is molded by her hands and heart."

Another laundry advertisement beckons: "Time for sale! Will you buy? Where can you buy back a single yesterday? Nowhere, of course. Yet, right in your city, you can purchase tomorrows. Time for youth and beauty! Time for club work, for church and community activities. Time for books and plays and concerts. Time for home and children."

Telephones came to Middletown in the early eighties, and by 1890 there were only seventy-one subscribers. So slowly did their use spread, that as late as 1900 the local press finds it necessary to warn the public that, when using the telephone, they "should not ask for a name but refer to the number list." "It was not until about the time of the Spanish War," according to a woman of the business class, "that 'phones became common enough so that you thought of feeling apologetic at not having one." In April, 1924, 4,348 Middletown homes had a telephone, leaving about 5,000 without this tool.[28] All forty of the business class families interviewed had telephones, while only 55 per cent. of the 122 workers' families for whom this point was checked had 'phones.[29] Around the telephone have grown up such time-savers for the housewife as the general delivery system for everything from groceries to a spool of thread.

The weekly washing and ironing of family clothing tends to make markedly different inroads upon the time of housewives

[28] The large number of families without 'phones is due in part to lack of plant facilities in certain newer sections of the city. The Commercial Department of the Bell Telephone Company anticipates connecting "at least 2,000 telephones within the next five years."

[29] This differential diffusion is important in view of the different emphasis upon vicinage in the social habits of the two groups noted in Ch. XIX.

The Commercial Department of the Bell Telephone Company classifies the houses and lots in Middletown into four groups: Class A (most expensive) 936; B, 1,822; C, 3,910; D (cheapest), 2,491; and these are rated as telephone prospects as follows: (Class A was reduced somewhat and the difference redistributed between B and C, on the advice of local

of the working class and of the business class, with somewhat less marked shifts between the generations:

	No. of Working Class Families		No. of Business Class Families	
	1924	*1890*	*1924*	*1890*
Total number of families 120		94	40	40
Spend no time on washing or ironing	0	5	24	20
Spend some time but less than two hours weekly..................	2	1	0	1
Spend two to four hours weekly...	24	3	6	4
Spend five to eight hours weekly...	65	28	8	6
Spend nine or more hours weekly...	29	57	2	9

Commercial laundries play a larger rôle today than in the nineties. Practically all family laundry in Middletown was done then either in the home or "taken out" by a "washwoman," and, according to the proprietor of a local laundry, local commercial laundries did almost exclusively hotel and boarding-house work and men's bundle work, i.e., "boiled" shirts, collars and cuffs, underwear, and handkerchiefs; about the only women's clothing done by them was some shirtwaists.[30] The advent of individually owned electric washing machines and electric irons has, however, slowed up the trend of laundry work—following baking, canning, sewing, and other items of household activity—out of the home to large-scale commercial agencies; Middletown laundries report some loss of trade.[31]

The rapid and uncontrolled spread of such new devices as labor-saving machinery under a system of free competition makes the housekeeping of Middletown present a crazy-quilt

real estate experts, in the revised distribution according to valuation given above, but the original distribution is retained here).

(A) Prospects for single party 'phones.

(B) Ten per cent. of these are prospects for single-party 'phones and 90 per cent. two-party.

(C) Two-thirds of these are telephone prospects, 10 to 15 per cent. one-party, 25 to 30 per cent. two-party, and the remaining four-party.

(D) Sixteen and two-thirds per cent. of these are prospects for 'phones; of these 20 per cent. are prospects for two-party 'phones and the balance for four-party 'phones. "Probably about 7 per cent. of the total class 'D' families now have service." (Letter of October 20, 1924.)

[30] Local laundries today offer six types of service, of which "rough dry," introduced about 1900 and consisting in washing everything, ironing flat pieces, and starching those requiring it, is most popular for family work, while next most popular comes "wet wash," catering to poorer families and consisting merely in washing.

[31] This is an example of the way in which a useful new invention vig-

appearance. A single home may be operated in the twentieth century when it comes to ownership of automobile and vacuum cleaner, while its lack of a bathtub may throw it back into another era and its lack of sewer connection and custom of pumping drinking-water from a well in the same back yard with the family "privy" put it on a par with life in the Middle Ages. Side by side in the same block one observes families using in one case a broom, in another a carpet sweeper, and in a third a vacuum cleaner for an identical use, or such widely varying methods of getting clothes clean as using a scrub board, a hand washing machine, an electric machine, having a woman come to the house to wash, sending the clothes out to a woman, or sending them to a laundry for any one of six kinds of laundry service. Even as late as 1890, when housework was still a craft passed down from mother to daughter—what Bentham scornfully called "ancestor wisdom"—the Middletown housewife moved in a narrower world of "either-or." "There is a right way and a wrong way to do everything, and you might just as well learn the right way as the wrong way," was the household gospel on which many present-day housewives were brought up. Likewise the routine of housework was more important in a day when "women's work [was] never done" and the twin concerns of home work and church work constituted the "right" activity for a woman and all the slow hand processes of the household had somehow to be crowded within the waking hours of the six working days of the week with little

orously pushed on the market by effective advertising may serve to slow up a secular trend. The heavy investment by the individual family in an electric washing machine costing from $60 to $200 tends to perpetuate a questionable institutional set-up—whereby many individual homes repeat common tasks day after day in isolated units—by forcing back into the individual home a process that was following belatedly the trend in industry towards centralized operation. The whole procedure follows the customary, haphazard practice of social change: the issue is not settled on its merits; each Middletown household stands an isolated unit in the midst of a baffling battery of diffusion from personally interested agencies: the manufacturers of laundry machinery spray the thinking of the housewife through her magazines with a shower of "educational" copy about the mistake of a woman's neglecting her children for mere laundry work, and she lays down her magazine to answer the door-bell and finds there the suave installment payment salesman ready to install a $155 washer on trial, to be paid for in weekly installments if satisfactory. The whole question has to be settled in terms of such immediate and often incidental considerations by each isolated family unit.

room for leisure. Today the various alternatives in the performance of her housework combine with the new alternative things to do in her newly-acquired leisure time to turn the either-or world of her mother into an array of multiple choices.

Here, as in the case of clothing, advertising and the multiplication of alternative standardized brands and methods are thrusting new illiteracies upon the home and its members more rapidly than the traditional "liberal education" and the deeply-grooved patterns under which "vocational education" is conceived are shifting to meet them. As more and more of the things utilized as food, clothing, and shelter are being shifted from home production to purchase for a price and as standards of living shift and varieties multiply, the housewife must distinguish among various grades of milk, bread, vacuum cleaners, and fireless cookers, must decide against rugs made of grass for another type or vice versa, and so on through hundreds of items. At a time when growing industrial units in Middletown are more and more entrusting their purchasing to specialized skill, the home is, as in 1890, an essentially isolated, small-unit purchaser served by an untrained, amateur purchasing agent exposed to competing, highly organized channels of diffusion.

Here again in these social sprains apparent in the activities of housekeeping one simply glimpses the enforced re-orientation taking place in every department of making a home. In the agricultural life so recently behind Middletown,

"the family was an economic institution as well as a biological and affectional. It possessed recreational, educational, and protective, as well as economic functions. The woman's duties of spinning, weaving, sewing, preparing food, and gardening were as economically important as the man's different activities. Divorce was infrequent in part because it disturbed these fundamental and supplementary economic and industrial activities. The family afforded a place where boys learned their trade, and where girls were trained to be skilled housewives. Owing to its settled abode and definite resources in property, dependent kin were easily protected and cared for." [32]

New cultural demands pressing upon this earlier compact home and family are altering its form: geographical vicinage and permanence of abode apparently play a weaker part in family

[32] F. S. Chapin, "The Lag of Family Mores in Social Culture," *Journal of Applied Sociology*, Vol. IX, No. 4, p. 245.

life; there are fewer children and other dependents in the home to hold husband and wife together; activities adapted to the age, sex, and temperament of its members are replacing many whole-family activities; with the growth of these extra-home activities involving money expenditure comes an increased emphasis upon the money nexus between members of the family; the impetus toward higher education, sending an increasing proportion of boys into lines of work not shared by their fathers, is likewise tending to widen the gap between the generations in standards of living and habits of thought; such new tools as the telephone and the automobile, while helping to keep members of the family in touch with each other, are also serving to make separate activities easier.

Such changes go on in the midst of protest; fresh encroachments tend to be met by a reassertion of the traditional *noli tangere* attitude toward the "sacred institution" of the home: [33]

"Our children should have a growing conception of the family as a sacred organization," says the Course of Study for Elementary Schools for 1924.

"There are four big, fine thoughts that must be thought in every community," said a prominent Middletown club woman in addressing a community meeting in 1924, "thoughts of the home, of the church, of the school, of the community. The home is the first institution that God made. He intended it to be the unit of society—a father, mother, and children."

"There are three notable words in the English language," said a paper given before one of the federated women's clubs in 1924, "mother, home, and heaven. . . . When women begin to realize the full value and importance of the profession of home-making, it grows dignified and glorified in their eyes and in the eyes of the community. It becomes an adequate career, and no girl need think she must leave her home in order to achieve her individuality. There is nothing like being a good home-maker to be able to raise up a child in the way he should go."

"Home life," according to a leading Middletown business man, "is the center of everything; if that is gone, everything is gone."

A speaker at the Middletown Chautauqua said, "We seem to be drifting away from the fundamentals in our home life. The

[33] The process of the secularization of man's several activities, involving "constant appeal from the new and clamorous economic interests of the day to the traditional Christian morality," has been described by R. H. Tawney in *Religion and the Rise of Capitalism* (New York; Harcourt, Brace, 1926).

home was once a sacred institution where the family spent most of its time. Now it is a physical service station except for the old and infirm."

Here, as elsewhere, one thing that determines whether a given change is welcomed or resisted appears to be the fact of its unsettling simply material traits or the personal attitudes, ideals, and values of the population. Changes, such as the adoption of electrical devices, refrigeration, and similar material factors, occur with relatively greater ease and speed than those touching marriage, the relation of parents and children, and other "sanctities" of family life. Thus Middletown faces in two directions as regards child-rearing: problems of child health, clothing, and diet are accepted as subjects for empirical study and matter-of-fact handling, whereas the "morals," "personalities," and "characters" of its children are heavily barricaded against any direct, matter-of-fact approach. In a forum called to discuss child-rearing, a mother prominent in the Mothers' Council and in the educational work of the city said, "I believe there is no boy problem. Children are not a problem, but a joy. We can trust them, and God will lead them." A leading minister, preaching on "The Bible as the Foundation of the Home," said, "We learn more by way of our heart than we do by our head." Many of the papers given before the Mothers' Council tend to deal in generalities rather than addressing themselves to specific situations, e.g., "The duty of parents is to make their children realize the sacred duty of every man to vote, to exercise the political sacrament of the ballot." In the case of some of the working class families there is a tendency to blanket questions of child-rearing under formulas; in answer to every question in regard to children these mothers would not answer in terms of the specific situation, but by some such blanket-explanation as, "The cause of that is sin" (or "the devil") and "the only remedy is salvation."

III. TRAINING THE YOUNG

Chapter XIII

WHO GO TO SCHOOL?

In an institutional world as seemingly elaborate and complex as that of Middletown, the orientation of the child presents an acute problem. Living goes on all about him at a brisk pace, speeded up at every point by the utilization of complex shorthand devices—ranging all the way from the alphabet to daily market quotations and automatic machinery—through which vast quantities of intricate social capital are made to serve the needs of the commonest member of the group. As already noted, the home operates as an important transfer point of civilization, mediating this surrounding institutional world to the uninitiated newcomer. Religious agencies take a limited part in the child's training at the option of his parents. *True Story* on the news-stand, *Flaming Youth* on the screen, books from the public library, the daily friction of life with playmates—all these make their casual though not insignificant contributions. But it is by yet another agency, the school, that the most formal and systematic training is imparted.

When the child is six the community for the first time concerns itself with his training, and his systematic, high-pressure orientation to life begins. He continues to live at home under the nominal supervision of his parents, but for four to six hours a day,[1] five days a week, nine months of the year, his life becomes almost as definitely routinized as his father's in shop or office, and even more so than his mother's at home; he "goes to school."

Prior to 1897 when the first state "compulsory education" law was passed, the child's orientation to life might continue throughout as casually as in the first six years. Even after the coming of compulsory schooling only twelve consecutive weeks'

[1] Ranging from three hours and fifty minutes, exclusive of recess periods, in the first and second years, to five hours and fifty minutes in every year above the seventh. In 1890 it was five hours daily for all years.

attendance each year between the ages of eight and fourteen was at first required. During the last thirty years, however, the tendency has been not only to require more constant attendance during each year,[2] but to extend the years that must be devoted to this formal, group-directed training both upward and downward. Today, no person may stop attending school until he is fourteen,[3] while by taking over and expanding in 1924 the kindergartens, hitherto private semi-charitable organizations, the community is now allowing children of five and even of four, if room permits, to receive training at public expense.

This solicitude on the part of Middletown that its young have "an education" is reflected in the fact that no less than 45 per cent. of all money expended by the city in 1925 was devoted to its schools. The fourteen school plants are valued at $1,600,-000—nearly nine times the value of the school equipment in 1890.[4] During 1923-24 nearly seven out of ten of all those in the city between the ages of six and twenty-one were going regularly to day school, while many others of all ages were attending night classes.

No records are kept in the Middletown schools of the ages and grades at which children withdraw from school. The lengthening average number of years during which each child remains in school today can only be inferred from the heavier attendance in high school and college. While the city's population has increased but three-and-one-half-fold since 1890, enrollment in the four grades of the high school has mounted nearly elevenfold, and the number of those graduating has increased nineteenfold. In 1889-90 there were 170 pupils in the

[2] The average daily attendance in the two school years 1889-91 was 66 per cent. of the school enrollment as against 83 per cent. for the two school years 1922-24.

[3] Cf. Ch. V. Children of fourteen who have completed the eighth grade in school may be given a certificate allowing them to start getting a living, provided they can prove that money is needed for the support of their families and that they attend school in special part-time classes at least five hours a week; children who have not finished the eighth grade may start getting a living at sixteen; the community supervises the conditions under which they shall work until they are eighteen. Until 1924 the upper age limit for required school attendance was fourteen. The state law allows a city to require the minimum school attendance (five hours a week) for all children up to eighteen, but Middletown does not do this.

[4] The annual expenditures of the state for elementary and secondary education, meanwhile, increased from $5,245,218 in 1890 to $63,358,807 in 1922.

high school, one for every sixty-seven persons in the city, and the high school enrollment was only 8 per cent. of the total school enrollment, whereas in 1923-24 there were 1,849 pupils in high school, one for every twenty-one persons in the city, and the high school enrollment was 25 per cent. of the total school enrollment. In other words, most of Middletown's children now extend their education past the elementary school into grades nine to twelve. In 1882, five graduated from high school, one for each 1,110 persons in the community;[5] in 1890 fourteen graduated, one for each 810 persons; in 1899 thirty-four graduated, "one of the largest graduating classes the city ever had," making one for each 588 persons;[6] in 1920 114 graduated, or one for each 320 persons; and in 1924, 236 graduated, or one for each 161.[7]

Equally striking is the pressure for training even beyond high school. Of those who continue their training for twelve years, long enough to graduate from high school, over a third prolong it still further in college or normal work. Two of the fourteen members of the high school graduating class of 1890 and nine of the thirty-two graduates of 1894 eventually entered a college or normal school, while by the middle of the October following graduation, a check of 153 of the 236 members of the class of 1924 revealed eighty as already in college, thirty-six of them in colleges other than the local college and forty-four taking either the four-year college course or normal training at the local college.[8] Between 1890 and 1924, while the population of the state increased only approximately 25 per cent., the number of students enrolled in the State University in-

[5] Population estimated at 5,550.

[6] Population estimated at 20,000.

[7] The 1920 Federal Census showed 76 per cent. of the city's population aged fourteen and fifteen and 30 per cent. of the group aged sixteen and seventeen as in attendance at school; the doubling of the high school graduating class between 1920 and 1924 suggests a substantial increase today over these 1920 percentages. According to the State Department of Public Instruction, high school attendance throughout the state increased 56 per cent. during the five years 1920 to 1924.

The high school in Middletown is used by the township, but the number of pupils from outside the city is small and the population of Middletown has therefore been used above as the basis in figuring.

[8] The 1924 data are from published lists in the high school paper and are not a sample, but probably include the majority of those who went to college. In addition to the eighty accounted for above, seven more were in business college, one in an art school at the state capitol, and three were taking post-graduate courses in high school.

creased nearly 700 per cent., and the number of those graduating nearly 800 per cent. During the same period the number of students enrolled in the state engineering and agricultural college increased 600 per cent., and the number of those graduating over 1,000 per cent.

Even among those who do not go on to college or do not finish high school the same leaven is working; there were, in the spring of 1925, 1,890 enrollments in evening courses in the local schools—719 of them in trade and industry courses,[9] 175 in commercial courses, and 996 in home-making courses.[10]

In addition to other forms of training, fifty to one hundred people of both sexes take correspondence courses annually in the city. The Middletown Business College has an annual enrollment of about 300 students, roughly half of them coming from Middletown.

So general is the drive towards education in Middletown today that, instead of explaining why those who continue in high school or even go on to college do so, as would have been appropriate a generation ago, it is simpler today to ask why those who do not continue their education fail to do so. Answers to this question were obtained from forty-two mothers who had a total of sixty-seven children, thirty-seven girls and thirty boys, who had left high school. Fourteen girls and six boys had left because their financial help was needed at home; three girls and twelve boys because they "wanted to work"; six girls because of "poor health," and one boy because of bad eyes; seven girls and six boys because they "didn't like high school"; three of each left to go to business college and one girl to study music; one girl and two boys "had to take so many

[9] Machine shop practice, carpentry, blue printing, drafting, pattern making, lathe and cabinet work, shop mathematics, chemistry.

[10] Sewing, dressmaking, millinery, applied design, basketry, planning and serving meals.

In general, attendance at evening courses of all kinds tends to be larger in "bad times." The director of this work attributes this to two factors: (1) people out of work have more time on their hands; (2) when competition for jobs is severe, workers realize the desirability of having education in addition to mere trade skill. A third factor, touching women only, is the increase in home sewing and the making of one's own hats when times are bad and the family pocket-book empty; these women may join a course to make one hat or dress and then drop out.

In 1923-24 the modal group among the men students were in their early twenties, while the modal group of women were in their thirties.

things of no use"; one girl was married; and one stayed home to help during her mother's illness.

Obviously such answers are superficial explanations, masking in most cases a cluster of underlying factors. The matter of mental endowment is, naturally, not mentioned, although, according to Terman, "The pupils who drop out [of high school] are in the main pupils of inferior mental ability." [11] And yet, important though this consideration undoubtedly is, it must not too easily be regarded as the prepotent factor in the case of many of those who drop out of high school and in that of perhaps most of those who complete high school but do not go on to college; [12] standards are relatively low both in the high school and in a number of near-by colleges. The formal, remote nature of much school work probably plays a larger rôle in discouraging children from continuing in school than the reference above to having "to take so many things of no use" indicates; save in the case of certain vocational courses, a Middletown boy or girl must take the immediate relevancy and value of the high school curriculum largely on faith.

Potent among the determining factors in this matter of continuance in school is the economic status of a child's family; here again, as in the case of the size of the house a given family occupies and in other significant accompaniments of living, we observe this extraneous pecuniary consideration dictating the course of the individual's life. The emphasis upon this financial consideration in the answers of the Middletown mothers cited above probably underestimates the importance of money. A number of mothers who said that a child had left

[11] Terman, *The Intelligence of School Children* (Boston; Houghton Mifflin, 1919), pp. 87-90.
Cf. in this connection the intelligence quotients of a cross-section of Middletown in Ch. V, and also, as pointed out in Ch. VI, the roughly identical proportions of working class and business class boys in the three upper years of the high school.

[12] In Middletown as in the rest of the state there seems to be little direct relation between the ability of high school seniors and the selection of those who go to college. Book says of the state, "Almost as many students possessing E and F grades of intelligence are going to college as merit a ranking of A-plus or A.

"Many of the brightest students graduating from our high schools are not planning to go to college at all. Of those rated A-plus, 22 per cent. stated that they never expected to attend a college or university. Of those rated A, 24 per cent. did not intend to continue their education beyond the high school. . . . Of those ranking D and E, 64 and 62 per cent. respectively stated they would attend college next year." (*Op. cit.,* pp. 39-40.)

school because he "didn't like it" finally explained with great reluctance, "We couldn't dress him like we'd ought to and he felt out of it," or, "The two boys and the oldest girl all quit because they hated Central High School. They all loved the Junior High School [13] down here, but up there they're so snobbish. If you don't dress right you haven't any friends." "My two girls and oldest boy have all stopped school," said another mother. "My oldest girl stopped because we couldn't give her no money for the right kind of clothes. The boy begged and begged to go on through high school, but his father wouldn't give him no help. Now the youngest girl has left 10B this year. She was doing just fine, but she was too proud to go to school unless she could have clothes like the other girls." The marked hesitation of mothers in mentioning these distasteful social distinctions only emphasizes the likelihood that the reasons for their children's leaving school summarized above understate the real situation in this respect.

This influential position of the family's financial status emerges again in the answers of the women interviewed regarding their plans for their children's future, although these answers cannot be satisfactorily tabulated as they tended to be vague in families where children were still below high school age. Every business class mother among the group of forty interviewed was planning to send her children through high school, and all but three of the forty were definitely planning to send their children to college; of these three, two were planning a musical education for their children after high school, and the third had children under eight. Eight of those planning to send their children to college added, "If we can afford it." Three were planning graduate work in addition to college. Two others said that musical study might be an alternative to college.

The answers of the working class wives were in terms of "hope to" or "want to"; in almost every case plans were contingent upon "if we can afford it." Forty of these 124 working class families had no plans for their children's education, eighteen of the forty having children in high school; the attitude of some of these mothers is expressed by the mother of nine chil-

[13] Working class children go to the Junior High School on the South Side until they have finished the ninth grade. For the last three years of the high school—tenth, eleventh, and twelfth grades—all children of the city go to the Central High School on the North Side.

dren who said wearily, "I don't know; we want them all to go as far as they can." Of those who had plans for their children's future, three were planning definitely to have their children stop school at sixteen, the legal age limit of compulsory attendance. Thirty-eight were planning if possible to have their children continue through high school. Five planned on the local Business College in addition to one or more years at high school; four on the local Normal School; twenty-eight on college following high school; one on musical training in addition to high school; five on music without high school; and one on Business College without high school.[14] The answer of one mother conveys the mood of many other families: "Our oldest boy is doing fine in high school and his father says he'd like to send him to some nice college. The others will go through high school anyhow. If children don't have a good education they'll never know anything except hard work. Their father wants them to have just as much schooling as he can afford." Over and over again one sees both parents working to keep their children in college. "I don't know how we're going to get the children through college, but we're *going* to. A boy without an education today just ain't *anywhere!*" was the emphatic assertion of one father.

If education is oftentimes taken for granted by the business class, it is no exaggeration to say that it evokes the fervor of a religion, a means of salvation, among a large section of the working class.[15] Add to this the further fact, pointed out below, that the high school has become the hub of the social life of the young of Middletown, and it is not surprising that high school attendance is almost as common today as it was rare a generation ago.

[14] These figures are given in terms of families, not children. In some cases the plans given apply to only one or two children in a family when the parents have no plans for the others.
[15] Cf. in Table XIV the greater emphasis of working class mothers on "getting good grades in school" and "making a name in the world" in training their children.

Chapter XIV

THE THINGS CHILDREN LEARN

The school, like the factory, is a thoroughly regimented world. Immovable seats in orderly rows fix the sphere of activity of each child. For all, from the timid six-year-old entering for the first time to the most assured high school senior, the general routine is much the same. Bells divide the day into periods. For the six-year-olds the periods are short (fifteen to twenty-five minutes) and varied; in some they leave their seats, play games, and act out make-believe stories, although in "recitation periods" all movement is prohibited. As they grow older the taboo upon physical activity becomes stricter, until by the third or fourth year practically all movement is forbidden except the marching from one set of seats to another between periods, a brief interval of prescribed exercise daily, and periods of manual training or home economics once or twice a week. There are "study-periods" in which children learn "lessons" from "text-books" prescribed by the state and "recitation-periods" in which they tell an adult teacher what the book has said; one hears children reciting the battles of the Civil War in one recitation period, the rivers of Africa in another, the "parts of speech" in a third; the method is much the same. With high school come some differences; more "vocational" and "laboratory" work varies the periods. But here again the lesson-text-book-recitation method is the chief characteristic of education. For nearly an hour a teacher asks questions and pupils answer, then a bell rings, on the instant books bang, powder and mirrors come out, there is a buzz of talk and laughter as all the urgent business of living resumes momentarily for the children, notes and "dates" are exchanged, five minutes pass, another bell, gradual sliding into seats, a final giggle, a last vanity case snapped shut, "In our last lesson we had just finished"—and another class is begun.

All this ordered industry of imparting and learning facts

and skills represents an effort on the part of this matter-of-fact community immersed in its daily activities to endow its young with certain essential supplements to the training received in the home. A quick epitome of the things adult Middletown has come to think it important for its children to learn in school, as well as some indication of regions of pressure and change, is afforded by the following summary of the work in Grades I and VII in 1890 and in 1924:

1890	1924
GRADE I	
Reading	Reading
Writing	Writing
Arithmetic	Arithmetic
Language	Language
Spelling	Spelling
Drawing	Drawing
Object Lessons (Science)	Geography
Music	Music
	Civic Training
	History and Civics
	Hygiene and Health
	Physical Education
GRADE VII	
Reading	Reading
Writing	Writing
Arithmetic	Arithmetic
Language	Language
Spelling	Spelling
Drawing	Drawing
Music	Music
Geography	Geography
Object Lessons (Science)	Civic Training
Compositions and Declamation	History and Civics
	Manual Arts (Boys)
	Home Economics (Girls)
	Physical Education

In the culture of thirty-five years ago it was deemed sufficient to teach during the first seven years of this extra-home training the following skills and facts, in rough order of importance: [1]

[1] The state law of 1865 upon which the public school system rests provided for instruction in "orthography, reading, writing, arithmetic, English, grammar, and good behavior," and the minutes of the Middletown School Board for 1882 (the only minutes for a decade on either side of

a. The various uses of language. (Overwhelmingly first in importance.)

b. The accurate manipulation of numerical symbols.

c. Familiarity with the physical surroundings of peoples.

d. A miscellaneous group of facts about familiar physical objects about the child—trees, sun, ice, food, and so on.

e. The leisure-time skills of singing and drawing.

Today the things for which all children are sent to school fall into the following rough order:

a. The same uses of language.

b. The same uses of numerical figures.

c. Training in patriotic citizenship.

d. The same familiarity with the physical surroundings of peoples.

e. Facts about how to keep well and some physical exercise.

f. The same leisure-time skills of singing and drawing.

g. Knowledge and skills useful in sewing, cooking and using tools about the home for the girls, and, for the boys, an introductory acquaintance with some of the manual skills by which the working class members get their living.

Both in its optional, non-compulsory character and also in its more limited scope the school training of a generation ago appears to have been a more casual adjunct of the main business of "bringing up" that went on day by day in the home. Today, however, the school is relied upon to carry a more direct, if at most points still vaguely defined, responsibility. This has in turn reacted upon the content of the teaching and encouraged a more utilitarian approach at certain points. A slow trend toward utilizing material more directly instrumental to the day-by-day urgencies of living appears clearly in such a course as that in hygiene and health, epitomized in the text-books used a generation ago and today, Jenkins' *Advanced Lessons in Human Physiology* in the one case and Emerson and Betts' *Physiology and Hygiene* in the other. The earlier book devoted twenty-one chapters, 287 of its 296 pages, to the structure and function of the body—"The Skeleton," "The Skin and the Kidneys," "The

1890 which describe the course of study in detail) affirm that "reading, writing, and arithmetic are the three principal studies of the public schools, and if nothing more is possible, pupils should be taught to read the newspapers, write a letter, and perform the ordinary operations of arithmetic."

Anatomy of the Nervous System," and so on, and a final chapter, eight and one-quarter pages, to "the laws of health"; a three-page appendix on "Poisons and Antidotes" gave the various remedies to be used to induce vomiting after poisoning by aconite, arsenic, and so on, as well as rules for treating asphyxia. The current book, on the other hand, is primarily concerned throughout with the care of the body, and its structure is treated incidentally. Examination questions in the two periods show the same shift. Characteristic questions of 1890 such as "Describe each of the two kinds of matter of the nervous system" and "Tell weight and shape of brain. Tell names of membranes around it" are being replaced by "Write a paragraph describing exactly the kind of shoe you should wear, stating all the good points and the reasons for them," and "What is the law of muscles and bones (regarding posture)? How should it guide you in your daily life?"

Geography, likewise, according to the printed courses of study for the two periods, is less concerned today with memorizing "at least one important fact about each city located" and more with the "presence of storm and sunshine and song of bird," "interests of the child"; but classes visited are preoccupied with learning of facts, and 1890 and 1924 examination questions are interchangeable. Reading, spelling, and arithmetic, also, exhibit at certain points less emphasis upon elaboration of symbols and formal drill and more on the "practical application" of these skills; thus in reading, somewhat less attention is being paid to "clear and distinct enunciation" and "proper emphasis and expression" and more to "silent reading," which stresses content. But, in general, these subjects which are "the backbone of the curriculum" show less flexibility than do the subjects on the periphery or the newcomers. Most of these changes are indeed relatively slight; the social values represented by an "elementary education" are changing slowly in Middletown.

When we approach the high school, however, the matter-of-fact tendency of the city to commandeer education as an aid in dealing with its own concerns becomes more apparent.[2] Caught less firmly than the elementary school in the cake of tradition and now forced to train children from a group not heretofore

2 Cf. in Ch. XXV the discussion of the dental and other health work in the schools as another manifestation of this same tendency.

reached by it, the high school has been more adaptable than the lower school. Here group training no longer means the same set of facts learned on the same days by all children of a given grade. The freshman entering high school may plan to spend his four years following any one of twelve different "courses of study"; [3] he may choose the sixteen different yearly courses which will make up his four years of training from a total of 102.[4] All this is something new, for the 170 students who were going to high school in the "bursting days of boom" of 1889-90 had to choose, as Middletown high school students had done for thirty years, between two four-year courses, the Latin and the English courses, the sole difference between them being whether one did or did not take "the language." The number of separate year courses open to them totaled but twenty.

The facts and skills constituting the present-day high school curriculum present a combination of the traditional learning reputed to be essential to an "educated" man or woman and

[3]
1. General Course	7. Applied Electricity Course
2. College Preparatory Course	8. Mechanical Drafting Course
3. Music Course	9. Printing Course
4. Art Course	10. Machine Shop Course
5. Shorthand Course	11. Manual Arts Course
6. Bookkeeping course	12. Home Economics Course

Courses Three to Twelve inclusive have a uniform first-year group of required and elective subjects. Four subjects are taken each half of each year, of which two or three are required and the rest selected from among a list offering from two to nine electives, according to the course and the year. The indispensables of secondary education required of every high school student are:

Four years of English for those taking Courses One through Six.
Three years of English for those taking Courses Seven through Twelve.
One year of algebra.
One year of general history.
One year of American history.
One-half year of civics.
One-half year of sociology.
One year of science.
One-half year of music.
One-half year of gymnasium.

This constitutes a total of ten required and six elective one-year courses or their equivalents during the four years for the academic department (Courses One through Six) and nine required and seven elective courses for those in the vocational department (Courses Seven through Twelve).

[4] The year unit rather than the term or semester unit is taken here as the measure of the number of courses, since it furnishes the only basis of comparison with 1890. When different subjects make up one year's course they are almost invariably related, e.g., civics and sociology, zoölogy and botany.

newer applied information or skills constantly being inserted into the curriculum to meet current immanent concerns. Here, too, English, the successor in its varied forms of the language work in the grades, far outdistances all competitors for student time, consuming 22 per cent. of all student hours.[5] It is no longer compulsory throughout the entire four years as it was a generation ago; instead, it is required of all students for the first two years, and thereafter the earlier literary emphasis disappears in seven of the twelve courses, being replaced in the third year by commercial English, while in the fourth year it disappears entirely in five courses save as an optional subject. Both teaching and learning appear at times to be ordeals from which teachers and pupils alike would apparently gladly escape: "Thank goodness, we've finished Chaucer's *Prologue!*" exclaimed one high school English teacher. "I am thankful and the children are, too. They think of it almost as if it were in a foreign language, and they *hate* it."

Latin, likewise, though still regarded by some parents of the business class as a vaguely significant earmark of the educated man or woman, is being rapidly attenuated in the training given the young. It is not required of any student for even one year, though in one of the twelve courses it or French is required for two years. Gone is the required course of the nineties taken by over half of the high school students for the entire four years and enticingly set forth in the course of study of the period as "Latin, Grammar, Harkness: Begun-Completed. Latin, Reader, Harkness: Begun-Completed. Latin, Caesar, Harkness: Begun-Completed. Latin, Virgil, Harkness: Begun-Completed." The "Virgil Club's" annual banquet and the "Latin Wedding" are, however, prominent high school social events today, and more than one pupil confessed that the lure of these in the senior year helped to keep him through four years of Latin. Although Latin is deader than last summer's straw hat to the men joshing each other about Middletown's Rotary luncheon table, tradition, the pressure of college entrance requirements, and such incidental social considerations as those just mentioned still manage to hold Latin to a place of prominence in the curriculum: 10 per cent. of all student hours are devoted to Latin, as against but 2 per cent. each to French

[5] See Table XVI for a complete distribution of student hours in the high school.

and Spanish;[6] only English, the combined vocational courses, mathematics, and history consume more student hours.

The most pronounced region of movement appears in the rush of courses that depart from the traditional dignified conception of what constitutes education and seek to train for specific tool and skill activities in factory, office, and home. A generation ago a solitary optional senior course in bookkeeping was the thin entering wedge of the trend that today controls eight of the twelve courses of the high school and claimed 17 per cent. of the total student hours during the first semester of 1923-24 and 21 per cent. during the second.[7] At no point has the training prescribed for the preparation of children for effective adulthood approached more nearly actual preparation for the dominant concerns in the daily lives of the people of Middletown. This pragmatic commandeering of education is frankly stated by the president of the School Board: "For a long time all boys were trained to be President. Then for a while we trained them all to be professional men. Now we are training boys to get jobs."

Unlike Latin, English, and mathematics in that they have no independent, honorific traditions of their own, these vocational courses have frankly adopted the canons of office and machine shop: they must change in step with the coming of new physical equipment in machine shops and offices, or become ineffective.[8]

[6] Since the World War German has not been taught in Middletown.

[7] See Table XVI for the distribution of these hours among commercial, domestic science, manual arts, and vocational work and also for their relation to other courses in point of student hours. In the case of English, the only subject or group of subjects to exceed the time spent on these non-academic courses, it should be borne in mind that in seven of the twelve courses of study offered by the high school one-third of the total English work required is a new vocational kind of English called commercial English, reflecting the workaday emphasis rather than the older academic emphasis in the curriculum.

It should also be recalled that, in addition to this high school work, manual arts is compulsory for all boys in Grades VI to VIII of the elementary school, and home economics is also compulsory for girls in Grades VII and VIII.

[8] This conformity to existing conditions is accentuated by the necessity of bidding for union support and falling in with current trade union practices. The attitude of the unions toward this school training varies all the way from that of the carpenters whose president attends the evening classes and who start a high school trained boy with a journeyman's card and corresponding wages to that of the bricklayers and plasterers who start a high school vocational graduate at exactly the same wage as an untrained boy.

A recently organized radio class shows the possibility of quick adaptability to new developments. More than any other part of the school training, these vocational courses consist in learning *how* rather than learning *about*. Actual conditions of work in the city's factories are imported into the school shops; boys bring repair work from their homes; they study auto mechanics by working on an old Ford car; they design, draft, and make patterns for lathes and drill presses, the actual casting being done by a Middletown foundry; they have designed and constructed a house, doing all the architectural, carpentry, wiring, metal work, and painting. A plan for providing work in a local machine shop, alternating two weeks of this with two weeks of study throughout the year, is under discussion.

Under the circumstances, it is not surprising that this vocational work for boys is the darling of Middletown's eye—if we except a group of teachers and of parents of the business class who protest that the city's preoccupation with vocational work tends to drag down standards in academic studies and to divert the future college student's attention from his preparatory courses.[9] Like the enthusiastically supported high school basket-ball team, these vocational courses have caught the imagination of the mass of male tax-payers; ask your neighbor at Rotary what kind of schools Middletown has and he will begin to tell you about these "live" courses. It is not without signifi-

[9] Many Middletown people maintain that the coming of vocational work to the high school has tended greatly to lower its standing as a college preparatory school. More than one mother shook her head over the fact that her daughter never does any studying at home and is out every evening but gets A's in all her work. It is generally recognized that a boy or girl graduating from the high school can scarcely enter an eastern college without a year of additional preparatory work elsewhere.

Leading nationally known universities in neighboring states gave the following reports of the work of graduates of the Middletown high school: In one, of eleven Middletown students over a period of fifteen years, one graduated, none of the others made good records, four were asked to withdraw because of poor scholarship; of the four in residence in 1924, two were on probation, one was on the warned list, and one was doing fair work. In another, of five Middletown students in the last five years, one did excellent work, one fair, two did very poor work and dropped out after the first term, one had a record below requirement at the time of withdrawal. In a third, of eight Middletown students in the last five years, one was an excellent student, four were fair, and three were on probation. The single Middletown student in a fourth university attended for only a year and was on probation the entire time.

cance that vocational supervisors are more highly paid than any other teachers in the school system.

Much of what has been said of the strictly vocational courses applies also to work in bookkeeping and stenography and in home economics. The last-named, entirely new since 1890, is devised to meet the functional needs of the major group of the girls, who will be home-makers. Beginning in the seventh and eighth years with the study of food, clothing, and house-planning, it continues as an optional course through the high school with work in dressmaking, millinery, hygiene and home nursing, household management, and selection of food and clothing. As in the boys' vocational work, these courses center in the more obvious, accepted group practices; much more of the work in home economics, for example, centers in the traditional household productive skills such as canning, baking, and sewing, than in the rapidly growing battery of skills involved in effective buying of ready-made articles. The optional half-year course for the future business girl in selection of food and clothing, equipping a girl "to be an intelligent consumer," marks, however, an emergent recognition of a need for training in effective consumption, as does also the class visiting of local stores to inspect and discuss various kinds of household articles. In 1925 a new course in child care and nutrition was offered in one of the grade schools; while it consists almost entirely in the study of child feeding rather than of the wider aspects of child care, it is highly significant as being the first and sole effort on the part of the community to train women for this fundamental child-rearing function. Standard women's magazines are resorted to in these courses for girls as freely as technical journals are employed in the courses for boys.

Second only in importance to the rise of these courses addressed to practical vocational activities is the new emphasis upon courses in history and civics. These represent yet another point at which Middletown is bending its schools to the immediate service of its institutions—in this case, bolstering community solidarity against sundry divisive tendencies. A generation ago a course in American history was given to those who survived until the eighth grade, a course in general history, "covering everything from the Creation to the present in one little book of a hundred or so pages," followed in the second year of the high school, and one in civil government in the third year. Today, separate courses in civic training and in history and civics begin with the first grade for all children and continue throughout the elementary school, while in high school the third-year course in American history and the fourth-year course in civics and sociology are, with the exception of the second-year English course, the only courses required of all students after the completion of the first year. Sixteen per cent. of the total student hours in the high school are devoted to these social studies—history, sociology, and civics—a total surpassed only by those of English and the combined cluster of vocational, domestic science, manual arts, and commercial courses.

Evidently Middletown has become concerned that no child shall be without this pattern of the group.[10] Precisely what this stamp is appears clearly in instructions to teachers:[11]

"The most fundamental impression a study of history should leave on the youth of the land when they have reached the period

[10] "Good citizenship as an aim in life is nothing new. . . . But good citizenship as a dominant aim of the American public school is something new. . . . For the first time in history, as I see it, a social democracy is attempting to shape the opinions and bias the judgment of oncoming generations." From the *Annual Report* of Dean James E. Russell of Teachers College for the year ending June, 1925.

In view of the manifest concern in Middletown to dictate the social attitudes of its young citizens, the concentration of college attendance of local high school graduates in local or near-by institutions is significant. As noted in the preceding chapter, forty-four of the eighty members of the high school class of 1924 who were attending college were enrolled in the small local college; twelve more were in the two state universities, ten more in other small colleges within the state, nine were in small colleges in adjoining states, two in nationally known state universities in adjoining states, two in prominent eastern colleges, and one in an eastern school giving specialized training—a total of sixty-six within the city or state, eleven in immediately adjoining states, and three in distant states. This, when coupled with the tendency already pointed out for from one-third to one-half of each high school graduating class, including almost certainly many of the most enterprising and original members, to migrate to other communities. and the further tendency of Middletown to favor teachers trained within the state, presents some interesting implications for the process of social change in Middletown.

[11] Descriptions of courses and instructions to teachers as set forth by the School Board or State Department of Education sometimes bear little relation to what children are actually being taught in the class-room. But they do show what those directing the training of the young think *ought* to be taught and what they believe the public thinks ought to be taught. As indicating major characteristics of this culture, therefore, they are, in one sense, even more significant than the things that actually go on in the class-room. And by and large they do, of course, indicate trends in teaching.

of citizenship," begins the section on history and civics of the Middletown Course of Study of the Elementary Schools, "is that they are their government's keepers as well as their brothers' keepers in a very true sense. This study should lead us to feel and will that sacrifice and service for our neighbor are the best fruits of life; that reverence for law, which means, also, reverence for God, is fundamental to citizenship; that private property, in the strictest sense, is a trust imposed upon us to be administered for the public good; that no man can safely live unto himself. . . ."

"History furnishes no parallel of national growth, national prosperity and national achievement like ours," asserts the State Manual for Secondary Schools for 1923. "Practically all of this has been accomplished since we adopted our present form of government, and we are justified in believing that our political philosophy is right, and that those who are today assailing it are wrong. To properly grasp the philosophy of this government of ours, requires a correct knowledge of its history."

The State Manual for Elementary Schools for 1921 instructs that "a sense of the greatness of their state and a pride in its history should be developed in the minds of children," and quotes as part of its directions to teachers of history: "The right of revolution does not exist in America. We had a revolution 140 years ago which made it unnecessary to have any other revolution in this country. . . . One of the many meanings of democracy is that it is a form of government in which the right of revolution has been lost. . . . No man can be a sound and sterling American who believes that force is necessary to effectuate the popular will. . . . Americanism . . . emphatically means . . . that we have repudiated old European methods of settling domestic questions, and have evolved for ourselves machinery by which revolution as a method of changing our life is outgrown, abandoned, outlawed."

The president of the Board of Education, addressing a meeting of Middletown parents in 1923, said that "many educators have failed to face the big problem of teaching patriotism. . . . We need to teach American children about American heroes and American ideals."

The other social studies resemble history in their announced aims: civic training, with its emphasis upon respect for private property, respect for public property, respect for law, respect for the home, appreciation of services of good men and women, and so on; economics, with its stressing of "common and fundamental principles," "the fundamental institutions of society: private property, guaranteed privileges, contracts, personal liberty, right to establish private enterprises"; and sociology.

Nearly thirty-five years ago the first high school annual summarized the fruits of four years of high school training as follows: "Many facts have been presented to us and thus more knowledge has been attained." Such a summary would be nearly as applicable today, and nowhere more so than in these social studies. Teaching varies from teacher to teacher, but with a few outstanding exceptions the social studies are taught with close reliance upon textbooks prescribed by the state and in large measure embodying its avowed aims. A leading teacher of history and civics in the high school explained:

"In class discussion I try to bring out minor points, two ways of looking at a thing and all that, but in examinations I try to emphasize important principles and group the main facts that they have to remember around them. I always ask simple fact questions in examinations. They get all mixed up and confused if we ask questions where they have to think, and write all over the place."

In the case of history, facts presented in the textbooks are, as in 1890, predominantly military and political, although military affairs occupy relatively less space than in the nineties. Facts concerning economic and industrial development receive more emphasis than in the earlier texts, although political development is still the core. Recent events as compared with the colonial period in colonial history are somewhat more prominent today.[12] Examination questions of the two periods indicate so little change in method and emphasis in teaching that it is almost impossible simply by reading a history examination to tell whether it is of 1890 or 1924 vintage.

[12] See W. C. Bagley and H. O. Rugg, *The Content of American History as Taught in the Seventh and Eighth Grades* (University of Illinois School of Education Bulletin No. 16, Vol. XIII, 1916), comparing textbooks from 1865 to 1911, with a supplementary study by Earle Rugg of *Eight Current Histories,* and Snyder's *An Analysis of the Content of Elementary High School History Texts* (University of Chicago Doctor's Dissertation, 1919). Montgomery's *The Leading Facts of American History,* used in the Middletown schools in the nineties, and Woodburn and Moran's *American History and Government,* used in 1924, were included in the Rugg-Bagley study. Fite's *History of the United States,* used in the Middletown schools in 1924, was included in Snyder's study.

It may be a commentary upon the vitality of this early and persistent teaching of American history that when pictures like the Yale Press historical series are brought to Middletown the children say they get enough history in school, the adults say they are too grown up for such things, and the attendance is so poor that the exhibitor says, "Never again!"

Further insight into the stamp of the group with which Middletown children complete their social studies courses is gained through the following summary of answers of 241 boys and 315 girls, comprising the social science classes of the last two years of the high school, to a questionnaire: [13]

Statement	Percentage answering "True"		Percentage answering "False"		Percentage answering "Uncertain"		Percentage not answering	
	Boys	Girls	Boys	Girls	Boys	Girls	Boys	Girls
The white race is the best race on earth..............	66	75	19	17	14	6	1	2
The United States is unquestionably the best country in the world............	77	88	10	6	11	5	2	1
Every good citizen should act according to the following statement: "My country —right or wrong!".........	47	56	40	29	9	10	4	5
A citizen of the United States should be allowed to say anything he pleases, even to advocate violent revolution, if he does no violent act himself..........	20	16	70	75	7	7	3	2
The recent labor government in England was a misfortune for England........	16	15	38	20	38	57	8	8
The United States was entirely right and England was entirely wrong in the American Revolution.......	30	33	55	40	13	25	2	2

[13] Students were requested to write "true," "false," or "uncertain" after each statement. No answers of Negroes are included in this summary. The greater conservatism of the girls in their answers to some of the questions is noteworthy. See Appendix on Method. See Index for other answers.

Statement	Percentage answering "True"		Percentage answering "False"		Percentage answering "Uncertain"		Percentage not answering	
	Boys	Girls	Boys	Girls	Boys	Girls	Boys	Girls
The Allied Governments in the World War were fighting for a wholly righteous cause	65	75	22	8	11	14	2	3
Germany and Austria were the only nations responsible for causing the World War	22	25	62	42	15	31	1	2
The Russian Bolshevist government should be recognized by the United States Government.........	8	5	73	67	17	24	2	4
A pacifist in war time is a "slacker" and should be prosecuted by the government	40	36	34	28	22	28	4	8
The fact that some men have so much more money than others shows that there is an unjust condition in this country which ought to be changed................	25	31	70	62	4	5	1	2

Other new emphases in the training given the young may be noted briefly. Natural sciences, taught in 1890 virtually without a laboratory [14] by a teacher trained in English and mathematics and by the high school principal who also taught all other junior and senior subjects, is today taught in well-equipped student laboratories by specially trained teachers. In the first and second semesters of 1923-24, 7 per cent. and 8 per cent. respectively of the student hours were devoted to the natural sciences.[15]

Although art and music appear to occupy a lesser place in

[14] Says the high school annual in 1894: "The laboratory is situated in what is known as the south office—a room six by four feet. On the east side of the room are a few shelves containing a half dozen bottles of chemicals. This is the extent of the chemical 'laboratory.' The physical laboratory will be found (with the aid of a microscope) in the closet adjoining the south office. Here will be found the remnants of an old electric outfit, and a few worn-out pieces of apparatus to illustrate the principles of natural philosophy."

[15] See Table XVI.

the spontaneous leisure-time life of Middletown than they did a generation ago,[16] both are more prominent in the training given the young. In 1890 both were unknown in the high school except for the informal high school choir; a lone music teacher taught three hours a day in the grades; and "drawing" was taught "as an aid to muscular coördination" on alternate days with writing. Today art is taught in all eight years of the grades, while in the high school a student may center his four years' work in either art or music, two of the twelve courses being built around these subjects. The high school art courses consist in creative work, art history, and art appreciation, while art exhibits and art contests reach far beyond formal class-room work.[17] Over and above the work in ear training and sight reading throughout the grades, there are today sixteen high school music courses in addition to classes in instrumental work. They include not only instruction in harmony, history of music, and music appreciation, but a chorus, four Glee Clubs, three orchestras, and two bands. Victrolas, now a necessary part of the equipment of all schools, and an annual music memory contest in the schools, further help to bring music within the reach of all children.[18]

Another innovation today is the more explicit recognition that education concerns bodies as well as minds. Gymnasium work, required of all students during the last year of the elementary school and the first year of the high school, replaces the earlier brief periods of "setting-up exercises" and seems likely to spread much more widely.

Abundant evidence has appeared throughout this chapter of the emphasis upon values and "right" attitudes in this busi-

[16] Cf. Ch. XVII.

[17] The class work itself reaches all groups of students. A barber commented proudly on the interest of his daughter, a high school junior, in her art work: "We have some friends that made fun of her for taking art—they thought it meant painting big pictures. My wife heard her talking art to some people the other day and says she could hold up her end with the best of 'em. I'm all for it. Now it has practical applications. When it comes to fixing up a house she'll know what things go good together."

[18] Cf. Ch. XVII. Neither this music work nor the art work in the schools appears to be as rooted in the present-day local life as the emphasis upon vocational education and the social studies. In fact, they represent a tradition less strong in the everyday life of the city today than a generation ago. Whether they will tend to increase spontaneous and active participation in music and art, as opposed to the passive enjoyment of them that predominates in this culture today, is problematical.

ness of passing along the lore of the elders to the young of Middletown. Since the religious attitudes and values are nominally held in this culture to overshadow all others, no account of the things taught in the schools would be adequate without a discussion of the relation of the schools to the religious beliefs and practices of the city. Getting a living, as we have observed, goes forward without any accompanying religious ceremonies or without any formal relation to the religious life of the city save that it "keeps the Sabbath day." Religion permeates the home at many points: marriage, birth, and death are usually accompanied or followed by religious rites, the eating of food is frequently preceded by its brief verbal blessing, most children are taught to say their prayers before retiring at night, a Bible is found in nearly every home, and the entire family traditionally prays together daily, though this last, as noted elsewhere, is becoming rare; the family itself is regarded as a sacred institution, though being secularized at many points. Leisure-time practices are less often today opened by prayer or hymns; though they have traditionally "observed the Sabbath," abundant testimony appears throughout this study of the attenuation of such observance of the "Lord's Day" by young and old. The common group affairs of the city, likewise, are increasingly carried on, like getting a living, without direct recourse to religious ceremonies and beliefs. In the midst of the medley of secularized and non-secularized ways of living in the city, education steers a devious course. One religious group, the Catholics, trains its children in a special school building under teachers who are professional religious devotees and wear a religious garb; this school adjoins the church and the children attend a church service as part of their day's schooling. For the great mass of children, however, separate Sunday Schools, in no way controlled by the secular schools, teach the accepted religious beliefs to those who choose to attend. The Y.M.C.A. and Y.W.C.A. serve as a liaison between church and school, teaching Bible classes in all elementary schools, for the most part on school time, and giving work in the high school for which credit is granted towards graduation. But while the public schools themselves do not teach the group's religious beliefs directly, these beliefs tacitly underlie much that goes on in the class-room, more particularly those classes concerned not with the manipulation of material tools but with the teaching of ideas, concepts, attitudes. The first paragraph in the "Course of Study of the Elementary Schools" enjoins upon the teachers that "all your children should join in opening the day with some exercise which will prepare them with thankful hearts and open minds for the work of the day. . . . The Bible should be heard and some sacred song sung." The School Board further instructs its teachers that geography should teach "the spirit of reverence and appreciation for the works of God— that these things have been created for [man's] joy and elevation . . . that the earth in its shape and movements, its mountains and valleys, its drought and flood, and in all things that grow upon it, is well planned for man in working out his destiny"; that history should teach "the earth as the field of man's spiritual existence"; that hygiene create interest in the care of the body "as a fit temple for the spirit"; finally that "the schools should lead the children, through their insight into the things of nature that they study, to appreciate the power, wisdom, and goodness of the Author of these things. They should see in the good things that have come out of man's struggle for a better life a guiding hand stronger than his own. . . . The pupils should learn to appreciate the Bible as a fountain of truth and beauty through the lessons to be gotten from it. . . ."

This emphasis was if anything even stronger in 1890. At the Teachers' Institute in 1890 botany was discussed as a subject in which "by the study of nature we are enabled to see the perfection of creation," and a resolution was passed that "the moral qualifications of the teachers should be of such a nature as to make them fit representatives to instruct for both time and eternity." "In morals, show the importance of building upon principles. Encourage the pupil to do right because it is right," said the School Board instructions for 1882. One gains a distinct impression that the religious basis of all education was more taken for granted if less talked about thirty-five years ago, when high school "chapel" was a religio-inspirational service with a "choir" instead of the "pep session" which it tends to become today.

Some inkling of the degree of dominance of religious ways of thinking at the end of ten or twelve years of education is afforded by the answers of the 241 boys and 315 girls in the social science classes of the last two years of the high school appraising the statement: "The theory of evolution offers a more accurate account of the origin and history of mankind than that offered by a literal interpretation of the first chapters of the Bible": 19 per cent. of them marked it "true," 48 per cent. "false," 26 per cent. were "uncertain," and 7 per cent. did not answer.

Chapter XV

THOSE WHO TRAIN THE YOUNG

In the school as in the home, child-training is largely left to the womenfolk. Four-fifths of Middletown's teachers are women, the majority of them unmarried women under forty.[1] This is not the result of a definite policy, although the general sentiment of the community probably accords with at least the first half of the local editorial statement in 1900 that teaching is "an occupation for which women seem to have a peculiar fitness and a greater adaptability than men; but whether from their qualities of gentleness or from superior mental endowment is open to question." Actually, however, here as at so many other points in the city's life, money seems to be the controlling factor; more money elsewhere draws men away from teaching, rather than special fitness attracting women. Middletown pays its teachers more than it did thirty-five years ago,[2] but even the $2,100 maximum paid to grade school principals and high school teachers, the $3,200 paid to the high school principal, and the $4,900 to the Superintendent of Schools are hardly enough to tempt many of the abler men away from business in a culture in which everything hinges on money.

One becomes a teacher by doing a certain amount of studying things in books. All teachers today must have graduated from high school and have spent at least nine months in a recognized teacher-training college, while at least two years more of normal

[1] Twenty-one per cent. of 273 teachers (247 academic teachers, 15 principals, 6 supervisors, and 5 physical directors) in 1923-24 were men. Only two of the thirty-four teachers in 1889-90 were men. The increase in the proportion of men is largely accounted for by the introduction of vocational work and the greater use of men as principals of grade schools.

Between 1880 and 1914 the percentage of male teachers in the United States was more than halved, dropping from 43 per cent. to 20 per cent. (Ernest C. Moore, *Fifty Years of American Education: 1867-1917*, Boston; Ginn, 1918, pp. 60-61.) The percentage of men teachers in Middletown at present corresponds closely to the national figure.

[2] Cf. Ch. VIII for rise in teachers' salaries since 1890 in relation to rise in cost of living.

training or a college degree are necessary to teach in the high school. In 1890 the common practice was for new teachers to secure a license simply by passing an examination and then to begin by teaching in rural schools, whence they eventually graduated to city grade schools, and from them in turn moved up to the high school; today it is not uncommon for teachers who have had no experience in teaching, save possibly "practice teaching" in connection with normal school work, to teach even in high school. For the most part, teachers in Middletown's schools today have more formal book training and less experience of dealing with children than those of a generation ago.[3] One sees in the high school young people, caught in the cross-currents of a period of rapid change in many deeply-lying institutional habits, being trained by teachers many of whom are only slightly older and less bewildered than their pupils. The whole situation is complicated by the fact that these young teachers go into teaching in many cases not primarily because of their ability or great personal interest in teaching; for very many of them teaching is just a job. The wistful remark of a high school teacher, "I just wasn't brought up to do anything interesting. So I'm teaching!" possibly represents the situation with many.

Cultural in-breeding is a dominant factor in the selection of teachers; for the most part, children are taught by teachers brought up in the same state. Preference tends to be given to Middletown's own teachers' training college and, failing that, to colleges in the state. A bill was even introduced into the legislature in 1924 providing that teachers' licenses be issued only to graduates of public high schools in the state. Of the 156 teachers and principals in 1923-24 who held a college or normal school diploma, degree, or certificate, only twenty-nine, less

[3] Of sixteen women teachers in the grades in 1889-90 of whose training some record could be found, ten had high school training, one had high school plus two years of normal training, three had high school plus six or twelve weeks of normal, one had a few weeks of normal but no high school, one had neither high school nor normal; of three high school teachers one, the principal, a man, was a graduate of the state university, and the other two of the high school. Of 268 teachers, principals, and supervisors in 1923-24 (excluding the five physical directors), fifty-six have college degrees; eight, college diplomas; three, college certificates; thirty have normal degrees; twenty-six, normal diplomas; thirty-three, normal certificates; 112 have no degree, diploma, or certificate from either college or normal, but are all high school graduates.

than a fifth, had received them from institutions outside the state.[4]

Two shifts in the general conditions surrounding teaching operate variously to lessen or to increase the handicaps of teachers selected and trained in the above fashion. On the one hand, class units are usually considerably smaller today; in 1923-24 there were thirty pupils per teacher as against fifty-eight in 1889-90.[5] Congestion in 1889-90 was greatest in the early years, some first-year classes numbering as many as eighty.

At the same time, a second trend operates to render it unnecessary for each teacher to spread herself over as much subject matter as formerly, but at the same time to allow her less contact with the individual student. From the fourth or fifth year on, a teacher becomes a specialized teacher of one subject to many groups instead of teaching a single group all their subjects for two years as was formerly the case; by the time a child reaches high school he has as many teachers as he has subjects. This becomes significant in view of the fact that in a period when the home is losing some of its close control of the child's activities to the school, the latter is likewise diminishing rather than increasing the former status of the teacher *in loco parentis,* through the teacher's having to concentrate more and more upon teaching subjects rather than upon teaching children. This situation is being met somewhat in the high school, as noted earlier, by the creation of a new type of teacher known as the "dean of girls"; the freeing of grade school principals of teaching responsibility operates, theoretically at least, to the same end. This growing impersonality of teacher-pupil relationships is responsible, in part, for such observed phenomena as the heavy relative drop in space devoted to the faculty in the high school annual, noted below; one no longer finds tributes to teachers such as those printed in 1894:

"The influence of her beautiful life will be felt by her pupils in after years, and make them stronger and better fitted for the realities of life."

[4] A few of the others, following their first degree, had received some summer school or other professional training outside the state.

[5] Figures on the basis of the number of pupils enrolled. The number of teachers in 1923-24 is taken as 252, i.e., all teachers exclusive of principals and special supervisors. Principals are included in the thirty-four teachers of 1889-90, since they all taught; special supervisors were then a thing unknown.

And "one of the brightest and dearest of our teachers . . . her persistent efforts have been rewarded in the love and respect which her pupils have for her, and her life is one which should be a criterion to all."

It seems unlikely that such appreciations could have been entirely perfunctory. Pupils today are apparently inclined to take their teachers more casually, unless the latter happen to be basket-ball enthusiasts or fraternity brothers or have some other extra-teaching hold upon them.

Indeed, few things about education in Middletown today are more noteworthy than the fact that the entire community treats its teachers casually. These more than 250 persons to whom this weighty responsibility of training the young is entrusted are not the wise, skilled, revered elders of the group. In terms of the concerns and activities that preoccupy the keenest interests of the city's leaders, they are for the most part nonentities; rarely does one run across a teacher at the weekly luncheons of the city's business men assembled in their civic clubs; nor are many of them likely to be present at the social functions over which the wives of these influential men preside. Middletown pays these people to whom it entrusts its children about what it pays a retail clerk, turns the whole business of running the schools over to a School Board of three business men appointed by the political machine, and rarely stumbles on the individual teacher thereafter save when a particularly interested mother pays a visit to the school "to find out how Ted is getting along." The often bitter comments of the teachers themselves upon their lack of status and recognition in the ordinary give and take of local life are not needed to make an observer realize that in this commercial culture the "teacher" and "professor" do not occupy the position they did even a generation ago.

Furthermore, as Middletown every year is confronted with an increasing number of children to be educated at public expense and as a growing number of more highly paid occupations are drawing men and women from teaching, emphasis almost inevitably comes to be laid upon the perfection of the system rather than upon the personality or qualifications of the individual teacher. The more completely a teacher is a part of the accepted school system in terms of his own training, the more he is valued and the more salary he receives. As in so many other aspects of Middletown's life, criticism of individuals or creakings in the system are met primarily not by changes in its foundations but by adding fresh stories to its superstructure. If teaching is poor, supervisors are employed and "critic teachers" are added; in 1890 the only person in the entire school system who did not teach was the superintendent, while between superintendent and teacher today is a galaxy of principals, assistant principals, supervisors of special subjects, directors of vocational education and home economics, deans, attendance officers, and clerks, who do no teaching but are concerned in one way or another with keeping the system going; in 1924 the office of superintendent itself was bifurcated into a superintendent of schools and a business director. Thus, in personnel as well as in textbooks and courses of study, strains or maladjustments in education are being met by further elaboration and standardization.

Chapter XVI

SCHOOL "LIFE"

Accompanying the formal training afforded by courses of study is another and informal kind of training, particularly during the high school years. The high school, with its athletics, clubs, sororities and fraternities, dances and parties, and other "extracurricular activities," is a fairly complete social cosmos in itself, and about this city within a city the social life of the intermediate generation centers. Here the social sifting devices of their elders—money, clothes, personal attractiveness, male physical prowess, exclusive clubs, election to positions of leadership—are all for the first time set going with a population as yet largely undifferentiated save as regards their business class and working class parents. This informal training is not a preparation for a vague future that must be taken on trust, as is the case with so much of the academic work; to many of the boys and girls in high school this is "the life," the thing they personally like best about going to school.

The school is taking over more and more of the child's waking life. Both high school and grades have departed from the attitude of fifty years ago, when the Board directed:

"Pupils shall not be permitted to remain on the school grounds after dismissal. The teachers shall often remind the pupils that the first duty when dismissed is to proceed quietly and directly home to render all needed assistance to their parents."

Today the school is becoming not a place to which children go from their homes for a few hours daily but a place from which they go home to eat and sleep.[1]

[1] This condition is deplored by some as indicative of the "break-up of the American home." Others welcome it as freeing the child earlier from the domination of parents and accustoming him to face adjustments upon the success of which adult behavior depends. In any event, the trend appears to be in the direction of an extension of the present tendency increasingly into the grades.

An index to this widening of the school's function appears in a comparison of the 1924 high school annual with the first annual, published thirty years before, though even this comparison does not reflect the full extent of the shift since 1890, for innovations had been so numerous in the years just preceding 1894 as to dwarf the extent of the 1890-1924 contrast. Next in importance to the pictures of the senior class and other class data in the earlier book, as measured by the percentage of space occupied, were the pages devoted to the faculty and the courses taught by them, while in the current book athletics shares the position of honor with the class data, and a faculty twelve times as large occupies relatively only half as much space. Interest in small selective group "activities" has increased at the expense of the earlier total class activities.[2] But such a numerical comparison can only faintly suggest the difference in tone of the two books. The description of academic work in the early annual beginning, "Among the various changes that have been effected in grade work are . . ." and ending, "regular monthly teachers' meetings have been inaugurated," seems as foreign to the present high school as does the early class motto "Deo Duce"; equally far from 1890 is the present dedication, "To the Bearcats."

This whole spontaneous life of the intermediate generation that clusters about the formal nucleus of school studies becomes focused, articulate, and even rendered important in the eyes of adults through the medium of the school athletic teams— the "Bearcats."[3] The business man may "lay down the law" to his adolescent son or daughter at home and patronize their friends, but in the basket-ball grandstand he is if anything a little less important than these youngsters of his who actually mingle daily with those five boys who wear the colors of

[2] The following shows the percentage of the pages of the annual occupied by the chief items in 1894 and 1924, the earlier year being in each case given first: Class data—39 per cent., 19 per cent.; faculty—16 per cent., 8 per cent. (brief biographies and pictures in 1894, list of names only and picture of principal in 1924); athletics—5 per cent., 19 per cent.; courses of study—6 per cent., 0.0 per cent.; class poems—13 per cent., 0.0 per cent.; activities other than athletics—5 per cent. (one literary society), 13 per cent. (thirteen *kinds* of clubs); jokes—5 per cent., 17 per cent.; advertisements and miscellaneous—11 per cent., 24 per cent.

[3] In the elementary grades athletics are still a minor interest, though a school baseball and basket-ball league have been formed of recent years and the pressure of inter-school leagues and games is being felt increasingly.

"Magic Middletown." There were no high school teams in 1890. Today, during the height of the basket-ball season when all the cities and towns of the state are fighting for the state championship amidst the delirious backing of the rival citizens, the dominance of this sport is as all-pervasive as football in a college like Dartmouth or Princeton the week of the "big game." [4] At other times dances, dramatics, and other interests may bulk larger, but it is the "Bearcats," particularly the basket-ball team, that dominate the life of the school. Says the prologue to the high school annual:

"The Bearcat spirit has permeated our high school in the last few years and pushed it into the prominence that it now holds. The '24 *Magician* has endeavored to catch, reflect and record this spirit because it has been so evident this year. We hope that after you have glanced at this book for the first time, this spirit will be evident to you.

"However, most of all, we hope that in perhaps twenty years, if you become tired of this old world, you will pick up this book and it will restore to you the spirit, pep, and enthusiasm of the old 'Bearcat Days' and will inspire in you better things."

Every issue of the high school weekly bears proudly the following "Platform":

"1. To support live school organizations.

"2. To recognize worth-while individual student achievements.

"3. Above all to foster the real 'Bearcat' spirit in all of Central High School."

Curricular and social interests tend to conform. Friday nights throughout the season are preëmpted for games; the Mothers' Council, recognizing that every Saturday night had its own social event, urged that other dances be held on Friday nights instead of school nights, but every request was met with the rejoinder that "Friday is basket-ball night."

This activity, so enthusiastically supported, is largely vicarious. The press complains that only about forty boys are prominent enough in athletics to win varsity sweaters. In the case of the girls it is almost 100 per cent. vicarious. Girls play some informal basket-ball and there is a Girls' Athletic Club which

---

[4] Cf. Ch. XXVIII for the discussion of basket-ball as a center of civic loyalty.

has a monogram and social meetings. But the interest of the girls in athletics is an interest in the activities of the young males. "My daughter plans to go to the University of ——," said one mother, "because she says, 'Mother, I just *couldn't* go to a college whose athletics I couldn't be proud of!'" The highest honor a senior boy can have is captaincy of the football or basket-ball team, although, as one senior girl explained, "Every member is almost as much admired."

Less spectacular than athletics but bulking even larger in time demands is the network of organizations that serve to break the nearly two thousand individuals composing the high school microcosm into the more intimate groups human beings demand. These groups are mainly of three kinds: the purely social clubs, in the main a stepping down of the social system of adults; a long distance behind in point of prestige, clubs formed around curriculum activities; and, even farther behind, a few groups sponsored by the religious systems of the adults.

In 1894 the high school boasted one club, the "Turemethian Literary Society." According to the early school yearbook:

"The Turemethian Society makes every individual feel that practically he is free to choose between good and evil; that he is not a mere straw thrown upon the water to mark the direction of the current, but that he has within himself the power of a strong swimmer and is capable of striking out for himself, of buffeting the waves, and directing, to a certain extent, his own independent course. Socrates said, 'Let him who would move the world move first himself.' . . . A paper called The Zetetic is prepared and read at each meeting. . . . Debates have created . . . a friendly rivalry. . . . Another very interesting feature of the Turemethian Society is the lectures delivered to us. . . . All of these lectures help to make our High School one of the first of its kind in the land. The Turemethian Society has slowly progressed in the last year. What the future has in store for it we can not tell, but must say as Mary Riley Smith said, 'God's plans, like lilies pure and white, unfold; we must not tear the close-shut leaves apart; time will reveal the calyxes of gold.'"

Six years later, at the turn of the century, clubs had increased to the point of arousing protest in a press editorial entitled "Barriers to Intellectual Progress." Today clubs and other extracurricular activities are more numerous than ever. Not only is the camel's head inside the tent but his hump as well; the first period of the school day, often running over into the next hour, has recently, at the request of the Mothers' Council, been set aside as a "convocation hour" dedicated to club and committee meetings.

The backbone of the purely social clubs is the series of unofficial branches of former high school fraternities and sororities; Middletown boasts four Alpha chapters. For a number of years a state law has banned these high school organizations, but the interest of active graduate chapters keeps them alive. The high school clubs have harmless names such as the Glendale Club; a boy is given a long, impressive initiation into his club but is not nominally a member of the fraternity of which his club is the undergraduate section until after he graduates, when it is said that by the uttering of a few hitherto unspoken words he comes into his heritage. Under this ambiguous status dances have been given with the club name on the front of the program and the fraternity name on the back. Two girls' clubs and two boys' clubs which every one wants to make are the leaders. Trailing down from them are a long list of lesser clubs. Informal meetings are usually in homes of members but the formal fall, spring, and Christmas functions are always elaborate hotel affairs.[5]

Extracurricular clubs have canons not dictated by academic standards of the world of teachers and textbooks. Since the adult world upon which the world of this intermediate generation is modeled tends to be dominated primarily by getting a living and "getting on" socially rather than by learning and "the things of the mind," the bifurcation of high school life is not surprising.

"When do you study?" some one asked a clever high school Senior who had just finished recounting her week of club meetings, committee meetings, and dances, ending with three parties the night before. "Oh, in civics I know more or less about politics, so it's easy to talk and I don't have to study that. In English we're reading plays and I can just look at the end of the play and know about that. Typewriting and chemistry I don't have to study outside anyway. Virgil is worst, but I've stuck out Latin four years for the Virgil banquet; I just sit next to —— and get it from her. Mother jumps on me for never studying, but I get A's all the time, so she can't say anything."

[5] Cf. Ch. XIX for fuller discussion of juvenile social life in Middletown.

The relative status of academic excellence and other qualities is fairly revealed in the candid rejoinder of one of the keenest and most popular girls in the school to the question, "What makes a girl eligible for a leading high school club?"

"The chief thing is if the boys like you and you can get them for the dances," she replied. "Then, if your mother belongs to a graduate chapter that's pretty sure to get you in. Good looks and clothes don't necessarily get you in, and being good in your studies doesn't necessarily keep you out unless you're a 'grind.' Same way with the boys—the big thing there is being on the basket-ball or football team. A fellow who's just a good student rates pretty low. Being good-looking, a good dancer, and your family owning a car all help."

The clubs allied to curricular activities today include the Dramatic Club—plays by sophomore, junior, and senior classes in a single spring have replaced the "programs of recitations, selections, declamations, and essays" of the old days; the Daubers, meeting weekly in school hours to sketch and in evening meetings with graduate members for special talks on art; the Science Club with its weekly talks by members and occasional lectures by well-known scientists; the Pickwick Club, open to members of English classes, meeting weekly for book reviews and one-act plays, with occasional social meetings; the Penmanship Club; and the Virgil Club, carrying with it some social prestige. Interest in the work of these clubs is keen among some students. All have their "pledges," making their rituals conform roughly to those of the more popular fraternities and sororities.

On the periphery of this high school activity are the church and Y.M.C.A. and Y.W.C.A. clubs. All these organizations frankly admit that the fifteen to twenty-one-year person is their hardest problem. The Hi-Y club appears to be most successful. The Y.M.C.A. controls the extracurricular activities of the grade school boys more than any other single agency, but it maintains itself with only moderate success in the form of this Hi-Y Club among the older boys. A Hi-Y medal is awarded each commencement to the boy in the graduating class who shows the best all-round record, both in point of scholarship and of character. The Y.W.C.A. likewise maintains clubs in the grades but has rough sledding when it comes to the busy, popular, influential group in high school. According to one representative senior girl:

"High School girls pay little attention to the Y.W. and the Girl Reserves. The boys go to the Y.M. and Hi-Y club because it has a supper meeting once a month, and that is one excuse for getting away from home evenings. There aren't any supper meetings for the girls at the Y.W. It's not much good to belong to a Y.W. club; *any one* can belong to them."

All manner of other clubs, such as the Hiking Club and the Boys' and Girls' Booster Club and the Boys' and Girls' Pep Club hover at the fringes or even occasionally take the center of the stage. Says the school paper:

"Pep Clubs are being organized in Central High School with a motive that wins recognition. Before, there has been a Pep Club in school, but this year we are more than fortunate in having two. Their business-like start this year predicts a good future. Let's support them!"

Pep week during the basket-ball season, engineered by these Pep Clubs, included:
"*Monday:* Speakers in each of the four assemblies. . . .
"*Tuesday:* Poster Day.
"*Wednesday:* Reverend Mr. —— in chapel. Booster pins and pep tags.
"*Thursday:* Practice on yells and songs.
"*Friday:* Final Chapel. Mr. —— speaks. Yells and songs.
"Pep chapel [6] for all students will be held in the auditorium the ninth period. Professor —— and his noisy cohorts will furnish the music for the occasion. Immediately following the chapel the students will parade through the business district."

With the growth of smaller competitive groups, class organization has also increased, reaching a crescendo of importance in the junior and senior years. In a community with such a strong political tradition it is not surprising that there should be an elaborate ritual in connection with the election of senior and other class officers. The senior officers are nominated early in the school year, after much wire-pulling by all parties. "The diplomatic agents of the candidates have been working for

[6] The evolution of the chapel to anything from a "Pep chapel" to a class rally is an interesting example of the change of custom while the label persists.

weeks on this election," commented the school paper. The election comes a week later so as to allow plenty of electioneering; the evening before election an "enthusiasm dinner" is held in the school cafeteria at which nominees and their "campaign managers" vie with each other in distributing attractive favors (menus, printed paper napkins, and so on), and each candidate states his platform.

Amid the round of athletics, clubs, committees, and class meetings there is always some contest or other to compete for the time of the pupils. Principals complain that hardly a week passes that they do not have to take time from class work in preparation for a contest, the special concern of some organization. In 1923-24 these included art and music memory contests, better speech and commercial department contests, a Latin contest, a contest on the constitution, essays on meat eating, tobacco, poster making, home lighting, and highways.

In this bustle of activity young Middletown swims along in a world as real and perhaps even more zestful than that in which its parents move. Small wonder that a local paper comments editorially, "It is a revelation to old-timers to learn that a genuine boy of the most boyish type nowadays likes to go to school." "Oh, yes, they have a much better time," rejoined the energetic father of a high school boy to a question asked informally of a tableful of men at a Kiwanis luncheon as to whether boys really have a better time in school than they did thirty-five years ago or whether they simply have more things. "No doubt about it!" added another. "When I graduated early in the nineties there weren't many boys—only two in our class, and a dozen girls. All our studies seemed very far away from real life, but today—they've got shop work and athletics, and it's all nearer what a boy's interested in."

The relative disregard of most people in Middletown for teachers and for the content of books, on the one hand, and the exalted position of the social and athletic activities of the schools, on the other, offer an interesting commentary on Middletown's attitude toward education. And yet Middletown places large faith in going to school. The heated opposition to compulsory education in the nineties [7] has virtually disappeared;

[7] The following, from editorials in the leading daily in 1891, reflect the virulence with which compulsory education was fought by many, and inci-

only three of the 124 working class families interviewed voiced even the mildest impatience at it. Parents insist upon more and more education as part of their children's birthright; editors and lecturers point to education as a solution for every kind of social ill; the local press proclaims, "Public Schools of [Middletown] Are the City's Pride"; woman's club papers speak of the home, the church, and the school as the "foundations" of Middletown's culture. Education is a faith, a religion, to Middletown. And yet when one looks more closely at this dominant belief in the magic of formal schooling, it appears that it is not what actually goes on in the schoolroom that these many voices laud. Literacy, yes, they want their children to be able to "read the newspapers, write a letter, and perform the ordinary operations of arithmetic," but, beyond that, many of them are little interested in what the schools teach. This thing, education, appears to be desired frequently not for its specific content but as a symbol—by the working class as an open sesame that will

dentally exhibit a pattern of opposition to social change that bobs up from time to time today as innovations appear:

"Taxpayers of this county are upset by the state and county teachers' resolutions favoring compulsory education. . . . The teachers in our schools are not well versed in political economy. The most of them are young, and have had little time to study anything other than textbooks and their reports and programs. The idea of compulsion is detestable to the average American citizen. Men do not become good under compulsion. Two classes of men are clamoring for compulsory education: those who are depending upon school work for a living and for place and power, and those who are afraid of the Catholic Church. . . . The school system has not done what was expected of it. Immorality and crime are actually on the increase. . . . The states that have the greatest percentage of illiteracy have the smallest percentage of crime. . . . Compulsory education has failed wherever tried on American soil."

"The danger to the country today is through too many educated scoundrels. Boys and girls learn to cheat and defraud in copying papers for graduation essays. . . . A law compelling a child seven years of age to sit in a poorly ventilated school room and inhale the nauseous exhalations from the bodies of his mates for six hours a day for three or four months at a time, is a wicked and inhuman law. . . . Children forced into schools are morally tainted—and neutralize the virtues of well-bred children. It is a great mistake for the state to undertake to carry forward the evolution of the race from such bad material when there is so much good material at hand. Every movement that tends to relieve the father or mother of the moral responsibility of developing, training and directing the moral and intellectual forces of their own children, tends to reduce marriage and the home to a mere institution for the propagation of our species."

The press of 1900 noted that "the problem of securing boy labor is still worrying [state] manufacturers. The truancy law, they say, is detrimental to their business."

mysteriously admit their children to a world closed to them, and by the business class as a heavily sanctioned aid in getting on further economically or socially in the world.

Rarely does one hear a talk addressed to school children by a Middletown citizen that does not contain in some form the idea, "Of course, you won't remember much of the history or other things they teach you here. Why, I haven't thought of Latin or algebra in thirty years! But . . ." And here the speaker goes on to enumerate what *are* to his mind the enduring values of education which every child should seize as his great opportunity: "habits of industry," "friendships formed," "the great ideals of our nation." Almost never is the essential of education defined in terms of the subjects taught in the class-room. One member of Rotary spoke with pitying sympathy of his son who "even brought along a history book to read on the train when he came home for his Christmas vacation—the poor overworked kid!"

Furthermore, in Middletown's traditional philosophy it is not primarily learning, or even intelligence, as much as character and good will which are exalted. Says Edgar Guest, whose daily message in Middletown's leading paper is widely read and much quoted:

> "God won't ask you if you were clever,
>   For I think he'll little care,
>   When your toil is done forever
>   He may question: 'Were you square?'"

"You know the smarter the man the more dissatisfied he is," says Will Rogers in a Middletown paper, "so cheer up, let us be happy in our ignorance." "I wanted my son to go to a different school in the East," said a business class mother, "because it's more cultured. But then I think you can have too much culture. It's all right if you're living in the East—or even in California—but it unfits you for living in the Middle West." [8] Every one lauds education in general, but relatively few people in Middletown seem to be sure just how they have ever used their own education beyond such commonplaces as the three R's and an occasional odd fact, or to value greatly its specific outcome in others.

[8] Cf. Ch. XXIV for discussion of this same emphasis upon character rather than ability in the choosing of public officials.

Some clew to these anomalies of the universal lauding of education but the disparagement of many of the particular things taught, and of the universal praise of the schools but the almost equally general apathy towards the people entrusted with the teaching, may be found in the disparity that exists at many points between the daily activities of Middletown adults and the things taught in the schools. Square root, algebra, French, the battles of the Civil War, the presidents of the United States before Grover Cleveland, the boundaries of the state of Arizona, whether Rangoon is on the Yangtze or Ganges or neither, the nature or location of the Japan Current, the ability to write compositions or to use semicolons, sonnets, free verse, and the Victorian novel—all these and many other things that constitute the core of education simply do not operate in life as Middletown adults live it. And yet, the world says education is important; and certainly educated men seem to have something that brings them to the top—just look at the way the college boys walked off with the commissions during the war. The upshot is, with Middletown reasoning thus, that a phenomenon common in human culture has appeared: a value divorced from current, tangible existence in the world all about men and largely without commerce with these concrete existential realities has become an ideal to which independent existence is attributed. Hence the anomaly of Middletown's regard for the symbol of education and its disregard for the concrete procedure of the school-room.

But the pressure and accidents of local life are prompting Middletown to lay hands upon its schools at certain points, as we have observed, and to use them instrumentally to foster patriotism, teach hand skills, and serve its needs in other ways. This change, again characteristically, is taking place not so much through the direct challenging of the old as through the setting up of new alternate procedures, e.g., the adding to the traditional high school, offering only a Latin and an English course in 1890, of ten complete alternate courses ranging all the way from shorthand to home economics and mechanical drafting. The indications seem to be that the optional newcomers may in time displace more and more of the traditional education and thus the training given the young will approach more nearly the methodically practical concerns of the group.

Lest this trend of education overtaking the life of Middletown appear too simple, however, it should be borne in mind that even while Middletown prides itself on its "up-to-date" schools with their vocational training, the local institutional life is creating fresh strains and maladjustments heretofore unknown: the city boasts of the fact that only 2.5 per cent. of its population ten years of age or older cannot read and write, and meanwhile the massed weight of advertising and professional publicity are creating, as pointed out above, new forms of social illiteracy, and the invention of the motion picture is introducing the city's population, young and old, week after week, into types of vivid experience which they come to take for granted as parts of their lives, yet have no training to handle. Another type of social illiteracy is being bred by the stifling of self-appraisal and self-criticism under the heavily diffused habit of local solidarity in which the schools coöperate. An organized, professional type of city-boosting, even more forceful than the largely spontaneous, amateur enthusiasm of the gas boom days, has grown up in the shelter of national propaganda during the war. Fostered particularly by the civic clubs, backed by the Chamber of Commerce and business interests, as noted elsewhere, it insists that the city must be kept to the fore and its shortcomings blanketed under the din of local boosting—or new business will not come to town. The result of this is the muzzling of self-criticism by hurling the term "knocker" at the head of a critic and the drowning of incipient social problems under a public mood of everything being "fine and dandy." Thus, while education slowly pushes its tents closer to the practical concerns of the local life, the latter are forever striking camp and removing deeper into the forest.

IV. USING LEISURE

Chapter XVII

TRADITIONAL WAYS OF SPENDING LEISURE

Some of Middletown's waking hours escape the routinization of getting a living, home-making, receiving training in school, or carrying on religious or communal practices; in contrast to more strictly marshaled pursuits, such hours are called "leisure time," and this precious time, quite characteristically in a pecuniary society, is "spent."

The manner of spending leisure is perforce conditioned by the physical environment of the city and by the rest of its culture. Its location in the flat ex-prairie known as the Corn Belt precludes such variety of activity as cities adjacent to mountains, lakes, or forests know; Middletown, according to a local editorial, "is unfortunate in not having many natural beauty spots." There is rolling country to the south, but no real hills nearer than one hundred miles. Equally distant to the north are "the lakes," large prairie ponds scattered through flat farming country. A small river wanders through Middletown, and in 1890 when timber still stood on its banks, White River was a pleasant stream for picnics, fishing, and boating, but it has shrunk today to a creek discolored by industrial chemicals and malodorous with the city's sewage. The local chapter of the Isaak Walton League aspires to "Make White River white." "This Corn Belt . . . is not a land to thrill one who loves hills, wild landscape, mountain panorama, waterfalls, babbling brooks, and nature undisturbed. In this flat land of food crops and murky streams rich with silt, man must find thrills in other things, perhaps in travel, print, radio, or movie." [1]

Middletown people today enjoy a greater variety of these alternate other things than their parents knew a generation ago. The lessening of the number of hours spent daily in getting a living and in home-making and the almost universal habit of the Saturday half-holiday combine with these new possibilities

[1] Smith, *op. cit.*, pp. 298-9.

for spending an extra hour to make leisure a more generally expected part of every day rather than a more sporadic, semi-occasional event. The characteristic leisure-time pursuits of the city tend to be things done with others rather than by individuals alone;[2] and except for the young, particularly the young males, they are largely passive, i.e., looking at or listening to something or talking or playing cards or riding in an auto; the leisure of virtually all women and of most of the men over thirty is mainly spent sitting down. Its more striking aspects relate to the coming of inventions, the automobile, the movies, the radio, that have swept through the community since 1890, dragging the life of the city in their wake. Yet these newer forms of leisure must be viewed against an underlying groundwork of folk-play and folk-talk that makes up a relatively less changing human tradition.

Middletown has always delighted in talk. The operation of its business as well as of many of its professional institutions depends upon talk; honored among those who get its living are those puissant in talking. The axis upon which the training of its children turns is teaching them to use language according to the rules of the group and to understand the talk of others, whether spoken or written. Talking is the chief feature of its religious services. Much of its leisure time it spends in talking or listening to talk.

The habit of thinking no occasion, from an ice cream social to the burial of the dead, complete without a speech, is nearly as strong as in the nineties when, on a characteristic occasion, it took no less than eight speakers to dedicate a public building:

"The evening's exercises were begun by placing Rev. O. M. T—— in the chair. Rev G—— then delivered a very fine prayer and was followed by the regular address of the evening by Mr. T——. He was succeeded by Mr. J. W. R——. . . . Then came Glenn M——, Charley K——, Charley M——, and George M——. Mrs. P—— then delivered a short address, after which the meeting adjourned."

The dedication of two new buildings at the local college in 1925 included in its morning, afternoon, and evening programs six formal "addresses" and five other "talks." Indeed, as the author

[2] Cf. Ch. XIX for a discussion of the basis of person-to-person association around which these activities are built.

of *The American Commonwealth* pointed out forty years ago, "there is scarcely an occasion in life which brings forty or fifty people together on which a prominent citizen or a stranger . . . is not called upon 'to offer a few remarks.' "

And today, as in the nineties, the oratory of the speaker is nearly, if not quite, as important as the subject of his speech. "No matter what it's about, there's nothing I like better than a real good speech," remarked a leading citizen in 1924. In 1890 it was not necessary to announce the subject of a "lecture" to draw a crowd; "Rev. C. R. Bacon of W—— will deliver a free lecture at the Methodist Church Wednesday night. . . . Everybody invited" ran a characteristic announcement in the 1890 press. Another minister was invited to repeat his lecture on "Sunshine." "The lecture has been delivered here before," said the press notice, "and yet so well pleased was the audience that the church was well filled to hear the eminent divine a second time." [3] The relative unimportance of lecture subjects today appears in the civic clubs which are kept alive week after week by an endless succession of speeches on almost every subject from Gandhi to the manufacture of a local brand of gas burners for coffee roasters. One of the most popular speakers frequently paid by Middletown to talk to it is a woman travel lecturer described by the local press as one who "delights in superlatives and whose fluency of expression has won for her an enthusiastic group of admirers in this city." "No subject is prescribed for him," continues Lord Bryce in the passage cited above; ". . . he is simply put on his legs to talk upon anything in heaven or earth which may rise to his mind."

If the subject of the address is one with which the hearers are unfamiliar or upon which they have no fixed views, they frequently adopt bodily not only the speaker's opinion but its

[3] "Judge —— of Rushville will speak on the political issues of the day at the opera house Saturday at 2 P.M.," read the vague announcement of yet another lecture. "The judge is an able and eloquent speaker and you will be well entertained if you hear him."

Nothing is more characteristic of the early interest in speeches than the commencement exercises of the high school in 1890. Every graduate wrote and delivered an essay and people discussed certain essays for years afterwards. On the great night "everybody in town" gathered to hear these talks on such subjects as "We Sinais Climb and Know It Not"; "Whence, What, and Whither"; "Timon of Athens"; "Pandora's Box"; "Flowers," discussed as "aids in making life pleasant for rich and poor"; and a new type of "Germs" filling the body, whose "action on mind and heart causes the impulses for good and bad."

weighting of emotion. It is not uncommon to hear a final judgment on "the Philippine problem," "economic fundamentals,"
"the cause of cancer," or "the future of the white race," delivered with the preamble, "Well, I heard —— say at Chautauqua
[or at Rotary] two years ago. . . ." [4] Heckling is unknown;
people think with the speaker; rarely do they challenge his
thought.[5]

Changes are, however, apparent in this complex of speech
habits. Speeches are getting shorter; the long, general public
lecture bringing its "message" is disappearing as a form of entertainment. At the Farmers and Knights of Labor picnic in
1890, the feature of the afternoon was "an address lasting two
hours" to "a great crowd." This sort of thing would not draw
a crowd today. "A large and cultured audience" no longer
"crowds the Opera House" to hear "a polished gentleman of
pleasant presence and happy manner, thoroughly at ease before
an audience," deliver a lecture at once "eloquent and humorous,
logical, and pathetic" on the subject of "Nicknames of Prominent Americans," "Milton as an Educator," or "The Uses of
Ugliness." The humorous lecture, so popular in 1890 when the
great Riley-Nye combination rocked the Middletowns of
America, has almost disappeared today. Likewise have all but
vanished the heavy crop of moral and religious lectures by visiting ministers and denominational college presidents on "That
Boy," "Strange Things and Funny People," "Backbone," "The

[4] One such popular speaker told Rotarians: "My friends, we've been
asleep as a nation! But we are waking up. We always have waked up in
time. We've been in just as bad holes as any of the nations of Europe, but
there is always this difference, we always wake in time. . . . That's a
characteristic of us Anglo-Saxons. Babylonia went into a hole—and stayed
there. Rome went down—and never came up. Greece—swept away. Spain
went down, and we don't see her getting out. But we somehow always do
and always will!" And the Rotary Club cheered him to the echo and went
home to bring its wives to his evening lecture, the school authorities sending special word for all teachers to turn out. At the evening lecture, speaking on "The Eagle and the Oyster," the speaker lauded the American
business individualist as the eagle and decried the radical and socialist as
the "colony-hugging non-individualist." According to an enthusiastic business class citizen, "He showed how some things we don't think much
about are really socialism creeping in. He said that all these attempts to
regulate wages and hours are a mistake—getting away from the law of supply and demand. It is just the sort of sound logic that puts you back on
your feet again!"

[5] The prominence of oratory as well as this docility of the audience is
probably not unrelated to the authority of the evangelical Protestant preaching tradition in the community.

Trials of Jesus." The secularization of lectures and lecturers is marked and includes the increasing supplanting of such lectures as the above by short talks to club groups, more and more of them talks on specific subjects to specialized groups such as the Advertising Club, Poultry Raisers, Bar Association, and Medical Association. The one-time popular money-raising device of Sunday School classes and Young People's Societies of sponsoring a public lecture or winter lyceum is almost unknown today.[6] The lecture at which an admission fee is charged is in general a losing proposition in these days of radio and movies. The teachers have abandoned their effort to conduct a winter lyceum, and the lyceum conducted by the Ministerial Association, after strenuous city-wide efforts to drum up audiences, lost $6.00 on a course of five lectures in 1923 and made $15.00 in 1924. The local Chautauqua, lasting less than a week, is rapidly ceasing to be popular with the business group and achieves only a precarious support, likewise after hard pushing by the churches. A cleavage between business and working class groups is apparent in the greater tendency of the latter to support the earlier type of general discourse.

Among the activities tending to displace listening to talk is another form of this complex of speech habits, the reading of printed matter. Most of Middletown's reading matter originates elsewhere.[7] Through the development of devices for producing and distributing this material, the city now has access to a range and variety of reading matter unknown to its parents.[8] Book

[6] Only one in six of the public lectures in Middletown during 1924 was delivered in a church auditorium, as against more than half in 1890. This is in part due to the development of available auditoriums outside of churches, but it carries with it the incidental loss by the religious agencies of their former close place-association with this phase of group activity.

[7] See Ch. XXVII for discussion of newspapers, which are, however, only in part written in Middletown.

[8] "In 1887 typesetting was essentially the same art as in the sixteenth century. Since 1890 machine composition has been rapidly supplanting typesetting by hand. The average rate of composition on the linotype is estimated . . . at between 4,000 and 5,000 ems per hour. The rate of hand composition does not exceed 1,000 ems per hour on the average." George E. Barnett, "The Introduction of the Linotype" (*Yale Review,* November, 1904). The first linotype machine was introduced into Middletown in the late nineties.

Book production in the United States has virtually doubled in the last generation, increasing from 4,559 in 1890 to 8,863 in 1923. (See the files of *The Publishers' Weekly.*)

reading in Middletown today means overwhelmingly, if we exclude school-books and Bibles, the reading of public library books.[9] Over 40,000 volumes are available in the library, roughly fifteen volumes for every one to be had in the early nineties. Middletown drew out approximately 6,500 public library books for each thousand of its population during 1924, as against 850 for each thousand of population during 1890. Four hundred and fifty-eight persons in each 1,000 were library card-holders in 1923, whereas even as late as 1910 only 199 people in each 1,000 had cards.[10]

The buying of current books is almost entirely confined to a limited number of the business class. The rest of the population buy few books, chiefly religious books, children's books, and Christmas gifts, in the order indicated. Only twenty-four housewives out of the 100 working class families from whom family expenditures were secured reported expenditures for books other than school-books by members of their families during the past twelve months. The totals for the year ranged from $0.50 to $52.50; twelve of the twenty-four had each spent less than $5.00, six between $5.00 and $10.00, and six $10.00 or more.[11]

[9] The religious organizations of the city, many of which maintained small separate libraries in 1890 which are said to have been a boon, have for the most part ceased to perform this service, though free Sunday School papers are the rule today. On the other hand, the public library has entered the schools of the city with sixty libraries and maintains a book truck carrying books to the outlying sections. Seven full-time librarians and a part-time assistant have displaced the single untrained librarian who received $45.00 a month for conducting the public library tucked away in upstairs rooms in 1890.

[10] Figures on card-holders in 1890 are not available. Such figures as these must be used carefully. Any literate resident can obtain a card today by merely asking for one, without references or delay of any kind, whereas cards were issued in 1890 only to those ten years of age and over and after more red tape than today. Branch deposits in school buildings, the use of the book truck, and the "supplementary reading" required of children, particularly in the high school, tend to diffuse the card-holding habit. The number of books withdrawn is influenced by the fact that more books may be taken out on a single card today, thereby allowing the borrower to take more books home and read the one that turns out to be most interesting. There is no way of estimating the circulation of Sunday School library books a generation ago.

[11] The following books were bought by these twenty-four families, the purchases of each family being set off by semicolons: *Lives of Great Men* and *University Encyclopædia* (total $5.00) ; a Bible for daughter; a $5.00 book on Sunday School work; a Bible and, for little son, A B C Books, and Bible stories; a fifty-cent *History of the Methodist Church;*

Even more marked than the greater availability of books is the increase in the number of weekly or monthly periodicals since the days of *"The Pansy* for Sunday and weekday reading" and *"The Household* sent free to every newly-married couple upon receipt of ten cents in stamps." Today the Middletown library offers 225 periodicals as against nineteen periodicals in 1890. Heavy, likewise, has been the increase in the number of magazines coming into Middletown homes.[12] Into the 9,200 homes of the city, there came in 1923, at a rough estimate, 20,000 copies of each issue of commercially published weekly and monthly periodicals, excluding denominational church papers, Sunday School papers distributed free weekly to most of the 6–7,000 attending Sunday School, and lodge and civic club magazines.[13] Forty-seven of the 122 working class families and one of the thirty-nine business class families giving information on this point subscribe to or purchase regularly no periodical; thirty-seven of the former and four of the latter subscribe to or purchase regularly only one or two periodicals; and thirty-eight (three in ten) of the workers' families and thirty-four (nine in ten) of this business group take three

Fox's *Book of Martyrs,* Hurlburt's *Story of the Bible;* technical books, ten-cent little leather library books for the children; a family doctor book; a New Testament and *Four Thousand Questions and Answers on the Bible;* two Bibles, and five or six Christmas books; a set of the *World's Wonder Books* ($57.00) ; Williams' *Tinsmith's Helper and Pattern Book* ($3.00—a book for the husband's trade) ; Bible stories; family doctor books; religious books ($14.00) ; a Bible with encyclopædia and concordance ($7.00) and a Prayer Book; boys' books for Christmas; the set of books studied by the Delphian Chapter of local club women, though the wife was not a member, and also some sociology books in connection with her club work; a Bible and a pamphlet; story books at Christmas time; *Human Interest Library* ($29.00, five volumes) ; *Beautiful Story of the Bible;* story books at Christmas time; *Human Interest Library.*

Data were not secured on the book purchases of the business group.

[12] Of a given issue of the *Literary Digest* 939 copies reached Middletown in 1923, as over against thirty-one in 1900. Three hundred and fifty-five copies of the *National Geographic* went into local homes in 1923 as against twenty-five in 1910. Both the number of different periodicals and their efficient distribution have increased notably. As many as seventy different current periodicals may be seen displayed in a single drug store window in Middletown today.

There were no national circulations in 1890 like the 2,000,000 circulations of today. The *Atlantic Monthly* had a circulation of only about 10,000 in the entire United States in 1890, as against twelve times that today. In 1898 the *Saturday Evening Post* had a national circulation of 33,069, whereas nearly 1,500 copies of each issue go to Middletown alone today.

[13] Based upon subscription and news-stand totals. The circulation of one issue, whether weekly or monthly, is here taken as the unit.

or more periodicals. In both groups additional periodicals are bought from the news-stand sporadically by certain families.

The significance of such a ceaseless torrent of printed matter in the process of diffusing new tools and habits of thought can scarcely be overstated. Does this greater accessibility and wider diffusion of books and periodicals today mean, however, that Middletown is spending more time in reading? Library and periodical records suggest a marked increase, but other evidence necessitates qualification of this conclusion. The type of intellectual life that brought anywhere from two dozen to a hundred people, chiefly men, together Sunday after Sunday for an afternoon of discussing every subject from "Books, What to Read and How to Read Them" to the *Origin of Species* and the "Nature of God" has almost disappeared among the males; men are almost never heard discussing books in Middletown today.[14] The impulse in the local labor movement represented in the statement in trade union constitutions, "Each labor union should found libraries, [and] hold lectures," and which eventuated in 1900 in the organization of an independent Workingmen's Library, has gone.[15] The "reading circles" of the nineties have all but disappeared, but the women's clubs of today fill much the same place they occupied as stimuli to reading. Although groups of ten or twelve women no longer meet weekly to gain "the college outlook" by following the four-year cycle of Chautauqua readings, at least one business class woman is still reading for her "diploma" and more than one points to the rows of her mother's "Chautauqua books," saying that they are her chief help in preparing her club programs.[16] When teachers come together today it is not to read

[14] Few male leaders in Middletown read except in a desultory manner. Social workers and ministers complain that their outstanding problem is lack of leadership. "There is no group of intelligentsia here," said a leading minister, "only the bourgeoisie of the Rotary Club." One never hears book-talk around the tables at civic club luncheons. Even the ministers, as noted elsewhere, have little time to read.

[15] The working men employed one of their number as librarian at $600.00 a year. Among the purchases for the library recorded in 1900 are the *American Statesmen Series,* John B. Clark's *Distribution of Wealth,* Charlotte Perkins Stetson's *Women and Economics,* David A. Wells' *Recent Economic Changes,* "seven books on religion," and 210 volumes of fiction, including Thackeray and Dickens. This was before the day of the automobile and movies. The library has long since disappeared.

[16] See Ch. XIX for discussion of the Delphian Chapter, which probably corresponds most closely to these early reading circles, and for an account

and discuss books as in the old state reading circles; nor do young people meet to read books suggested by the state for Young People's Reading Circles, although according to some of the teachers, State Young People's Reading Circles were never organized very extensively in Middletown and the required supplementary reading now being done in connection with English classes in the high school more than makes up for their decline.[17] No longer do a Young Ladies' Reading Circle, a Christian Literary Society (of fifty), a Literary League, a Literary Home Circle, a Literary Fireside Club meet weekly or bi-weekly as in 1890, nor are reading circles formed in various sections of the city, nor does a group of young women meet to study the classics.[18] The young people's societies of the various churches do not form Dickens Clubs or have "literary evenings," as, for example, "an evening with Robert Burns," with the singing and recitation of Burns' poems and the reading of the poet's biography—a program of eleven numbers, concluding with "a discussion of Burns and his writings." "Young ladies' " clubs in Middletown are more likely now to play bridge; the attenuated church young people's societies follow mission study or other programs sent out

of the other women's study clubs. In these days of multiplied periodicals, public libraries, motion pictures at every cross-road, and no farm house or village too isolated to "tune in" on a metropolitan lecture or symphony concert, it is difficult to appreciate the vigor and enthusiasm aroused by the Chautauqua and Bay View circles. By 1903, says the Chautauqua booklet on *Literature and the Larger Life*, "more than 11,000 Chautauqua circles" had been conducted in "about 6,000" different localities. To these eager groups that met "around the study lamp" week after week, "Mehr Licht," the motto of the Bay View Circle, and Bishop Vincent's quotation of "Knowledge is power" at the opening of the Chautauqua Circle, were no mere mottoes but promises of fuller life. In Middletown a number of women, after having followed the "readings" on German literature or Greek life, one year on each country through four years, took examinations and received their "diplomas," a few going on to Chautauqua, New York, for the graduation exercises.

[17] State Teachers' Reading Circles were active in the nineties. Earlier figures are not available, but in 1902-3, 180 of 286 teachers in the county are listed as members of the state reading circles, and of the 10,566 children enrolled in the schools of the county, 1,817 belonged to state reading circles owning 3,412 books.

[18] Such a group as this last did not meet in 1890, but in 1895 the press reported a class of ten young women organized two years earlier which "meets Saturday afternoon, and has read Ruskin's *Essays,* Dante's *Vision,* Pope's Homer's *Iliad,* and is now reading the *Odyssey,* a new prose translation by George Herbert Palmer, professor of philosophy in Harvard College. This little society is a class of students, not a club, and enjoys the distinction of having no officers."

from denominational headquarters; men do not talk books; chiefly in women's clubs does the earlier tradition persist.

Middletown papers do not now carry such book advertisements as were familiar in the nineties: advertisements of the *Franklin Square Library,* Stanley's *In Darkest Africa,* the life of Dwight L. Moody, the life of Barnum, "New Books this week—Star Drug Co.: *A Laggard in Love* by Bettany, *The Tale of Chloe* by Meredith," or literary notes on *"Black Beauty,* the *Uncle Tom's Cabin* of the horse (price 12c.)," *John Ward, Preacher,* or a new edition of *The Autocrat of the Breakfast Table.* So widespread was the reading of certain books that, when the census of 1890 was being taken, a local paper suggested as additional questions for the census taker: "What is your opinion of the evolution theory? and, Have you read *Robert Elsmere?"* Even a young baker noted in his diary the reading of *Robert Elsmere* between his hilarious evenings of "banging about town." A newspaper in 1892 offered "a set of Dickens to the person sending in the best record of the number, contents, and firms inserting advertisements in the *Tribune"*—a procedure hardly apt to elicit popular response today.

Concrete evidence of sorts as to the amount of time Middletown spends reading today is afforded by the statements of the women interviewed as to the amount of time they and their families spend in reading of all kinds, including newspapers, periodicals, and books. Of the forty business class and 117 working class male heads of families for whom data were secured from their wives, only one of the former and nineteen of the latter were said to spend no time in reading of any kind.[19] None of these business class wives said she reads none at all, as against twenty-one of the workers' wives; twenty (half) of the former and thirty (a fourth) of the latter average six or more hours of reading a week.

To working class women reading apparently presents itself as a less urgent leisure-time pursuit than to women of the busi-

[19] "When do you get time to read?" a rising young lawyer was asked. "I don't—much. It's next to impossible to get any regular reading or study done. We know an awful lot of people and at least two or three evenings every week go out with them. Now Monday evening we went over to a little club we belong to where we play cards every week, Tuesday we went to that lecture for the benefit of the Day Nursery, Wednesday I forget what we did, but we went somewhere or other—and so the week goes."

ness group. Although eighteen of the former mentioned reading as the chief thing they would like to do with an extra hour in the day,[20] answers of these women were given only after some urging and tended to be vague, as "Oh, I don't know—read, I guess." Considerably more frustration was reflected in the more pointed answers of the fourteen business class women who mentioned reading:

"I would read if I only had the energy and quiet."

"I have tried three times to get into *The New Decalogue of Science*, but I never have time to give to it."

"I would take up some definite study. I am really interested in intellectual work, but I have so little time for it. I have tried to get one of the librarians at the public library to make me out a list of books on history but it doesn't work very well. My reading is so scattered, due to my children being small, and I want to organize it."

"I just read magazines in my scraps of time. I should so like to do more consecutive reading but I don't know of any reading course or how to make one out."

According to statements of the women interviewed as to the time they and their husbands spend in reading as compared with the time spent by the wife's parents in the nineties, business class men would appear to read somewhat less, business class women about the same amount, and working class men and women somewhat more than did the wife's parents a generation ago.[21] The trends possibly suggested by these answers are supported by the observation of teachers and others who have

---

[20] For answers to this question see Ch. XIX.

[21] Some working men, however, found time for a good deal of reading in the nineties. A partial list of the books recorded in the diary of the baker quoted above as read during the year 1889-91 includes: *Barriers Burned Away, Robert Elsmere, Dinna Forget, Dr. Jekyll and Mr. Hyde* ("made my hair stand on end"), *Camille, or the Fate of a Coquette* ("a tale of fast life"), *The Rock or the Rye, Broken Vows, The Kreutzer Sonata, Love's Conflict, Linda Newton: or Life's Discipline, Huckleberry Finn,* and *History of the United States in Our Time* (Pt. I). The diary of this young blood, whose comment after seeing *Faust* at the local Opera House was: "Very good. Makes a person think of the hereafter," and who had only eighteen evenings in a six-month stretch when he was not "on the go" "out having a time," records his managing to read one book every six weeks or two months.

known Middletown since the nineties. Any increase in book reading indicated by heavier library circulation is certainly not among business class men and only to a limited extent among their wives.[22] It may reflect an increase in the amount of reading done by the working class and particularly by children of this group, as habits of prolonged schooling and increased use of the library are spreading among them.[23] More buying of cheap paper-covered books in the nineties and the reading of books from the meager Sunday School libraries undoubtedly offset to some extent the increased use of the public library and the far wider periodical circulations of today. Testimony of many local people suggests that more things are skimmed today but that there is less of the satisfaction of "a good evening of reading." There appears to be considerably less reading aloud by the entire family.

As Middletown reads it is participating in other worlds, being subjected to other ways of living. The nature of this culture stream from without is worth scrutinizing. A catalogue of the

[22] The predominance of women among users of the library appears in the fact that of 486 adults taking out new cards in a sample period immediately following January 1, 1922, 64 per cent. were women, and of 492 adults taking out cards in a similar period following September 1, 1923, 61 per cent. were women.

[23] Eleven per cent. of 275 boys and 8 per cent. of 341 girls in the three upper years of the high school answered "reading" to the question asked in mid-November, "In what thing that you are doing at home this fall are you most interested?"

Three hundred and six boys and 379 girls in the last three years of the high school listed their reading of books in a "usual month," and 315 boys and 395 girls, largely the same group, listed their reading of magazines in a "usual month" as follows:

	Percentage of Boys	Percentage of Girls		Percentage of Boys	Percentage of Girls
Total	100	100		100	100
Read no books outside of school reading	16	12	Read no magazines	7	5
Read 1 book	19	14	Read 1 magazine...	25	20
Read 2 books	19	26	Read 2 magazines..	27	28
Read 3-5 books	28	29	Read 3-5 magazines	35	42
Read 6-8 books	9	7	Read 6-8 magazines	5	5
Read 9 or more books	9	12	Read 9 or more magazines	1	0

The boys of this group averaged 2.4 magazines read regularly, and the girls 2 6. The boys averaged 3.2 books read monthly and the girls 3.1.

books available in the adult department of the Middletown library, issued in 1893, showed in useful arts (technology, advertising, salesmanship, etc.) 91 books as against 1,617 in the adult department of the library in 1924; in fine arts 45 as against 1,166; in history 348 as against 2,867; in biography 132 as against 1,396; in sociology 106 as against 1,937; in literature 164 as against 2,777; in science 89 as against 585; and so on through the other classifications. During the twenty years between 1903, the first year for which library circulation figures are available, and 1923, while the population less than doubled and the reading of library books in the adult department of the library increased more than fourfold, reading of library books on useful arts increased sixty-two-fold;[24] on the fine arts twenty-eight-fold; on philosophy, psychology, etc., twenty-six-fold; on religion, particularly the religion of this group, elevenfold; on the institutional devices involved in group life, sociology, economics, etc., ninefold; on history eightfold; on science sixfold; of fiction less than fourfold; and so on.[25] To Middletown adults, reading a book means overwhelmingly what story-telling means to primitive man— the vicarious entry into other, imagined kinds of living; in 1903, 92 per cent. of all the library books read by adults were fiction, as against 83 per cent. in 1923. Under a trained children's librarian the reading of fiction by children has decreased from 90 per cent. of the total of books read in 1903 to 67 per cent. in 1923.[26] This interest in imaginative narratives, like the constant movie attendance to be noted later and the prime popularity of comedy and society films, obviously assumes greater significance when viewed against the background of the day-long preoccupation with getting a living and other routinized activities in this prairie city. Says one of Middletown's

[24] The librarian says that books on business, particularly salesmanship and advertising, and technical books, particularly those on automobiles and radio, are in such demand that they are "never on the shelves."

[25] See Table XVII.

A distinct trend towards the secularization of the books in the library on natural science is observable when the current catalogue is compared with that of thirty-five years ago. In the 1893 catalogue the scientific books commonly bore such titles as *The Wonders of Creation, The Wonders of Bodily Strength, The Wonders of God's Universe,* as over against such titles in the 1924 catalogue as *Introduction to Geology.*

[26] In 1903 the children of Middletown read 12,224 library books, of which 11,048 were fiction, as against a total of 93,873 in 1923, of which 63,307 were fiction.

prominent citizens in an editorial in the local press under the caption "Kicking Over the Traces":

"Most of us, perhaps fortunately, take it out in wishing. We can't rush off to Timbuctoo when it is necessary for us to stay home and provide food, coal, and shoes 'for the missus and the kids.' If we were to do so, we wouldn't have a good time.

"And so we remain at home, go to the office at eight in the morning and depart from it at six at night, and we attend committee meetings, and drive the old family bus over the streets that we have traversed a thousand times before, and in general continue the life of the so-called model citizen.

"But these conditions need not fetter our fancy. In that realm we can scale the lofty Matterhorn, sail the sleepy Indian Sea, mine glittering gold in the snow-clad mountains of Alaska, tramp the Valley of the Moon, and idle along the majestic Amazon." [27]

And so, as in primitive story-telling, the social function of these forays into the realm of fancy demands that the experiences thus vicariously shared be happy or valorous ones. "There's enough trouble in the world all about one, so why should people have to put it in books?" is an opinion frequently heard in connection with the prevailing demand for "happy endings—or at least endings that if not exactly happy still exalt you and make you feel that the world is coming out all right." Many people in Middletown would agree with their favorite poet, Edgar Guest,[28] in condemning people who condone "sin or unhappiness" in fiction by saying, "The book is sordid, but it's art!" He concludes his syndicated message in the local press:

[27] The position of this reading of books, predominantly fiction, as a marginal leisure-time activity with many people, is suggested by the fact that, if the circulation to students of the local normal college be not included, the monthly totals of library books read tend to be from 50 to 100 per cent. heavier in winter than during the summer. The compulsory supplementary reading of high school students figures somewhat in this increase.

[28] "Eddie" Guest is more widely read in Middletown than any other poet, with Riley as runner-up in popularity. Rotary has tried to secure him as a speaker, as has the Men's Club in a leading church. In a group of college-trained men prominent in local life, one said that "Eddie" Guest and Riley were his favorite poets, "That man Guest certainly gets to my heart"; one liked Kipling, "never could get Burns, and Byron always seemed a dirty fellow dressed up in poetic form"; while a third prefers Kipling and "never could get Browning. Why didn't he say it in prose instead of the awful way he did?"

"But should you by chance be cheerful, using people not so fearful,
   Should your characters go smiling down the street;
Should your fiction girl or man do just the very best they can do
   With the obstacles and trials they must meet;
Should they come to sane conclusions about life and its illusions,
   Should they keep their marriage vow 'till death do part,'
Should they find a thrill in duty and in life some joy and beauty,
   They will say: 'The story's pretty, but not art!'"

Any detailed analysis of the contents of the periodicals which flood Middletown weekly and monthly is impossible. Of the 225 periodicals offered by the Middletown library in 1924, twenty-five were trade and technical journals, twenty-three religious, fifteen educational, nine each women's, public administration, and juvenile magazines, eight each business and financial, scientific, and arts and decoration, seven nature, six economic, five musical, five garden, four household arts, four travel, three theater, etc.; of the nineteen periodicals in the 1890 library, seven were children's papers, six general monthlies, three weeklies, two scientific journals, one agricultural. Some further indication of the way periodicals operate, probably even more powerfully than books, to shape the habits and outlook of the city, may be gathered from the distribution of certain national periodicals. Approximately one in each five of the 9,200 homes in the city receives the *American Magazine* and one in each six the *Saturday Evening Post*. Each of the following goes regularly into from one in each five to one in each ten of the homes: *Delineator, Ladies' Home Journal, McCall's, Physical Culture, True Story, Woman's Home Companion*. Two hundred to 500 homes receive one or more of *Adventure, Argosy, Collier's, College Humor, Cosmopolitan, Country Gentleman, Dream World, Good Housekeeping, Hearst's International, Modern Marriage, Motion Picture Magazine, National Geographic, Pictorial Review, Popular Science, Red Book, True Romance,* and others. One hundred to 200 receive *College Comics, Flynn's Magazine, Triple X, True Confessions, True Detective Magazine, Whiz Bang,* and many others. Approximately sixty receive *Vogue* and the same number *Vanity Fair,* about thirty-five the *Atlantic Monthly,* about twenty each *Harper's* and the *Century,* about fifteen the

New Republic, five the *Living Age,* four the *Survey,* three the *Dial.*[29]

The different levels of diffusion within the city appear in the fact that fifty-four periodicals drawing 115 subscriptions from the thirty-nine business class families giving information on this point have not one from the 122 workers' families, while forty-eight periodicals drawing ninety-six subscriptions from seventy-five working class families have none of this group of business class families as a subscriber; in between is a narrow group of twenty periodicals with 128 subscriptions from thirty-eight business class families and 105 from seventy-five workers' families. Nine of the 122 workers and seventeen of this business class group take the *Literary Digest;* seven of the former and twenty of the latter the *Saturday Evening Post;* forty-four of the workers' wives subscribe to women's magazines, scattering a total of 101 subscriptions among twenty-one different women's magazines, while the twenty-seven of the business class wives who take women's magazines bunch their forty-eight subscriptions among only nine magazines, almost entirely recognized leading magazines;[30] thirteen of the workers and nineteen of this business group subscribe to the *American Magazine;* there are only seven subscriptions to juvenile magazines among 122 workers' families, as against twenty-six among the thirty-nine business class families; none of the sample of workers takes a magazine of the *Atlantic, Harper's, World's Work* type, as against a total of twenty-two such subscriptions among less than one-third as many of the business group.

A cleavage between the reading habits of the two sexes is possibly suggested by the answers of 310 boys and 391 girls in

---

[29] All figures include both subscriptions and news-stand sales. Beyond this count the list of periodicals received in various homes of the city extends on indefinitely—*Hot Dog, Sporting News, Secrets, Psychology, Movie Thriller, Mind Power Plus, Art and Life, Art Lovers, Boxing Blade, Experience, Health and Life, Correct Eating, Ace-High, Action Stories, Black Mask, Breezy Stories, Droll Stories, Hunting and Fishing, Jim Jam Jems, La Vie Parisienne, Live Stories, Love Story Magazine, Real Life, Motor, Cartoons, Golf, Red Pepper, Screenland, Sea Stories,* specialized business and technical magazines, etc.

[30] The import of this fact that, by and large, the poorer grades of women's magazines go to the workers and the better grades to the business group cannot be overlooked in its significance for the differential rate of diffusion of modern habits of making a home to the two groups.

the three upper years of the high school to the question, "What magazines other than assigned school magazines do you usually read every month?" Forty-four boys and 367 girls read women's magazines;[31] ninety-five boys and fifteen girls read scientific magazines; 114 boys and seventy-six girls read the *Saturday Evening Post, Collier's* and *Liberty* group of week-lies; thirty-five boys and two girls read outdoor magazines; seventy-two boys and sixteen girls read juvenile magazines; other smaller groups were more evenly balanced.

Although, according to the city librarian, increased interest in business and technical journals has been marked, as in its reading of books Middletown appears to read magazines pri-marily for the vicarious living in fictional form they contain. Such reading centers about the idea of romance underlying the institution of marriage; since 1890 there has been a trend toward franker "sex adventure" fiction. It is noteworthy that a culture which traditionally taboos any discussion of sex in its systems of both religious and secular training and even until recently in the home training of children should be receiving such heavy diffusion of this material through its periodical reading matter. The aim of these sex adventure magazines, dif-fusing roughly 3,500 to 4,000 copies monthly throughout the city, is succinctly stated in the printed rejection slip received by a Middletown author from the New Fiction Publishing Corporation:

"*Live Stories* is interested in what we call 'sex adventure' stories told in the first person. The stories should embody pic-turesque settings for action; they should also present situations of high emotional character, rich in sentiment. A moral conclu-sion is essential."

"Until five years ago," said a full-page advertisement in a Mid-dletown paper in 1924, "there was nowhere men and women, boys and girls could turn to get a knowledge of the rules of life. They were sent out into the world totally unprepared to cope with life.

[31] The groups given here are, of course, not mutually exclusive. The predominance of the reading of magazines on making a home among girls of this age and the fact that, aside from this negligible fringe of boys, the males of the group neither here nor elsewhere come in contact with any discussion of home-making problems is suggestive in connection with the concentration of the males upon matters divorced from the home and the fact that habits of management are still in vogue in the homes of Middle-town that are becoming obsolete in Middletown's industries.

. . . Then came *True Story*, a magazine that is different from any ever published. Its foundation is the solid rock of truth. . . . It will help you, too. In five years it has reached the unheard-of circulation of two million copies monthly, and is read by five million or more appreciative men and women."

In these magazines Middletown reads "The Primitive Lover" ("She wanted a caveman husband"), "Her Life Secret," "Can a Wife Win with the Other Woman's Weapons?" "How to Keep the Thrill in Marriage," "What I Told My Daughter the Night Before Her Marriage" ("Every girl on the eve of her marriage becomes again a little frightened child").

While four leading motion picture houses were featuring synchronously four sex adventure films, *Telling Tales* on the Middletown news-stands was featuring on its cover four stories, "Indolent Kisses," "Primitive Love," "Watch Your Step-Ins!" ("Irene didn't, and you should have seen what happened!") and "Innocents Astray." The way Middletown absorbs this culture about (to quote the advertisement of a local film) "things you've always wanted to do and never DARED" was suggested by the coverless, thumb-marked condition of the January, 1925, *Motion Picture Magazine* in the Public Library a fortnight after its arrival. One page, captioned "Under the Mistletoe," depicted seven "movie kisses" with such captions as:

"Do you recognize your little friend, Mae Busch? She's had lots of kisses, but never seems to grow blasé. At least, you'll agree that she's giving a good imitation of a person enjoying this one," and "If some one should catch you beneath the mistletoe and hold you there like this, what would you do? Struggle? But making love divinely is one of the best things Monte Blue does. Can't you just hear Marie Prevost's heart going pitty-pat?"

And a Middletown mother complained to the interviewer, "Children weren't bold like they are today when we were young!"

Music, like literature, is a traditional leisure activity regarded as of sufficient importance to be made compulsory for the young. The emergence of music to a prominent place in the school curriculum has already been described. It seems probable from informal local testimony that the taking of music lessons is a generally accepted essential in a child's home training among a wider group of the city's families than in the nineties. In forty-one of the 124 working class families, all of whom, it will be recalled, have children of school age, one or more children had taken music lessons during the preceding year; in twenty-seven of the group of forty business class families interviewed, one or more children had taken lessons.[32] Of fifty-four workers' wives reporting on the amount of time their children spend on music, forty-four said that they themselves spent less time on music as children than do their children today, while only five spent more time, and five "about the same." Among the business class, sixteen reported that they had spent less time on music than do their children, three that they spent more time, and three "about the same." In answer to the question to the three upper classes in the high school, "In what thing that you are doing at home this fall are you most interested?" "music" led the list with the 341 girls, being named by 26 per cent. of them, with "sewing" next most often mentioned, by 15 per cent.; among the 274 boys "radio" led the list, being mentioned by 20 per cent., while "music" followed with 15 per cent. The current interest arises in part from the muscularity injected into music by jazz, the diffusion of instruments other than the piano, and the social and sometimes financial accompaniments of knowing how to "play." The one musical club among women of the business class maintains a Junior and a Juvenile section with social meetings at which children play for an audience of their mothers; in addition to the two high school bands and three orchestras, Middletown has a boys'

[32] Of the 100 working class families for whom income distribution was secured, twenty-four reported money spent on music lessons for children. The amounts ranged from $2.50 to $104.00, averaging $44.57. See Table VI.

Comments by various parents reveal how seriously many of them take this training in music; witness a working class family of five in which the mother teaches one son to play the piano and the father teaches another son the drums, the mother explaining that "there is so much bad in life to keep children away from that we've decided it's hopeless to try, and the only thing to do is to make them love so many good things that they'll never pay attention to the bad things." Another worker's wife said that her sister criticized them for spending so much on their son's music, "but I feel his talent should be developed and want him to have a musical education if that's what he wants." The ubiquitous clash between having more children and maintaining a higher standard for fewer children came out in the remark of a modest-salaried business man's wife, "Our oldest boy takes music lessons and they cost us a good deal—$1.85 a week. That's why you can't have more children when everything you do for them costs so much."

band, a girls' band, and a band of both boys and girls from nine to thirteen years.[33]

Mechanical inventions such as the phonograph and radio are further bringing to Middletown more contacts with more kinds of music than ever before. Thirty-five years ago diffusion of musical knowledge was entirely in the handicraft stage; today it has entered a machine stage. The first phonograph was exhibited locally in 1890 and was reported as "drawing large crowds. The Edison invention is undoubtedly the most wonderful of the age." [34] Now these phonographs have become so much a part of living that, for example, a family of three, when the father was laid off in the summer of 1923, "strapped a trunk on the running board of the Ford, put the Victrola in the back seat with the little girl, and went off job-hunting. Wherever we lived all summer we had our music with us." [35]

And yet, although more music is available to Middletown than ever before and children are taught music with more or-

[33] Those familiar with the local musical life express the belief that the orchestral and other musical work in the high schools is recruiting an entirely new crop of musicians rather than reducing the number playing the piano. This is probably less true in the case of girls than of boys. Boys are more attracted to other instrumental work than to the piano because of the prestige of playing in one of the high school bands or in the well-known local boys' band, and more particularly because of the money they can earn playing in small dance orchestras. The energetic jazz aggregation of four or five boys, featuring the easily learned saxophone, presents a new and relatively distinguished occupation by which sons of working class parents are seeking in some cases to escape from the industrial level. The city has several of these small groups seeking engagements playing for dances.

[34] As late as 1900 "graphophones" were still curiosities. "The graphophone is rapidly superseding the piano in Middletown saloons. Fully fifty are being used and they never fail to draw large crowds," says the press.

[35] The ownership of radios in Middletown is noted elsewhere. No check was made of the ownership of phonographs and of pianolas. Of 100 working class families from whom expenditures were secured, however, twenty-three had bought phonograph records during the last twelve months. Eleven had bought less than $5.00 worth each; the amounts ranged from $1.05 to $50.00, averaging $11.17. Three had spent money for pianola records, each less than $4.00 worth. More than twenty-three phonographs were owned by these 100 families, however, as a number of others spoke of owning them but of having no money to spend on records. There was a marked tendency to sacrifice phonograph records both to the cost of installing a radio and to the cost of children's music lessons.

In the study of Zanesville, Ohio, in 1925 there were discovered phonographs in 54 per cent. of the homes, pianos in 43 per cent., organs in 3 per cent. and other musical instruments in 8 per cent.; the figures for the thirty-six cities, including Middletown, compared with Zanesville were 59 per cent., 51 per cent., 1 per cent., and 11 per cent., respectively. (*Op. cit.*, p. 112.)

ganized zeal than formerly, the question arises, as in the case of reading, as to whether music actually bulks larger as a form of leisure-time enjoyment than in the nineties. If one boy in each six or seven in high school enjoys music more than any other leisure-time home activity, this enthusiasm evaporates between high school and his active life as one of those getting Middletown's living. Music, like poetry and the other arts, is almost non-existent among the men. As noted elsewhere, "having a love of music and poetry" was ranked ninth among the qualities desirable in a father by 369 high school boys and seventh by 415 girls, only 4 per cent. of the boys and 6 per cent. of the girls ranking it as one of two qualities in the list of ten that they considered most desirable.[36] Music for adults has almost ceased to be a matter of spontaneous, active participation and has become largely a passive matter of listening to others. The popular singing societies of the nineties have disappeared, with one working class exception. One such group in the nineties, composed of workers, met every Sunday afternoon and Thursday evening with a "keg of beer" and a hired "instructor." Another singing society celebrated its sixty-fourth anniversary in 1890. Still another group of forty of the city's male social leaders, calling themselves the Apollo Club and dubbed "dudes" by the others, met every Friday evening, "instructed by Professor B——" who came over from a neighboring city; its three concerts each year were widely attended and received enthusiastic reports in the press. Even schoolboys apparently enjoyed chorus singing, for as late as 1900, 300 of them gave a concert at the Opera House, the program including Gounod's "Praise Ye the Father," "The Lord's Prayer," and "Follow On" from *Der Freischütz*. In commenting upon the rehearsals, the local press said, "The boys are enjoying the practicing and are attending well despite the fine marble weather."

Even more characteristic of the nineties was spontaneous singing as a part of the fun of any and all gatherings. When a family reunion was held it began with prayer and ended with the inevitable address and singing; at the lawn fêtes of the day some of those present would sing or play while the others sat in the windows or on the porch rail and listened. "Lay awake awhile last night," says a local diary, "listening to sere-

[36] See Table XV.

naders." The diary of the young baker mentions music of all sorts as an informal part of his "banging around town" night after night:

"Went to L——s' and serenaded them." "Gang over at N——s'. Singing, guitar, mouth harp, piano, cake, bananas, oranges and lemonade. Had a time!" "Yesterday ——'s birthday, so he set up cigars and a keg at the union meeting. After the meeting we played cards and sang till eleven."

Even at an "elegant party" of "some of our society young men" of the Success Club in 1900 the press states that "an entertainment of vocal solos and readings was enjoyed by those present."

Solo singing or group singing to jazz accompaniment still appears occasionally at small parties but is far less common than a generation ago. Serenading is a thing of the past. Chorus choirs are disappearing in the churches most frequented by the business class. There is today no chorus of business class men. In the city of today, nearly three and one-half times as large as that of 1890, there are only two adult musical societies in which the earlier tradition survives, as over against four in 1890.[37] The first is a chorus of working class men. This, together with the chorus choirs in working class churches and the frequent appearance of songs and recitations in the 1890 manner in the "socials" of these churches, suggests the relatively greater place of singing and playing in the play life of working class adults. It suggests, too, the tendency noted elsewhere for many of the workers' habits to lag roughly a generation behind those of the business class.

A second group participating actively in music is composed of women of the business class. This group, responsible for most of the organized musical life of the city, began in 1889 with a membership of thirty of the city's leading women, each of whom appeared on the program at every third meeting. To-day it has 249 members, sixty-seven of them active, forty-eight professional, and ten chorus members, with many even of these taking no active part in the meetings. "The interest in the club and participation in its programs is not as great today as

[37] The war brought "community singing" and it survives today principally in the civic clubs. Here it has relatively little life *qua* singing, aside from the stunts or novelties that the pianist or leader injects into it.

in the nineties," lamented one member. "Now there are so many clubs and other diversions to occupy people's time." Those most active in the club complain of continual lack of interest on the part of the members. The sophistication of a few of the more privileged women in the city tends to make them impatient of a less cultivated group to whom the club affords a more satisfying form of social and artistic expression. There was a net drop in membership of fifty-nine from 1923-24 to 1924-25. A possible indication of what the mass of the members prefer is furnished by the fact that a recitation to music of Eugene Field's "the dear little boy, the sweet little boy, the pretty little bow-legged boy" will be greeted with more applause on a program of American music than characteristic examples of Negro and Indian music. Leaders complain of lack of support for the various concerts which they bring to the city, although they say that Middletown is "music hungry"; the concert of an organization like the Letz Quartet barely pays for itself, but some song recitals receive more support.

It is an open question whether the devotion of Middletown to music as a personal art, as opposed to listening to music, today is not more a part of tradition and the institutional relationships kept alive as part of the adult social system than of the spontaneous play life of the city. Music seems to serve in part as a symbol that one belongs, and much of the musical activity of the women appears as a rather self-conscious appendage of the city's club life. An incipient trend away from the ritual of music lessons for children may be apparent in the remark of a prominent mother: "My children are not interested in music and there are so many things children can be interested in today that we are not going to waste time and money on them until they really want it. I had five years of lessons as a girl and can't play a thing today. I'm not going to make this mistake with my children." The mothers of the present generation of children were brought up in a culture without Victrola and radio when the girl in the crowd who could play while the others sang or danced was in demand. In their insistence upon music lessons for their children they may be reliving a world that no longer exists. Today when great artists or dance orchestras are in the cabinet in the corner of one's living room or "on the air," the ability to "play a little" may be in increasingly less demand. It seems not unlikely that, within the next generation, this habit of taking music lessons may become more selective throughout the entire population as music is made available to all through instruction in the schools and wide diffusion of Victrolas, radios, and other instruments in the home, while other abilities supplant it as the ritualistic social grace it so often is today.

Like music, art as a leisure-time activity appears to be somewhat more a thing of passive knowledge than of creative enjoyment. It, too, has its national "week" celebrated locally and its place in the school curriculum, but even more than music it is among adults largely a social ritual of a small group of privileged women.

The art activity of the city in 1890 was largely confined to the business class. Early in the eighties a number of citizens lent a promising artist in the state capital a sum of money to go to Munich to study; in return, he undertook to copy celebrated pictures for their homes. In 1888 this artist, with an associate, set up an art school in Middletown. A studio was opened in a downtown business block, and larger studios had to be taken in each of the two succeeding years. The pupils were all local business class women. Annual exhibitions of the work of the school were held and enthusiastically reviewed in local papers, and many of the paintings found their way to the walls of the homes of the city. In 1892 the school was given up and the Art Students' League was organized. "The great thing in this world," according to the motto of the League, "is not so much where we stand as in which direction we are moving." The members rented a studio where some of them worked daily, having a class model one evening a week. Members brought sketches to each meeting, sketches made on the river bank or in their yards or homes.[38]

Creating "art" is no part of the program of the present Art Students' League. This fashionable club now meets to listen to papers by members or by traveled speakers giving "an ex-

[38] Typical of the prevailing neighborly simplicity of this concern for local art were the annual loan exhibitions of art objects from the homes of the city conducted by the League. In 1894, e.g., the exhibition consisted of thirty-two paintings in oil from local homes, twenty water colors, seven studies in charcoal and pastel, needlework and laces lent by fourteen people, bric-à-brac and antiques lent by twenty-one people, china painting lent by fourteen people, and wood carving lent by three people.

haustive description of modern excavations in North and Central America," or an account of "Renaissance Art," or "Japanese Prints." The programs for a given year have coherence but may skip in fourteen meetings from "The Character of the Early Christian, Byzantine, Romanesque, and Gothic Periods," through "A Historic Survey of the Middle Ages," "Medieval Sculpture and Painting," and various aspects of Renaissance history, architecture, sculpture, and painting. The Literature and Art Department of the local Woman's Club is a less socially exclusive group of women, engaged largely in writing and listening to papers on similar subjects. Art figures sporadically in the programs of other women's clubs. A few women in the city continue their interest in painting and sketching.

One is struck by the gap that exists between "art" as discussed in these groups and as observed in most of the homes. As noted in Chapter IX, although art is regarded, at least among the business class, as a thing of unusual merit, art in the homes is highly standardized and used almost entirely as furniture.[39] With the exception of a few wealthy families, people do not have collections of unframed prints as they do of books and Victrola records. It is noteworthy, too, that the buildings for religious worship are, with the exception of the Catholic church and an occasional Sunday School room, naked of art objects.

Art is diffused to Middletown by more channels today than in 1890. In place of the early art school, local loan exhibitions, and forty-five volumes on fine arts in the public library, there are today the Art Students' League, the Art and Literature

[39] It is to the point to recall how essentially primitive is this limitation of creative spontaneity that has gone forward in the art of Middletown since 1890. Speaking of the period of man's known history, Teggart says: "It is difficult for the modern man to realize that, in the earlier period, individuality did not exist. . . . So completely was the individual subordinated to the community that art was just the repetition of tribal designs, literature the repetition of tribal songs, and religion the repetition of tribal rites." *Processes of History* (New Haven; Yale University Press, 1918), p. 86.

It is also noteworthy that the tendency observed in Middletown art clubs to make the group leisure-time activities "merely imitations of ceremonial occupations" also has its roots in life among primitive man. Rivers points out, e.g.: "The artistic side of life among the Todas is but little developed. Their interest is so much absorbed in ceremony that little is left for the development of art, even of a primitive kind. . . . In their amusements again we shall find that the influence of ceremonial is so great, that many of the games are merely imitations of ceremonial occupations." (*The Todas*, p. 570.)

Department of the Women's Club, a collection of paintings in the high school, loan exhibitions from museums in other cities, 1,150 volumes on the fine arts in the public library, art periodicals such as did not exist in 1890 in the library and in a few homes, a high grade of art work in popular magazines, a wide contact with art in certain travel and other motion picture films, and finally the group training in art in the schools, particularly in the high school.

The most spontaneous artistic life of the city, aside from that exhibited in some of the clothing of the younger women, is to be found in the high school with its Daubers' Club made up of boys and girls from all sections of the city. Nine girls out of 341 in the three upper years of the high school listed "art" as the thing they were doing at home in which they were most interested. It is significant that although art was cultivated in 1890 only by a narrow group of business class women, only four of these nine girls came from the business class. More interesting still in view of the fact that art has always been an exclusively female accomplishment in Middletown is the fact that twelve boys listed "art" as the thing they are doing at home in which they are most interested, only six of them coming from the business class. But, like music, art seems somehow to drop out of the picture between the time boys and girls sketch in their high school classes and the time they become immersed in the usual activities of Middletown adults.

Chapter XVIII

INVENTIONS RE-MAKING LEISURE

Although lectures, reading, music, and art are strongly intrenched in Middletown's traditions, it is none of these that would first attract the attention of a newcomer watching Middletown at play.

"Why on earth do you need to study what's changing this country?" said a lifelong resident and shrewd observer of the Middle West. "I can tell you what's happening in just four letters: A-U-T-O!"

In 1890 the possession of a pony was the wildest flight of a Middletown boy's dreams. In 1924 a Bible class teacher in a Middletown school concluded her teaching of the Creation: "And now, children, is there any of these animals that God created that man could have got along without?" One after another of the animals from goat to mosquito was mentioned and for some reason rejected; finally, "The horse!" said one boy triumphantly, and the rest of the class agreed. Ten or twelve years ago a new horse fountain was installed at the corner of the Courthouse square; now it remains dry during most of the blazing heat of a Mid-Western summer, and no one cares. The "horse culture" of Middletown has almost disappeared.[1]

Nor was the horse culture in all the years of its undisputed sway ever as pervasive a part of the life of Middletown as is the cluster of habits that have grown up overnight around the automobile. A local carriage manufacturer of the early days estimates that about 125 families owned a horse and buggy in 1890, practically all of them business class folk. "A regular sight summer mornings was Mrs. Jim B—— [the wife of one of the city's leading men] with a friend out in her rig, shelling

[1] Two million horse-drawn carriages were manufactured in the United States in 1909 and 10,000 in 1923; 80,000 automobiles were manufactured in 1909 and 4,000,000 in 1923.

peas for dinner while her horse ambled along the road." As spring came on each year entries like these began to appear in the diaries:

"April 1, '88. Easter. A beautiful day, cloudy at times but very warm, and much walking and riding about town."

"May 19, '89. Considerable carriage riding today."

"July 16, '89. Considerable riding this evening. People out 'cooling off.'"

"Sept. 18, '87. Wife and myself went to the Cemetery this afternoon in the buggy. Quite a number of others were placing flowers upon the graves of their dear ones. . . ."

But if the few rode in carriages in 1890, the great mass walked. The Sunday afternoon stroll was the rule.

Meanwhile, in a Middletown machine shop a man was tinkering at a "steam wagon" which in September, 1890, was placed on the street for the first trial. . . .

"The vehicle has the appearance of an ordinary road wagon, when put in motion," said the newspaper, "though there is no tongue attached. It is run on the principle of a railroad locomotive, a lever in front which guides the vehicle being operated by the person driving. The power is a small engine placed under the running gears and the steam is made by a small gasoline flame beneath a fuel tank. Twenty-five miles an hour can be attained with this wonderful device. The wagon will carry any load that can be placed on it, climbing hills and passing over bad roads with the same ease as over a level road. The wagon complete cost nearly $1,000."

In other cities other men were also working at these "horseless wagons." [2] As late as 1895 Elwood Haynes of Kokomo, Indiana, one of the early tinkerers, was stopped by a policeman as he drove his horseless car into Chicago and ordered to take the thing off the streets. In 1896 the resplendent posters of the alert P. T. Barnum featured in the foreground a "horseless carriage to be seen every day in the new street parade"—with elephants, camels and all the rest of the circus lost in the background while the crowd cheers "the famous Duryea Motorwagon or Motorcycle."

[2] See the Silver Anniversary Number of the *Automobile Trade Journal*, Vol. XXIX, No. 6. Dec. 1, 1924, for the story of these adventurous days.

The first real automobile appeared in Middletown in 1900. About 1906 it was estimated that "there are probably 200 in the city and county." At the close of 1923 there were 6,221 passenger cars in the city, one for every 6.1 persons, or roughly two for every three families.[3] Of these 6,221 cars, 41 per cent. were Fords; 54 per cent. of the total were cars of models of 1920 or later, and 17 per cent. models earlier than 1917.[4] These cars average a bit over 5,000 miles a year.[5] For some of the workers and some of the business class, use of the automobile is a seasonal matter, but the increase in surfaced roads and in closed cars is rapidly making the car a year-round tool for leisure-time as well as getting-a-living activities. As, at the turn of the century, business class people began to feel apologetic if they did not have a telephone, so ownership of an automobile has now reached the point of being an accepted essential of normal living.

Into the equilibrium of habits which constitutes for each individual some integration in living has come this new habit, upsetting old adjustments, and blasting its way through such

[3] These numbers have undoubtedly increased greatly since the count was made.

As a matter of fact, by far the greater part of the wide diffusion of the automobile culture one observes today in Middletown has taken place within the last ten or fifteen years. There were less than 500,000 passenger automobiles registered in the entire United States in 1910 and only 5,500,000 in 1918, as over against 15,500,000 in 1924. (Cf. *Facts and Figures of the Automobile Industry, 1925 Edition,* published by the National Automobile Chamber of Commerce.)

[4] Some further idea of the spread of automobiles, involving different degrees of inroad into the family budgets of the city, is afforded by the following list in order of frequency: Ford, 2,578; Chevrolet, 590; Overland, 459; Dodge, 343; Maxwell, 309; Buick, 295; Studebaker, 264; Oakland, 88; Willys-Knight, 74; Nash, 73; Interstate, 73; Durant, 65; Star, 62; Oldsmobile, 59; Saxon, 53; Reo, 50; Chalmers, 47; Franklin, 45; Essex, 45; Hudson, 44; Cadillac, 36; Chandler, 32; Monroe, 31; Paige, 31; Haynes, 29; International, 26; Sheridan, 26; Hupmobile, 25. Sixty-nine other makes are represented by less than twenty-five cars each, including fifteen Marmons, fourteen Packards, one Pierce-Arrow, one Lincoln, but for the most part cheap, early models, many of them of discontinued makes.

The 6,221 cars owned in the city at the end of 1923 included models of the following years: 1924—13; 1923—901; 1922—1,053; 1921—633; 1920—746; 1919—585; 1918—447; 1917—756; 1916—517; 1915—294; 1914—154; 1913—85; earlier than 1913—37.

[5] This is a rough figure based upon the total of 11,660 passenger cars and 1,768 trucks registered in the county at the close of 1924, the gasoline tax paid during the year, an arbitrary assumption that a truck used three times the gas used by a passenger car, and upon an estimate of 17.5 miles per gallon. The number of motorcycles is negligible.

accustomed and unquestioned dicta as "Rain or shine, I never miss a Sunday morning at church"; "A high school boy does not need much spending money"; "I don't need exercise, walking to the office keeps me fit"; "I wouldn't think of moving out of town and being so far from my friends"; "Parents ought always to know where their children are." The newcomer is most quickly and amicably incorporated into those regions of behavior in which men are engaged in doing impersonal, matter-of-fact things; much more contested is its advent where emotionally charged sanctions and taboos are concerned. No one questions the use of the auto for transporting groceries, getting to one's place of work or to the golf course, or in place of the porch for "cooling off after supper" on a hot summer evening; however much the activities concerned with getting a living may be altered by the fact that a factory can draw from workmen within a radius of forty-five miles, or however much old labor union men resent the intrusion of this new alternate way of spending an evening,[6] these things are hardly major issues. But when auto riding tends to replace the traditional call in the family parlor as a way of approach between the unmarried, "the home is endangered," and all-day Sunday motor trips are a "threat against the church"; it is in the activities concerned with the home and religion that the automobile occasions the greatest emotional conflicts.

Group-sanctioned values are disturbed by the inroads of the automobile upon the family budget.[7] A case in point is the not uncommon practice of mortgaging a home to buy an automobile. Data on automobile ownership were secured from 123 working class families. Of these, sixty have cars. Forty-one of

[6] "The Ford car has done an awful lot of harm to the unions here and everywhere else," growled one man prominent in Middletown labor circles. "As long as men have enough money to buy a second-hand Ford and tires and gasoline, they'll be out on the road and paying no attention to union meetings."

[7] What a motor car means as an investment by Middletown families can be gathered from the following accepted rates of depreciation: 30 per cent. the first year, 20 per cent. more the second, 10 per cent. more each of the next three years. The operating cost of the lightest car of the Ford, Chevrolet, Overland type, including garage rent and depreciation, has been conservatively figured by a national automotive corporation for the country as a whole at $5.00 a week or $0.05 a mile for family use for 5,000 miles a year and replacement at the end of seven years. The cost of tires, gas, oil, and repairs of the forty-seven of the workers' families interviewed who gave expenditures on cars for the past year ranged from $8.90 to $192.00.

che sixty own their homes. Twenty-six of these forty-one families have mortgages on their homes. Forty of the sixty-three families who do not own a car own their homes. Twenty-nine of these have mortgages on their homes. Obviously other factors are involved in many of Middletown's mortgages. That the automobile does represent a real choice in the minds of some at least is suggested by the acid retort of one citizen to the question about car ownership: "No, sir, we've *not* got a car. *That's* why we've got a home." According to an officer of a Middletown automobile financing company, 75 to 90 per cent. of the cars purchased locally are bought on time payment, and a working man earning $35.00 a week frequently plans to use one week's pay each month as payment for his car.

The automobile has apparently unsettled the habit of careful saving for some families. "Part of the money we spend on the car would go to the bank, I suppose," said more than one working class wife. A business man explained his recent inviting of social oblivion by selling his car by saying: "My car, counting depreciation and everything, was costing mighty nearly $100.00 a month, and my wife and I sat down together the other night and just figured that we're getting along, and if we're to have anything later on, we've just got to begin to save." The "moral" aspect of the competition between the automobile and certain accepted expenditures appears in the remark of another business man, "An automobile is a luxury, and no one has a right to one if he can't afford it. I haven't the slightest sympathy for any one who is out of work if he owns a car."

Men in the clothing industry are convinced that automobiles are bought at the expense of clothing,[8] and the statements of a number of the working class wives bear this out:

"We'd rather do without clothes than give up the car," said one mother of nine children. "We used to go to his sister's to visit, but by the time we'd get the children shoed and dressed

---

[8] "The *National Retail Clothier* has been devoting space to trying to find out what is the matter with the clothing industry and has been inclined to blame it on the automobile. "In one city, to quote an example cited in the articles, a store 'put on a campaign that usually resulted in a business of 150 suits and overcoats on a Saturday afternoon. This season the campaign netted seventeen sales, while an automobile agency across the street sold twenty-five cars on the weekly payment plan.' In another, 'retail clothiers are unanimous in blaming the automobile for the admitted slump in the retail clothing trade.'" (*Chicago Evening Post,* December 28, 1923.)

there wasn't any money left for carfare. Now no matter how they look, we just poke 'em in the car and take 'em along."

"We don't have no fancy clothes when we have the car to pay for," said another. "The car is the only pleasure we have."

Even food may suffer:

"I'll go without food before I'll see us give up the car," said one woman emphatically, and several who were out of work were apparently making precisely this adjustment.

Twenty-one of the twenty-six families owning a car for whom data on bathroom facilities happened to be secured live in homes without bathtubs. Here we obviously have a new habit cutting in ahead of an older one and slowing down the diffusion of the latter.[9]

Meanwhile, advertisements pound away at Middletown people with the tempting advice to spend money for automobiles for the sake of their homes and families:

"Hit the trail to better times!" says one such advertisement.

Another depicts a gray-haired banker lending a young couple the money to buy a car and proffering the friendly advice: "Before you can save money, you first must make money. And to make it you must have health, contentment, and full command of all your resources. . . . I have often advised customers of mine to buy cars, as I felt that the increased stimulation and opportunity of observation would enable them to earn amounts equal to the cost of their cars."

[9] This low percentage of bathtubs would not hold for the entire car-owning group. The interviewers asked about bathtubs in these twenty-six cases out of curiosity, prompted by the run-down appearance of the homes.

While inroads upon savings and the re-allocation of items of home expenditure were the readjustments most often mentioned in connection with the financing of the family automobile, others also occur: "It's prohibition that's done it," according to an officer in the Middletown Trades Council: "drink money is going into cars." The same officer, in answering the question as to what he thought most of the men he comes in contact with are working for, guessed: "Twenty-five per cent. are fighting to keep their heads above water; 10 per cent. want to own their own homes; 65 per cent. are working to pay for cars." "All business is suffering," says a Middletown candy manufacturer and dealer. "The candy business is poor now to what it was before the war. There is no money in it any more. People just aren't buying candy so much now. How can they? Even laboring-men put all their money into cars, and every other branch of business feels it."[2]

Many families feel that an automobile is justified as an agency holding the family group together. "I never feel as close to my family as when we are all together in the car," said one business class mother, and one or two spoke of giving up Country Club membership or other recreations to get a car for this reason. "We don't spend anything on recreation except for the car. We save every place we can and put the money into the car. It keeps the family together," was an opinion voiced more than once. Sixty-one per cent. of 337 boys and 60 per cent. of 423 girls in the three upper years of the high school say that they motor more often with their parents than without them.[10]

But this centralizing tendency of the automobile may be only a passing phase; sets in the other direction are almost equally prominent. "Our daughters [eighteen and fifteen] don't use our car much because they are always with somebody else in their car when we go out motoring," lamented one business class mother. And another said, "The two older children [eighteen and sixteen] never go out when the family motors. They always have something else on." "In the nineties we were all much more together," said another wife. "People brought chairs and cushions out of the house and sat on the lawn evenings. We rolled out a strip of carpet and put cushions on the porch step to take care of the unlimited overflow of neighbors that dropped by. We'd sit out so all evening. The younger couples perhaps would wander off for half an hour to get a soda but come back to join in the informal singing or listen while somebody strummed a mandolin or guitar." "What on earth *do* you want me to do? Just sit around home all evening!" retorted a popular high school girl of today when her father discouraged her going out motoring for the evening with a young blade in a rakish car waiting at the curb. The fact that 348 boys and 382 girls in the three upper years of the high school placed "use of the automobile" fifth and fourth respectively in a list of twelve possible sources of disagreement between them and their parents

[10] As over against these answers regarding the automobile, 21 per cent. of the boys and 33 per cent. of the girls said that they go to the movies more often with their parents than without them, 25 per cent. and 22 per cent. respectively answered similarly as regards "listening to the radio," and 31 per cent. and 48 per cent. as regards "singing or playing a musical instrument." On the basis of these answers it would appear that the automobile is at present operating as a more active agency drawing Middletown families together than any of these other agencies.

suggests that this may be an increasing decentralizing agent.[11]

An earnest teacher in a Sunday School class of working class boys and girls in their late teens was winding up the lesson on the temptations of Jesus: "These three temptations summarize all the temptations we encounter today: physical comfort, fame, and wealth. Can you think of any temptation we have today that Jesus didn't have?" "Speed!" rejoined one boy. The unwanted interruption was quickly passed over. But the boy had mentioned a tendency underlying one of the four chief infringements of group laws in Middletown today, and the manifestations of Speed are not confined to "speeding." "Auto Polo next Sunday!!" shouts the display advertisement of an amusement park near the city. "It's motor insanity—too fast for the movies!" The boys who have cars "step on the gas," and those who haven't cars sometimes steal them: "The desire of youth to step on the gas when it has no machine of its own," said the local press, "is considered responsible for the theft of the greater part of the [154] automobiles stolen from [Middletown] during the past year." [12]

The threat which the automobile presents to some anxious parents is suggested by the fact that of thirty girls brought before the juvenile court in the twelve months preceding September 1, 1924, charged with "sex crimes," for whom the place where the offense occurred was given in the records, nineteen were listed as having committed the offense in an automobile.[13] Here again the automobile appears to some as an "enemy" of the home and society.

Sharp, also, is the resentment aroused by this elbowing new device when it interferes with old-established religious habits.

---

[11] See Table XIII.

[12] In any consideration of the devotion to "speed" that accompanies the coming of the automobile, it should be borne in mind that the increased monotony for the bulk of the workers involved in the shift from the large-muscled hand-trades, including farming, to the small-muscled high-speed machine-tending jobs and the disappearance of the saloon as an easy means of "tellin' the world to go to hell" have combined with the habit-cracking, eye-opening effect of service in the late war to set the stage for the automobile as a release. The fact that serviceable second-hand cars can be bought for $75.00 and up, the simplicity of installment payment, "the fact that everybody has one"—all unite to make ownership of a car relatively easy, even for boys. Cf. the incident cited above of the boy who wanted to "swap in his Ford for a Studebaker that will go seventy-five miles an hour."

[13] For ten others charged with sex offenses during this same period the scene of the offense was not given.

The minister trying to change people's behavior in desired directions through the spoken word must compete against the strong pull of the open road strengthened by endless printed "copy" inciting to travel. Preaching to 200 people on a hot, sunny Sunday in midsummer on "The Supreme Need of Today," a leading Middletown minister denounced "automobilitis—the thing those people have who go off motoring on Sunday instead of going to church. If you want to use your car on Sunday, take it out Sunday morning and bring some shut-ins to church and Sunday School; then in the afternoon, if you choose, go out and worship God in the beauty of nature—but don't neglect to worship Him indoors too." This same month there appeared in the *Saturday Evening Post,* reaching approximately one family in six in Middletown, a two-page spread on the automobile as an "enricher of life," quoting "a bank president in a Mid-Western city" as saying, "A man who works six days a week and spends the seventh on his own doorstep certainly will not pick up the extra dimes in the great thoroughfares of life." "Some sunny Sunday very soon," said another two-page spread in the *Post,* "just drive an Overland up to your door—tell the family to hurry the packing and get aboard —and be off with smiles down the nearest road—free, loose, and happy—bound for green wonderlands." Another such advertisement urged Middletown to "Increase Your Week-End Touring Radius." [14] If we except the concentrated group pressure of war time, never perhaps since the days of the camp-meeting have the citizens of this community been subjected to such a powerfully focused stream of habit diffusion. To get the full force of this appeal, one must remember that the near-est lakes or hills are one hundred miles from Middletown in either direction and that an afternoon's motoring brings only mile upon mile of level stretches like Middletown itself.

"We had a fine day yesterday," exclaimed an elderly pillar of a prominent church, by way of Monday morning greeting. "We left home at five in the morning. By seven we swept into ———. At

[14] Over against these appeals the Sunday of 1890 with its fewer alternatives should be borne in mind: as a Middletown plumber described it, "There wasn't anything to do but go to church or a saloon or walk uptown and look in the shop windows. You'd go about hunting saloons that were open, or maybe, if you were a *hot* sport, rent a rig for $1.50 for the afternoon and take your girl out riding."

eight we had breakfast at ——, eighty miles from home. From there we went on to Lake ——, the longest in the state. I had never seen it before, and I've lived here all my life, but I sure do want to go again. Then we went to —— [the Y.M.C.A. camp] and had our chicken dinner. It's a fine thing for people to get out that way on Sundays. No question about it. They see different things and get a larger outlook."

"Did you miss church?" he was asked.

"Yes, I did, but you can't do both. I never missed church or Sunday School for thirteen years and I kind of feel as if I'd done my share. The ministers ought not to rail against people's driving on Sunday. They ought just to realize that they won't be there every Sunday during the summer, and make church interesting enough so they'll want to come."

But if the automobile touches the rest of Middletown's living at many points, it has revolutionized its leisure; more, perhaps, than the movies or any other intrusion new to Middletown since the nineties, it is making leisure-time enjoyment a regularly expected part of every day and week rather than an occasional event. The readily available leisure-time options of even the working class have been multiplied many-fold. As one working class housewife remarked, "We just go to lots of things we couldn't go to if we didn't have a car." Beefsteak and watermelon picnics in a park or a near-by wood can be a matter of a moment's decision on a hot afternoon.

Not only has walking for pleasure become practically extinct,[15] but the occasional event such as a parade on a holiday attracts far less attention now.

"Lots of noise on the street preparing for the 4th," reports the diary of a Middletown merchant on July 3, 1891. And on the 4th: "The town full of people—grand parade with representatives of different trades, an ox roasted whole, four bands, fire-

[15] Sunday afternoon motorists today report only an occasional anachronism of a person seen walking about town. A newcomer to Middletown remarked, "I was never in a city where people walk so little. They ride to work, to a picture show or to dances, even if they are going only a few blocks." One business class woman says that her friends laugh at her for walking to club meetings or downtown.

In the early nineties street car rides were somewhat of an event and "trolley parties" were popular. "Saturday afternoon out with ma and Trese car riding and walking," writes the young baker in his diary, and a news note in June, 1890, states that "tickets for free street car ride were given out Sunday morning at the First Christian Church for Sunday School children."

works, races, greased pig, dancing all day, etc." An account in '93 reports: "Quite a stir in town. Firecrackers going off all night and all this day—big horse racing at the Fair Ground. Stores all closed this afternoon. Fireworks at the Fair Ground this evening."

Today the week before the Fourth brings a pale edition of the earlier din, continuing until the night before. But the Fourth dawns quietly on an empty city; Middletown has taken to the road. Memorial Day and Labor Day are likewise shorn of their earlier glory.

Use of the automobile has apparently been influential in spreading the "vacation" habit. The custom of having each summer a respite, usually of two weeks, from getting-a-living activities, with pay unabated, is increasingly common among the business class, but it is as yet very uncommon among the workers.[16] "Vacations in 1890?" echoed one substantial citizen. "Why, the word wasn't in the dictionary!" "Executives of the 1890 period *never* took a vacation," said another man of a type common in Middletown thirty-five years ago, who used to announce proudly that they had "not missed a day's work in twenty years." Vacations there were in the nineties, nevertheless, particularly for the wives and children of those business folk who had most financial leeway. Put-In Bay, Chautauqua, country boarding-houses where the rates were $5.00 a week for adults and $3.00 for children, the annual conference of the State Baptist Association, the Annual National Christian Endeavor Convention, the annual G.A.R. encampment, all drew people from Middletown. But these affected almost entirely business class people. A check of the habits of the parents of the 124 working class wives shows that summer vacations were almost unknown among this large section of the population in the nineties. In lieu of vacations both for workers and many of the business class there were excursions: those crowded, grimy, exuberant, banana-smelling affairs on which one sat up nights in a day coach, or, if a "dude," took a sleeper, from Sat-

[16] Not all of the business class are paid while on vacation, e.g., many retail clerks are not paid, but the custom is usual.

The growth of the vacation habit is reflected in the fact that the Woman's Club met with unabated vigor all through the summer in 1890. In 1900 it took the first vacation in its history for July and August. Commencing with 1914 it has closed earlier and earlier, and since 1919-20 has closed for the three months from June 1.

urday till Monday morning, and went back to work a bit seedy from loss of sleep but full of the glamour of Petoskey, or the ball game at Chicago. Two hundred and twelve people from Middletown went to Chicago in one week-end on one such excursion. One hundred and fifty journeyed to the state capital to see the unveiling of a monument to an ex-governor—"a statesman," as they called them in those days. Even train excursions to towns fifteen, twenty, and forty miles away were great events, and people reported having "seen the sights" of these other Middletowns with much enthusiasm.

Today a few plants close for one or two weeks each summer, allowing their workers an annual "vacation" without pay. Others do not close down, but workers "can usually take not over two weeks off without pay and have their jobs back when they return." Foremen in many plants get one or two weeks with pay. Of the 122 working class families giving information on this point, five families took one week off in 1923 and again in 1924, seven others took something over a week in each year, twelve took a week or more in only one of the two years. No others had as extensive vacations as these twenty-four, although other entire families took less than a week in one or both years, and in other cases some members of the families took vacations of varying lengths. Of the 100 families for whom income distribution was secured, thirty-four reported money spent on vacations; the amounts ranged from $1.49 to $175.00, averaging $24.12.

But even short trips are still beyond the horizon of many workers' families, as such comments as the following show:

"We haven't had a vacation in five years. He got a day off to paint the house, and another year they gave him two hours off to get the deed to the house signed."

"Never had a vacation in my life, honey!"

"Can't afford one this year because we're repairing the house."

"I don't know what a vacation is—I haven't had one for so long."

"We like to get out in the car each week for half a day but can't afford a longer vacation."

But the automobile is extending the radius of those who are allowed vacations with pay and is putting short trips within the reach of some for whom such vacations are still "not in the dictionary."

"The only vacation we've had in twenty years was three days we took off last year to go to Benton Harbor with my brother-in-law," said one woman, proudly recounting her trip. "We had two Fords. The women slept in the cars, the men on boards between the two running boards. Here's a picture of the two cars, taken just as the sun was coming up. See the shadows? And there's a *hill* back of them."

Like the automobile, the motion picture is more to Middletown than simply a new way of doing an old thing; it has added new dimensions to the city's leisure. To be sure, the spectacle-watching habit was strong upon Middletown in the nineties. Whenever they had a chance people turned out to a "show," but chances were relatively fewer. Fourteen times during January, 1890, for instance, the Opera House was opened for performances ranging from *Uncle Tom's Cabin* to *The Black Crook,* before the paper announced that "there will not be any more attractions at the Opera House for nearly two weeks." In July there were no "attractions"; a half dozen were scattered through August and September; there were twelve in October.[17]

Today nine motion picture theaters operate from 1 to 11 P.M. seven days a week summer and winter; four of the nine give three different programs a week, the other five having two a week; thus twenty-two different programs with a total of over 300 performances are available to Middletown every week in the year. In addition, during January, 1923, there were three plays in Middletown and four motion pictures in other places than the regular theaters, in July three plays and one additional movie, in October two plays and one movie.

About two and three-fourths times the city's entire population attended the nine motion picture theaters during the month of July, 1923, the "valley" month of the year, and four and one-half times the total population in the "peak" month of December.[18] Of 395 boys and 457 girls in the three upper years

[17] Exact counts were made for only January, July, and October. There were less than 125 performances, including matinées, for the entire year.
[18] These figures are rough estimates based upon the following data: The total Federal amusement tax paid by Middletown theaters in July was $3,002.04 and in December $4,781.47. The average tax paid per admission

of the high school who stated how many times they had attended the movies in "the last seven days," a characteristic week in mid-November, 30 per cent. of the boys and 39 per cent. of the girls had not attended, 31 and 29 per cent. respectively had been only once, 22 and 21 per cent. respectively two times, 10 and 7 per cent. three times, and 7 and 4 per cent. four or more times. According to the housewives interviewed regarding the custom in their own families, in three of the forty business class families interviewed and in thirty-eight of the 122 working class families no member "goes at all" to the movies.[19] One family in ten in each group goes as an entire family once a week or oftener; the two parents go together without their children once a week or oftener in four business class families (one in ten), and in two working class families (one in sixty); in fifteen business class families and in thirty-eight working class families the children were said by their mothers to go without their parents one or more times weekly.

In short, the frequency of movie attendance of high school boys and girls is about equal, business class families tend to go more often than do working class families, and children of both groups attend more often without their parents than do all

is about $0.0325, and the population in 1923 about 38,000. Attendance estimates secured in this way were raised by one-sixth to account for children under twelve who are tax-free. The proprietor of three representative houses said that he had seven admissions over twelve years to one aged twelve or less, and the proprietor of another house drawing many children has four over twelve to one aged twelve or less.

These attendance figures include, however, farmers and others from outlying districts.

[19] The question was asked in terms of frequency of attendance "in an average month" and was checked in each case by attendance during the month just past.

Lack of money and young children needing care in the home are probably two factors influencing these families that do not attend at all; of the forty-one working class families in which all the children are twelve years or under, eighteen never go to the movies, while of the eighty-one working class families in which one or more of the children is twelve or older, only twenty reported that no member of the family ever attends.

"I haven't been anywhere in two years," said a working class wife of thirty-three, the mother of six children, the youngest twenty months. "I went to the movies once two years ago. I was over to see Mrs. —— and she says, 'Come on, let's go to the movies.' I didn't believe her. She is always ragging the men and I thought she was joking. 'Come on,' she says, 'put your things on and we'll see a show.' I thought, well, if she wanted to rag the men, I'd help her, so I got up and put my things on. And, you know, she really meant it. She paid my carfare uptown and paid my way into the movies. I was never so surprised in my life. I haven't been anywhere since."

the individuals or other combinations of family members put together. The decentralizing tendency of the movies upon the family, suggested by this last, is further indicated by the fact that only 21 per cent. of 337 boys and 33 per cent. of 423 girls in the three upper years of the high school go to the movies more often with their parents than without them. On the other hand, the comment is frequently heard in Middletown that movies have cut into lodge attendance, and it is probable that time formerly spent in lodges, saloons, and unions is now being spent in part at the movies, at least occasionally with other members of the family.[20] Like the automobile and radio, the movies, by breaking up leisure time into an individual, family, or small group affair, represent a counter movement to the trend toward organization so marked in clubs and other leisure-time pursuits.

How is life being quickened by the movies for the youngsters who bulk so large in the audiences, for the punch press operator at the end of his working day, for the wife who goes to a "picture" every week or so "while he stays home with the children," for those business class families who habitually attend?

"Go to a motion picture . . . and let yourself go," Middletown reads in a *Saturday Evening Post* advertisement. "Before you know it you are *living* the story—laughing, loving, hating, struggling, winning! All the adventure, all the romance, all the excitement you lack in your daily life are in —— Pictures. They take you completely out of yourself into a wonderful new world. . . . Out of the cage of everyday existence! If only for an afternoon or an evening—escape!"

The program of the five cheaper houses is usually a "Wild West" feature, and a comedy; of the four better houses, one feature film, usually a "society" film but frequently Wild West or comedy, one short comedy, or if the feature is a comedy, an educational film (e.g., *Laying an Ocean Cable* or *Making a Telephone*), and a news film. In general, people do not go to the movies to be instructed; the Yale Press series of historical films, as noted earlier, were a flat failure and the local exhibitor discontinued them after the second picture.

[20] Cf. N. 10 above.

The ex-proprietor of one of the largest saloons in the city said, "The movies killed the saloon. They cut our business in half overnight."

As in the case of the books it reads, comedy, heart interest, and adventure compose the great bulk of what Middletown enjoys in the movies. Its heroes, according to the manager of the leading theater, are, in the order named, Harold Lloyd, comedian; Gloria Swanson, heroine in modern society films; Thomas Meighan, hero in modern society films; Colleen Moore, ingénue; Douglas Fairbanks, comedian and adventurer; Mary Pickford, ingénue; and Norma Talmadge, heroine in modern society films. Harold Lloyd comedies draw the largest crowds. "Middletown is amusement hungry," says the opening sentence in a local editorial; at the comedies Middletown lives for an hour in a happy sophisticated make-believe world that leaves it, according to the advertisement of one film, "happily convinced that Life is very well worth living."

Next largest are the crowds which come to see the sensational society films. The kind of vicarious living brought to Middletown by these films may be inferred from such titles as: *"Alimony*—brilliant men, beautiful jazz babies, champagne baths, midnight revels, petting parties in the purple dawn, all ending in one terrific smashing climax that makes you gasp"; *"Married Flirts—Husbands:* Do you flirt? Does your wife always know where you are? Are you faithful to your vows? *Wives:* What's your hubby doing? Do you know? Do you worry? Watch out for *Married Flirts."* So fast do these flow across the silver screen that, e.g., at one time *The Daring Years, Sinners in Silk, Women Who Give,* and *The Price She Paid* were all running synchronously, and at another *"Name the Man* —a story of betrayed womanhood," *Rouged Lips,* and *The Queen of Sin.*[21] While Western "action" films and a million-dollar spectacle like *The Covered Wagon* or *The Hunchback of Notre Dame* draw heavy houses, and while managers lament that there are too few of the popular comedy films, it is the film with burning "heart interest," that packs Middletown's motion picture houses week after week. Young Middletown enters eagerly into the vivid experience of *Flaming Youth:* "neckers, petters, white kisses, red kisses, pleasure-mad daughters, sensation-craving mothers, by an author who didn't dare sign his name; the truth bold, naked, sensational"—so ran the press

[21] It happens frequently that the title overplays the element of "sex adventure" in a picture. On the other hand, films less luridly advertised frequently portray more "raw situations."

advertisement—under the spell of the powerful conditioning medium of pictures presented with music and all possible heightening of the emotional content, and the added factor of sharing this experience with a "date" in a darkened room. Meanwhile, *Down to the Sea in Ships,* a costly spectacle of whaling adventure, failed at the leading theater "because," the exhibitor explained, "the whale is really the hero in the film and there wasn't enough 'heart interest' for the women."

Over against these spectacles which Middletown watches today stand the pale "sensations" of the nineties, when *Sappho* was the apogee of daring at the Opera House: *"The Telephone Girl*—Hurricane hits, breezy dialogue, gorgeous stage setting, dazzling dancing, spirited repartee, superb music, opulent costumes," *Over the Garden Wall, Edith's Burglar, East Lynne, La Belle Maria, or Women's Revenge, The Convict's Daughter, Joe, a Mountain Fairy, The Vagabond Heroine, Guilty Without Crime, The World Against Her* (which the baker pronounced in his diary, "good, but too solemn"), *Love Will Find a Way, Si. Plankard.* These, it must be recalled, were the great days when *Uncle Tom's Cabin,* with "fifty men, women, and children, a pack of genuine bloodhounds, grandest street parade ever given, and two bands," packed the Opera House to capacity.

Actual changes of habits resulting from the week-after-week witnessing of these films can only be inferred. Young Middletown is finding discussion of problems of mating in this new agency that boasts in large illustrated advertisements, "Girls! You will learn how to handle 'em!" and "Is it true that marriage kills love? If you want to know what love really means, its exquisite torture, its overwhelming raptures, see ———."

"Sheiks and their 'shebas,'" according to the press account of the Sunday opening of one film, ". . . sat without a movement or a whisper through the presentation. . . . It was a real exhibition of love-making and the youths and maidens of [Middletown] who thought that they knew something about the art found that they still had a great deal to learn."

Some high school teachers are convinced that the movies are a powerful factor in bringing about the "early sophistication" of the young and the relaxing of social taboos. One working class mother frankly welcomes the movies as an aid in child-rearing, saying, "I send my daughter because a girl has to learn the ways of the world somehow and the movies are a good safe way." The judge of the juvenile court lists the movies as one of the "big four" causes of local juvenile delinquency,[22] believing that the disregard of group mores by the young is definitely related to the witnessing week after week of fictitious behavior sequences that habitually link the taking of long chances and the happy ending. While the community attempts to safeguard its schools from commercially intent private hands, this powerful new educational instrument, which has taken Middletown unawares, remains in the hands of a group of men—an ex-peanut-stand proprietor, an ex-bicycle racer and race promoter, and so on—whose primary concern is making money.[23]

Middletown in 1890 was not hesitant in criticizing poor shows at the Opera House. The "morning after" reviews of 1890 bristle with frank adjectives: "Their version of the play is incomplete. Their scenery is limited to one drop. The women are ancient, the costumes dingy and old. Outside of a few specialties, the show was very 'bum.'" When *Sappho* struck town in 1900, the press roasted it roundly, concluding, "[Middletown] has had enough of naughtiness of the stage. . . . Manager W—— will do well to fumigate his pretty playhouse before one of the clean, instructive, entertaining plays he has billed comes before the footlights." The newspapers of today keep their hands off the movies, save for running free publicity stories and cuts furnished by the exhibitors who adver-

[22] Cf. Ch. XI.
Miriam Van Waters, referee of the juvenile court of Los Angeles and author of *Youth in Conflict,* says in a review of Cyril Burt's *The Young Delinquent:* "The cinema is recognized for what it is, the main source of excitement and of moral education for city children. Burt finds that only mental defectives take the movies seriously enough to imitate the criminal exploits portrayed therein, and only a small proportion of thefts can be traced to stealing to gain money for admittance. In no such direct way does the moving picture commonly demoralize youth. It is in the subtle way of picturing the standards of adult life, action and emotion, cheapening, debasing, distorting adults until they appear in the eyes of the young people perpetually bathed in a moral atmosphere of intrigue, jealousy, wild emotionalism, and cheap sentimentality. Burt realizes that these exhibitions stimulate children prematurely." (*The Survey,* April 15, 1926.)
[23] One exhibitor in Middletown is a college-trained man interested in bringing "good films" to the city. He, like the others, however, is caught in the competitive game and matches his competitors' sensational advertisements.

tise. Save for some efforts among certain of the women's clubs to "clean up the movies" and the opposition of the Ministerial Association to "Sunday movies," Middletown appears content in the main to take the movies at their face value—"a darned good show"—and largely disregard their educational or habit-forming aspects.

Though less widely diffused as yet than automobile owning or movie attendance, the radio nevertheless is rapidly crowding its way in among the necessities in the family standard of living. Not the least remarkable feature of this new invention is its accessibility. Here skill and ingenuity can in part offset money as an open sesame to swift sharing of the enjoyments of the wealthy. With but little equipment one can call the life of the rest of the world from the air, and this equipment can be purchased piecemeal at the ten-cent store. Far from being simply one more means of passive enjoyment, the radio has given rise to much ingenious manipulative activity. In a count of representative sections of Middletown, it was found that, of 303 homes in twenty-eight blocks in the "best section" of town, inhabited almost entirely by the business class, 12 per cent. had radios; of 518 workers' homes in sixty-four blocks, 6 per cent. had radios.[24]

As this new tool is rolling back the horizons of Middletown for the bank clerk or the mechanic sitting at home and listening to a Philharmonic concert or a sermon by Dr. Fosdick, or to President Coolidge bidding his father good night on the eve of election,[25] and as it is wedging its way with the movie, the automobile, and other new tools into the twisted mass of habits that are living for the 38,000 people of Middletown,

[24] Both percentages have undoubtedly increased notably since 1924, when the counts were made.

[25] In 1890 the local press spoke of an occasional citizen's visiting "Paris, France," and "London, England," and even in 1924 a note in one of the papers recording the accident of some Middletown people finding themselves in a box at a New York theater with a group of Englishmen was captioned "Lucky they weren't Chinese!" The rest of the world is still a long way from Middletown, but movies and radio are doing much to break down this isolation: "I've got 120 stations on my radio," gleefully announced a local working man. Meanwhile, the president of the Radio Corporation of America proclaims an era at hand when "the oldest and newest civilizations will throb together at the same intellectual appeal, and to the same artistic emotions."

readjustments necessarily occur. Such comments as the following suggest their nature:

"I use time evenings listening in that I used to spend in reading."

"The radio is hurting movie going, especially Sunday evening." (From a leading movie exhibitor.)

"I don't use my car so much any more. The heavy traffic makes it less fun. But I spend seven nights a week on my radio. We hear fine music from Boston." (From a shabby man of fifty.)

"Sundays I take the boy to Sunday School and come straight home and tune in. I get first an eastern service, then a Cincinnati one. Then there's nothing doing till about two-thirty, when I pick up an eastern service again and follow 'em across the country till I wind up with California about ten-thirty. Last night I heard a ripping sermon from Westminster Church somewhere in California. We've no preachers here that can compare with any of them."

"One of the bad features of radio," according to a teacher, "is that children stay up late at night and are not fit for school next day."

"We've spent close on to $100 on our radio, and we built it ourselves at that," commented one of the worker's wives. "Where'd we get the money? Oh, out of our savings, like everybody else."

In the flux of competing habits that are oscillating the members of the family now towards and now away from the home, radio occupies an intermediate position. Twenty-five per cent. of 337 high school boys and 22 per cent. of 423 high school girls said that they listen more often to the radio with their parents than without them,[26] and, as pointed out above, 20 per cent. of 274 boys in the three upper years of the high school answered "radio" to the question, "In what thing that you are doing at home this fall are you most interested?"—more than gave any other answer.[27] More than one mother said that her family used to scatter in the evening—"but now we all sit around and listen to the radio."

Likewise the place of the radio in relation to Middletown's

[26] Cf. N. 10 above.
[27] Les than 1 per cent. of the 341 girls answered "radio."

other leisure habits is not wholly clear. As it becomes more perfected, cheaper, and a more accepted part of life, it may cease to call forth so much active, constructive ingenuity and become one more form of passive enjoyment. Doubtless it will continue to play a mighty rôle in lifting Middletown out of the humdrum of every day; it is beginning to take over that function of the great political rallies or the trips by the trainload to the state capital to hear a noted speaker or to see a monument dedicated that a generation ago helped to set the average man in a wide place. But it seems not unlikely that, while furnishing a new means of diversified enjoyment, it will at the same time operate, with national advertising, syndicated newspapers, and other means of large-scale diffusion, as yet another means of standardizing many of Middletown's habits. Indeed, at no point is one brought up more sharply against the impossibility of studying Middletown as a self-contained, self-starting community than when one watches these space-binding leisure-time inventions imported from without—automobile, motion picture, and radio—reshaping the city.

Chapter XIX

THE ORGANIZATION OF LEISURE

In the main these leisure activities are carried on by people in groups rather than singly; one plays with one's family or friends. At the same time that the family is declining as a unit of leisure-time pursuits, the basis of other associations is shifting; many of the earlier informal ties are being displaced by more rigid lines of union and demarcation.

Only one of the thirty-eight wives of business men who gave information on this point said that she had no friends whatever in Middletown, as against fifteen of the 118 wives of the working class; of the former group an additional four said that they had no "intimate" friends but only "casual" acquaintances, as against an additional twenty-five of the latter—making a total of one in eight of this business group and one in three of the working class who had either no friends at all in Middletown or no intimate friends. This degree of social isolation in the case of a third of the women of the dominant numerical group may be expected to have wide implications.

Such answers as the following are characteristic of the forty of the 118 workers' wives who had no friends or no intimate friends:

"It doesn't pay to be too friendly."

"I never chum with any one; it's dangerous."

"I have no best friends. In town you never know who is your friend."

"Even your best friend will do you dirt. I never run around with people. I let every one alone."

"I haven't any friends in the city. I see plenty of people at church and clubs, but I treat them all alike."

"Our neighbors used to be good friends and we had lots of good times together, but in the last seven or eight years all that's gone. People don't pay much attention to each other any more."

"I don't even know the names of the people next door and we have lived here a year."

Among the wives of the group of business men interviewed, the isolation is not so marked, possibly in part because these people have moved about somewhat less,[1] and partly, no doubt, due to their more extensively developed system of social clubs. Here, too, however, there is indication of shallowing friendships in such remarks as:

"I haven't many intimate friends. There is really only one person in town whom I regard as a really close friend."

"I have no intimate friends; it is difficult and too involving to become intimate with anybody but a few close relatives."

"We've let all our friends slip away as our children have taken up more and more of our time."

"I have no intimate friends. I just haven't had time for friends while I've been bringing up my children."

Vicinage plays a part in the forming of friendships, but, although living "next door" or "on the same block" still operates prominently among the working class, it appears to be less controlling among the business class than a generation ago. Eighty-nine (more than half) of the 173 "best friends" of the present working class wives and seventy-three (nearly two-thirds) of 116 friends of their mothers were reported to have been met first "around the neighborhood"; while seven of the seventy-five friends of the business group and twenty-five of their mothers' seventy-one friends were said to have been met in the "neighborhood." More than two-thirds of the best friends of the mothers of both working class and business class wives interviewed were said to have lived within six blocks of them, while slightly over one-half of the best friends of the present generation of working class wives and less than one in five of the best friends of the business class women interviewed live within six blocks of them.[2] Friends living eleven

[1] Cf. Ch. IX.

[2] Nearly a fifth of the best friends of the present working class women live on the same block with them, as against only two friends in the case of the business class.

Each housewife was asked to report regarding her two best friends, but in cases where the women said they had no friends or no close friends they

or more blocks distant were again almost identically infrequent in the case of the mothers of both groups—less than one in five in each case; today, however, more than a third of the best friends of working men's wives and nearly two-thirds of those of this business group live eleven blocks or more away.

The neighborhood appears likewise to have declined as a place of most constant association of friends.[3] According to almost universal working class testimony:

"Neighbors used to be in each other's houses much more than they are now."

"Mother couldn't understand when she came to live with us why people didn't run in more and neighbor as they used to."

"We ain't got neighbors any more. People ain't so friendly as they used to be. There's less neighborhood visiting. You have to go places to see them."

"My friends and I see each other most at each other's houses and at the five-and-ten-cent store—generally when we go to the store. If you go to people's houses you aren't sure you're welcome."

Women of the business class were even more emphatic:

"I like this new way of living in a neighborhood where you can be friendly with people but not intimate and dependent."

"People used to have more neighbors and every one else knew what you were doing and commented on it. I like it much better

reported on the one or two acquaintances whom they saw most frequently. The same procedure was followed in the case of their mothers.

Data of this sort based upon small groups must be regarded as suggestive only.

[3] The closeness of Middletown to its farm background is important in this connection. In his study of *The Rural Primary Groups and Their Discovery* in Wisconsin, J. H. Kolb speaks of this break-up of "neighboring": "The people who made these groups fifty years ago surely knew what it was to have a group social life. . . . The group bonds were strong of necessity because of the type of life which the settlers led. There was the visiting of neighbors. Complaint was often heard that now with the good roads and automobiles, less 'neighboring' was done. 'The young people go miles away,' some one said, 'but fail to get well acquainted with those near by.'" (*Wisconsin Research Bulletin 51,* December, 1921, p. 28.)

Cf. the decline of the habit, common in Middletown a generation ago, as noted in Ch. VII, for workers in a given plant to live together close about the factory in which they work, with their other major life-activities interlacing back and forth.

as things are now, when you can be independent and do things without thinking what others are going to say about it."

"My friends and I don't go back and forth to each other's houses much except for definite social engagements. Then, too, the old-time call with cards and white kid gloves has completely gone out. No one ever comes to call."

"Clubs have done away with calling. They have spoiled that old spirit of friendliness. People used to call on a bride just after she was married and she would go promptly with her mother to return the calls. There were calls when some one had died and church people always called on a new person in the neighborhood."

"People just don't call in Middletown. I have lived here four years and I have had practically no calls."

"I don't see my friends at all. That is really true—I never see them unless I run into them somewhere occasionally or they come over to dinner. It was different with my mother. She and her friends were always in each other's homes."

"I do very little visiting—mostly keep in touch with my friends by telephone." [4]

Like the neighborhood, the church is one of the recognized agencies acquainting members of the community with each other. "Affiliate yourself with some church if you want to get acquainted," newcomers to Middletown are told. "Our first winter when we were so lonely we met people through the church," said more than one woman.[5] Forty-four of the 173

[4] The attenuation of visiting to telephone visiting is one of the phenomena that has appeared since 1890. A worker who had been injured in an industrial accident complained, "Radios and telephones make people farther apart. Instead of going to see a person as folks used to, you just telephone nowadays." A woman living in one of the larger homes of the city, when asked where she sees her best friend, replied, "I do a great deal of telephone visiting and then I see her at evening parties." In 1890 there was much "dropping in," a ritual that frequently extended from the original intention of a simple errand to a leisurely half-morning's visit: "When the 'phone came," according to another business class woman, "it took up a lot of time, since you were within reach of so many more people, but it saved all the time formerly spent with women who 'ran in' on you while you were trying to do your morning's work."

[5] It is not unusual to hear such an erstwhile stranger add, "But these people aren't our best friends now." According to local testimony, the tendency to employ the church's social function instrumentally is growing, and it is more common today than formerly for strangers deliberately to

friends of the working class women interviewed and sixteen of the seventy-five friends of the business class women were said to have been met first either at church or at church and some other place at about the same time. Thereafter, the church, or the church linked with some other agency, was the most frequent place of seeing fifty of these friends in the case of the working class and eleven in the case of the business class.[6] Women of both groups said that their mothers saw their friends much more frequently at church, though they did not give detailed information on this point.

But if the neighborhood and the church appear to have declined somewhat as bases of association, organized club groups appear from the very rough data available to have become more important. Clubs figure as a most constant place of meeting with ten of the 173 friends of the working class group interviewed, as against two of 116 friends of their mothers, and with twenty-six of the seventy-five friends of the business class wives interviewed as against six of seventy-one friends of their mothers.

Turning from the women of Middletown to their husbands, somewhat similar rough trends appear.[7] The church furnishes a place of most constant meeting of friends among the business men on whom data were secured in the case of but one man out of a total of thirty-eight—and even then it is linked with a civic club; among the men of the working class group, eleven out of a total of ninety-nine friends are said to be seen most often in church. Lodges exhibit about the same relative frequency as churches for both classes, affording a most frequent place of meeting for none of the business group, but for eleven of the friends of the working class; and on the other

"shop about" among leading churches, appraising congregations as well as ministers.

On the other hand barriers are appearing within business class congregations; a rather plain young couple, members of a large, fashionable church, complained of the coldness of Middletown: "It's a hard town to get acquainted in," said the young wife. "I go to the church Flower Mission and other meetings and my husband to the Men's Club, but the people you see there don't see much of you outside."

[6] If only those answers stating "church" alone (including "church work" and "Sunday School") be included, these last figures drop to thirty-seven and three respectively.

[7] These data on the basis of men's associations must be handled even more tentatively than those on their wives; not only are samples small, but data are second-hand in nearly all cases, i.e., secured from the wives.

hand clubs other than lodges were not once mentioned for the working class, while they account for the most constant places of meeting in the case of twelve of the thirty-eight friends of this business group.

One gains an impression that the women, especially those of the business class, actively cultivate friendships more than do the men. In fact, as pointed out in an earlier chapter, this social contact activity of the female partner in marriage is increasingly prominent today. One business class woman described a common situation when she said, "Most of my husband's friends are the husbands of my friends. I met the women through my club and he met their husbands through me." Another prevailing aspect of men's friendships was pointed out in the remark, "My husband has so many acquaintances that he has time for few friends. It's a modern tendency, I guess. Relationships are more artificial since people don't drop in as they used to." Among the working class, the isolated man is apparently more common than formerly, as the decline of unions and lodges as social agencies, the disappearance of the corner saloon,[8] greater mobility, and similar factors have combined to scatter the working personnel of a given factory, and no new organizations such as the business class civic clubs have arisen. Such statements as the following by some of the working class wives interviewed appear to represent a considerable group:

"He don't go to the lodge any more. The picture show has killed the lodge. He just stays home and don't see any one."

"He liked the man who lived next to us, but he's only seen him on the street once since we moved."

"He just sees the men at work. He don't go to the lodge any more. The auto has ruined the lodges and everything else."

"He says he's just a lone wolf."

[8] Drinking as a convivial activity has not been common among the business men of Middletown since the coming of prohibition, due largely apparently to the abstinence of a group of men powerful in the industrial, social, and civic life both in 1890 and today. Among the workers the abolition of the saloon has removed a place of frequent association with their fellows, which is only in part taken by the "speak-easies." There were forty saloons in Middletown in November, 1891.

It is a shortcoming of this study that it did not consider more directly the drinking habits of Middletown before and after the coming of Federal prohibition.

Never was there more pressure in the business world for solidarity, conformity, and wide personal acquaintance than exists today under the current credit economy. But among the working class certain factors operate to make more tolerable than formerly the position of the "queer cuss" or the "lone dog": no particular expression of sociability is necessary for or evoked by operating a machine; if the position of the individual tool-worker has become more precarious as he has been increasingly reduced to the status of one of the plant's raw materials, his isolation is being compensated for at certain new points. Workmen's compensation steps in when he is hurt in getting his living, the Visiting Nurses' Association may replace the neighbors who "run in to help" when his wife has a baby, life insurance may be provided by his employers or is available on easy weekly terms through the agent at the door, and such new inventions as the movie, the automobile, the radio, make him less dependent upon his friends in his leisure. It will not be surprising if we find the worker's leisure time less closely organized than that of the business man. But his increasing isolation and a rising standard of living fostered by the habit of leaning the present against the future through time payments are constantly exposing the worker at new points more rapidly than the organized agencies bolstering him in his emergencies can develop; the lone-wolf worker in Middletown has his flanks somewhat protected, but he follows a precarious trail.

Somewhat less informal than the chance associations growing out of neighborhood and church, are parties, which bring a certain group together for a single occasion. The leading daily paper in Middletown in 1923 reported for the months of January, July, and October respectively eighty-two, 104, and 155 parties of all kinds, including picnics, dinner parties, and so on, as over against eight, thirty-one, and fourteen in the corresponding three months in 1890. While the population has increased a little less than three and one-half fold, the parties reported in January and October increased over tenfold each, while in July the increase was a little less than three and one-half fold.[9] The relatively greater number of parties represents prob-

[9] Such press reports are a dubious source for statistically usable data, but the attention paid to "personals" in the press of the smaller Middle-

ably not so much an increase in occasions for association as greater organization and formality. "People used just to drop over in the evening," as one working class wife put it, "but now they invite them 'way ahead of the date and make a party of it."

The prominence of the informal "dropping in" type of social contacts in 1890 is reflected in the day-by-day diary of the young baker for the years 1888-95. Four, five, and six evenings a week the items run as follows:

"I picked up the bunch. Went to H——'s and I set up ice cream for the crowd. Then we all took a ride round town on the street-cars. Stopped in at F——'s awhile, then meandered to K——'s awhile, then home."

"Last night I got K—— and we went to church. Then picked up a crowd of fourteen and all went down to the glass factory to watch glass blown. Back uptown at ten and all got ice cream. Stopped in at N——'s; gang stayed there and I took my two girls home. Had an elegant evening."

"Bunch over at M——'s. Pulled candy and sang. Had a time!"

"Ice cream social at church. Then to K——'s. Set up a keg, Had a time!"

Among the business class, social intercourse appears to have been scarcely more formal.

The wider range of alternatives in Middletown today necessitates more organization; parties tend increasingly to center about a core of organized club groups—"pledge parties" of sororities, the Country Club Halloween party, the "husbands' dinner" of the Sew We Do Club.[10] Even when not definitely a part of a club organization, parties, particularly among the business class, are not infrequently today given in some public

town of the earlier day offsets somewhat the activity of the Sunday "society editor" of today. In fact, testimony of people who knew Middletown in both periods indicates that the parties of 1890 were as thoroughly covered as those of today.

[10] The leading paper reported 127 parties given by clubs in Middletown between December 11, 1924, and January 3, 1925, as against ninety-four parties given by all other organizations and individuals, including churches, business or industrial units, small home affairs, and so forth. It is probable, however, that the more thorough reporting of club parties than of small or home parties and the special celebration of the Christmas season by the clubs affect this ratio.

or semi-public building outside of the home; a woman may entertain a table of guests for luncheon as part of a large Country Club luncheon; or two women may hold a bridge party or reception in the ball room of a hotel or the parlors of the Elks' Club elaborately transformed for the occasion into a Japanese garden; or four wives and husbands may jointly entertain a dozen mutual friends at dinner at a hotel or the Elks' Club and then repair to one of their homes for an evening of bridge.

Accompanying this incipient decline in home parties is the almost total disappearance of whole-family parties before the specialized parties for each age group and the self-sufficient social system of the high school. Almost unknown today is such a list of guests as appeared frequently in the papers of the nineties:

"Among the many present at the surprise party were Grandma Walker, Mrs. C. P—— and family, S. C—— and family, John W—— and family, Isaac B—— and family, James W—— and family and S. H—— and family."

Likewise the home-made entertainment that enlivened the more spontaneous parties of the nineties is tending to narrow, especially among the business group, to a few correct variants. Such an occasion as the following reported in the 1890 paper would hardly be so acceptable as a party in these days of movies, dancing at a near-by resort to "the finest jazz orchestra in the state," and the ever-vocal radio:

"A pleasant surprise was held last night at the elegant residence of Oliver J—— in honor of the fortieth anniversary of the birth of Mrs. Ella J——. Besides the neighbors and friends to the number of forty, there were present . . . Every face was beaming with delight, and happiness flowed from heart to heart. . . . After dinner a season of song and prayer was had, after which the house was made to ring with music. . . . Mr. McC—— favored us with a song, *A Thousand Years My Own Columbia.*"

Nor would a Young Ladies' Cooking Club composed of "our young social leaders" entertain their friends by "thirteen tableaux" followed by a sumptuous dinner cooked by their own hands and a recitation of *"Curfew Shall Not Ring Tonight* in a most pleasing manner." Nor would a surprise party of six leading citizens and their wives meet "at a near-by house," hold "a council of war until their proposed victims had retired for the night, when with an abundance of 'taffy' sugar, all made a bold dash for Mr. C——'s." Gone, or nearly so among the business class, though to a far less extent among the working class, are such "jolly affairs" as "trolley parties," "progressive tiddledy-winks," "shoe socials" where one secured one's partner by seeing her shoes under a sheet pinned up at the end of the room, "lemon squeezes," "going to Jerusalem," "cobweb parties," "pin the tale on a mule," conundrums, charades, evenings of "euchre, whist, pedro, and crocono" or "parchesi, authors, and checkers," or evenings when "all had a sing." Waggish tricks do not form such a large part of an evening's diversion. The young baker's diary related every so often during the nineties, "They went home with my hat," or "Attended a wake at ——'s. Somebody hid my hat."

The growing rigidity of the social system today is centering parties more and more upon cards, pedro among the workers and bridge among the others.[11] Cards and dancing are the standard entertainment of Middletown. In general, dancing holds the position of preëminence with the younger group prior to marriage, while from marriage on the more sedentary activity predominates. In 1924 the sectional state conference of the numerically most powerful religious denomination in Middletown renewed its traditional prohibition upon card playing, and in some of the more religious working class families the ban is still maintained not only upon cards but upon checkers and other games as well, but among the business group there are virtually no people who debar card playing.[12] The local press

[11] Mah jong was a furious alternate to bridge for a while and then disappeared.

It must be borne in mind that in leisure-time pursuits as in so many other activities, the workers of Middletown still do many of the things the business group did a generation ago. Thus a men's chorus of a church whose membership is made up largely of working men gave a party in 1925 at which there was singing of sacred hymns, a long recitation in the rhetorical manner of 1890, a "humorous selection," "a few words from our pastor," and copious food.

[12] The extent of the relaxing of the ban upon cards is witnessed by the maintenance by at least three of the local semi-religious, benevolent lodges of regular gambling at cards in their club-houses, a percentage of the winnings going to the lodges for their charity.

Playing cards on Sunday is still tabooed by many people, though less so than in 1890.

reported thirty "card parties" for the three months of January, July, and October, 1923, with card playing a part at the entertainment of many others, as against only one for the corresponding months of 1890. This does not mean that cards were not played in 1890, but probably that they were not such an inevitable and formalized feature of social intercourse.

Dancing is today a universal skill among the young; their social life, particularly among the high school group, is increasingly built about it. The dance apparently held no such prominent place in the leisure activities of 1890. Dancing there was, to be sure: great balls by the policemen, cab-drivers, clerks, nail-makers, green glass workers, and other occupational groups—usually for charity; the grand ball of the Amalgamated Association of Iron and Steel Workers in 1890 was proclaimed "the largest event of its kind ever given in [Middletown] or the Gas Belt, with 1,200-1,500 present." Among the business class small dances in fashionable homes on New Year's Eve and other holidays were not uncommon; this group, moreover, patronized Professor Daisy's fortnightly dancing lessons at the skating rink, culminating in a "ball and German with fifty society couples and seventy-five spectators, the latter watching with interest the fancy dancing, heel and toe polka, then the German." But dances were not mentioned in the leading paper in January or July, 1890, and were mentioned only five times in October, as against ten times in January, 1923, twenty-two times in July, and twenty-four times in October. More significant of the shift is the fact that the leading fashionable young men's social club of 1890 gave parties but no dances. Not until 1900 does the press speak of dancing as a "local craze."

Today the social pace is set for the unmarried group by the elaborate formality of club, fraternity, and sorority dances in hotels, each costing $150-$300 and involving keen rivalry in decorations, music, partners, dress, and number of invitations. These reach chiefly the business group but tend to include a wider range as high school attendance grows. The old round of informal Christmas holiday pleasantries has been largely crowded out by a rigid ritual of fourteen annual formal dances; the principal public celebration of Thanksgiving Day consists in three dances—an annual matinée dance by one of the fraternities and two evening dances. High school commencement no longer means the program of essays, the solitary ball and faculty-student party of the nineties, but a dizzy week of junior-senior dance, "a fitting climax to the social affairs of the junior class"; senior formal dance, "one of the most elaborate affairs of the season"; banquets, picnics, and receptions—all carefully planned in April, two months before the events.

Among the working class home dances, like those of the business class thirty-five years ago, still survive, though the better music of public dance halls has greater attraction. So exacting has the public taste in dance music become that a local church was forced to abandon the effort to hold dances in its parish house "because the young people demanded better music than we could afford." Such new customs as the replacement of boys' and girls' walking to and from dances in a crowd by the almost universal custom of going by couples in an automobile, and the disappearance of "odd" girls at dances, the pairing off of boys and girls being emphasized by the full press reports of those who attend by couples, tend to emphasize the rigidity of the social ritual of the dance.[13]

Just as the unorganized social associations of the neighborhood in 1890 have given way increasingly to semi-organized dances and clubs, so the more active leisure-time pursuits involving physical exertion in various sports exhibit a similar recession of the unorganized before the organized. Few people today walk for pleasure in Middletown, the river is now too polluted for fishing, and the small boys of the city are wont to call out in a disgusted tone to a stray bicycler, "Aw, why don't you buy a machine!" Instead of these unorganized, one-man types of physical recreation, the city affords facilities for organized sport undreamed of in 1890: Y.M.C.A., Y.W.C.A., high school gymnasium, municipal golf course and Country Club course, and grade school, high school, factory, lodge, and Y.M.C.A. leagues in various sports. The sporadic factory

[13] It should be noted in passing, particularly in view of the fact that nine of the fourteen formal Christmas dances are given by girls, that the dance has apparently, whether consciously or not, been seized upon by the unmarried girls as a device whereby the newer aggressiveness of the females can assume overt form despite the persistence of the male ban upon it. The extent to which these dances exist as an appendage of the mating ritual of the younger set is reflected in the frank remark of a popular high school senior girl, "The girls in each club are awfully catty about the extra invitations to their club dances they are allowed, and it's sure some test of a girl's popularity if she gets invited to all nine of the girls' formal Christmas dances."

base ball teams of the nineties represent one of the few fore-runners of the present tendency.

Organized sports appear, from a brief check, to exhibit a greater increase since 1890 in the relative amount of news space devoted to them in the Middletown press than any other department of news—from 4 per cent. of the total news content of the leading paper in 1890 to 16 per cent. in 1923.[14] For the three sample months of January, July, and October, 1923, the leading paper mentioned 169, seventy, and ninety-eight organized and unorganized sporting or athletic events in the city, as against six, fifteen, and seven for the corresponding months of 1890.[15]

The athletic activity of the city today culminates in basket-ball in the high school. The high schools of the entire state are organized into a state-wide league involving each year "regionals," "sectionals," and "finals," during which the city's civic pride is deeply involved: leading citizens give " 'Bearcat' parties" prior to attending the final games, hundreds of people unable to secure tickets stand in the street cheering a score board, classes are virtually suspended in the high school, and the children who are unable to go to the state capital to see the game meet in the school in a chapel service of cheers and songs and sometimes prayers for victory. In the series of games leading up to the "finals" the city turns out week after week to fill to the doors the largest auditorium available.[16] In con-

[14] See Table XXIII. Counts could be made only of one representative week—the first week in March—for both periods. Counts based upon such a short period must obviously be used only tentatively.

[15] These figures apparently represent increases both in actual participation in sports and in occasions for watching others play. To a certain extent they are misleading in that they do not include the extensive unorganized "vacant lot" sports of 1890, but on the other hand they do not include much of the day-by-day activity for men and boys at the Y.M.C.A. in 1923. The heavy increase in organized sports is shown by the following distribution of sports mentioned for the month of October in each year: 1890, three competitive shoots by the local gun club with clubs from other towns, two fishing trips, and two announcements of hunting trips. 1923, three baseball games, seventy-three bowling matches, three basket-ball games, three weekly shoots by the gun club, four golf tournaments, nine football games, one bicycling party, one prize fight, one Y.M.C.A. track meet.

[16] Basket-ball sweeps all before it. Witness the jubilant voting of the city council to spend $100,000 for a new gymnasium for the "Bearcats" at a time when the cry on all sides was for retrenchment in city expenditures and the public library was understaffed because it could secure no assistant for less than $1,800 when a cut in its funds made only $1,500 available. The bond issue for the new gymnasium was finally overruled through appeal

trast to this is the complete absence of high school athletic teams of any kind in 1890, noted in an earlier chapter.

It is notably characteristic of this culture that active physical sports tend to drop off sharply among all classes after high school.[17] The automobile encourages this physical "settling down," while on the other hand the rise of golf, including the launching of a municipal course, hand-ball, and business men's gymnasium classes at the Y.M.C.A., bowling alleys in certain lodge buildings, and shorter working hours and Saturday half-holidays, all pull in the opposite direction. Women, too, have begun to engage in athletic sports, though not so commonly as men.[18]

This trend toward greater organization appearing in so many leisure pursuits culminates in the proliferating system of clubs which touches the life of the city in all its major activities. A total of 458 active clubs was discovered in Middletown after an exhaustive canvass during the spring and summer of 1924,

by a small and unpopular group of citizens to the state authorities having ultimate supervision over such fiscal matters.

It is widely reported that "the chief thing that got [the new superintendent of schools, a young man elected after the preceding superintendent, a veteran school executive, was dropped for alleged political reasons] his job was that he put [Middletown] on the map as a basket-ball town."

[17] Gillin says of Cleveland, "In the course of the successive age periods conventional spare-time interests turn less and less to outdoor, athletic activities; as people grow older their pursuits tend to diverge more and more from those forms which have been established in the history of the race. This is especially true of the activities of men. Their activities in the later periods follow the lead of the women's, emphasizing the trend away from the more direct and simply organized and physically active pursuits." John L. Gillin, *Wholesome Citizens and Spare Time* (Cleveland; Cleveland Recreation Survey, 1918), p. 18.

[18] There was considerable opposition from some older citizens to the proposal to put a swimming pool in the new Y.W.C.A. building.

A "doctor book" of the nineties warned the women of a generation ago against the current sedentary, indoor life of women, which "besides hurting their figure and complexion, relaxes their solids, weakens their minds, and disorders all the functions of the body." But Marion Harland in her *Talks upon Practical Subjects* warns that "the *fin de siècle* girl and her bicycle have hardly been acquainted long enough for the passage of correct judgment upon the consequence of the intimacy."

The girls entering Vassar College in 1916-20 engaged in an average of 9.2 sports, as against 2.0 for the incoming freshmen of 1896-1900; 0.6 per cent. of the former reported no sports at all, as against 26.5 per cent. of the earlier girls. Mabel Newcomer, "Physical Development of Vassar College Students, 1884-1920" (*Quarterly Publication of the American Statistical Association*, New Series No. 136, Vol. XVII.)

roughly one for each eighty people. This probably includes four-fifths of all active organized club groups. A canvass of the city of 1890, believed by the staff to be roughly as comprehensive as the 1924 count, revealed ninety-two clubs, or one for every 125 people.[19] While the city has grown less than three and one-half fold, adult social clubs have increased from twenty-one to 129, church adult social clubs from eight to 101, adult benevolent groups, trade unions, and the group of literary, artistic, and musical clubs have each doubled, business and professional groups have increased from one to nine, and civic clubs (most of them with a strong business flavor as well) from one to eleven. The current of organization has apparently run even more rapidly in the region of the more formal type of juvenile clubs, as national organizations such as the Boy Scouts and Girl Reserves have increased in Middletown from zero to ten groups, as the church has organized its children in the effort to hold them against outside competition, and as clubs have sprung up to sift the 1,600 high school students into the smaller groupings. Organized juvenile clubs of all kinds have increased from six (all church clubs) in 1890 to ninety-five, although obviously many transitory neighborhood "clubs" of children were omitted in the counts of both periods.

Although there has been a growing tendency among the business class to make club life serve other than recreational ends, notably those of getting a living, most Middletown clubs apparently offer people not an extension of their customary activities but a way of escape from them. The city is dotted with social clubs, chiefly women's clubs, but in a limited number of cases including husbands as well: the Kill Kare Club, Jolly Eight, Best of All Club, Happy Twelve, Bitter Sweet Club, and

---

[19] See Table XVIII for distribution of clubs by classes in both periods.

The number of formal and informal social groupings in the city is, of course, almost indefinite. Only those organizations which have a definite and regular social meeting monthly or oftener are included here as "clubs." The figures given are based upon a careful count of clubs mentioned in the two leading papers in 1890 and in 1924 from January 1 to October 1, checked and augmented by the city directories and by reports of individuals for both periods. The 1890 count was further supplemented by two diaries, one of a working man and one of a business man. Other informal sources of information were used for both periods. The chief types of clubs that are missing from these totals are certain informal neighborhood groups (luncheon, bridge, sewing, children's clubs, etc.) that are not reported in the press. The reporting in both periods appears, however, from all available evidence, to have been very comprehensive.

so on. Here cards, games among the working class, music, or dancing, and always "refreshments" offer Middletown an alternation from routine duties of life. Among the business class the Country Club, bridge clubs, and so on, and, among the group from fifteen to thirty years of age, fraternities and sororities, tend to supplement or displace the smaller, less formal neighborhood gatherings of a generation ago.[20]

The value which Middletown places upon education is reflected in its clubs, but, although working class families press toward schooling for their children and to some extent avail themselves of technical training in evening classes, it is the more leisured women of the business class who compose the literary and artistic study clubs of the city.[21] These nineteen groups, fifteen of them forming a part of the county "Federated Club of Clubs," vie with the men's civic clubs in local prominence. The total membership of all nineteen clubs is approximately 700.[22]

"Mutual mental improvement" was the stated aim of the earliest study club to be organized; many others sprang up in the nineties for "the social and intellectual advancement of its members," "general education in art, science, literature, and

[20] Of the eighty-eight women's "social" clubs on which data were secured, forty-five, chiefly of the working class, play games or have "contests," twenty-six, largely business class, play cards (this number would probably be greatly increased if more of the informal clubs could have been found and included in this enumeration), twelve sew, three are luncheon and dinner clubs, one a dancing club, and one a bowling club.

Thirteen of these eighty-eight clubs were originally formed with a purpose which included the mutual saving of money ("Christmas savings clubs"). Seven have some kind of devotional exercises at their meetings. Most of them have occasional parties and picnics. Two do some regular charitable work, though many more help specific needy cases.

Thirty-eight of these eighty-eight clubs are composed of women between twenty-one and forty years of age, fourteen of women over forty, twenty-nine of both groups, seven of both of the above groups and also of members under twenty-one.

[21] Efforts have been made to form study organizations among working class women, but with relatively little success. The tightening social lines of the city appear in the feeling of these women that they are being patronized by the business class women.

[22] There is some overlapping in membership of the various clubs.

Age distribution of members was secured for seventeen of these nineteen clubs: three of them are made up of women between twenty-one and forty years of age, two of women over forty, and twelve of both groups. By a comparison with N. 20 above it will be observed that age is apparently more of a factor influencing the coming together of women in social clubs than in these study clubs.

music," or "work in literature, music, art, needlework, and philanthropy." Each club has some symbol of its work; characteristically, all are verbal symbols; again characteristically, all relate to the master symbol, "Progress," the word standing alone as the motto of one club. "That what we have done already is but the earnest of what we shall do," says another, while the Federated Club of Clubs sets forth, "Our motto, 'The Actual and the Ideal,' means that from the actual we will grow into the ideal, that is, we have imagined an ideal woman, and we wish to grow toward that perfection in womanly beauty, grace, and culture." Bound up with the symbol of progress are strong religious as well as educational traditions. One club sets at the outset of its year's printed program:

> "On the threshold of our task,
> Let us light and guidance ask,
> Let us pause in silent prayer." [23]

"Progress" in mental improvement, today as a generation ago, is sought largely through writing and listening to papers and speeches. These are supplemented variously in the different clubs by devotions, music, and "responses" in which each woman says a sentence or two on "Forget-Me-Nots of My Summer," "Current Events," "Literary Gems," "Household Hints," "Bible Verses," "Who's Who," "Famous Sayings of Great Soldiers," "Prominent Women of the Civil War," "Short Accounts of New Reforms," "Wise and Foolish Women of the Bible," "American Industries," etc. A comparison of club programs of the nineties with those of today

---

[23] Six of these federated clubs read from the Bible at their meetings and six repeat at each meeting the club woman's creed:

COLLECT FOR CLUB WOMEN

"Keep us, O God, from pettiness; let us be large in thought, in word, in deed.
Let us be done with fault-finding and leave off self-seeking.
May we put off all pretense and meet each other face to face without self-pity and without prejudice.
May we never be hasty in judgment and always generous.
Teach us to put into action our better impulses, straightforward and unafraid.
Let us take time for all things; make us grow calm, serene and gentle.
Grant that we may realize it is the little things that create differences; that in the big things of life we are as one.
And may we strive to touch and to know the great common woman's heart of us all; and, O Lord God, let us not forget to be kind."

suggests that active study in connection with the programs is in general somewhat less consecutive than formerly, although only those clubs are admitted to the federation which do some definite "work." [24] Most of the programs tend to oscillate somewhat from subject to subject in an effort to be as comprehensive as possible in each year's work. The program of one characteristic federated club took up within one recent year "Prophets of the Bible," "Wonders of the Radio," "What Do Colleges for Women in the Orient Accomplish?" and "The Life of Paul." Another club proceeded within one winter from "Recent Religious Movements: Christian Science and New Thought," to "The Dictograph," "Mural Paintings," "The Panama Canal," "The Drama," "Hull House," and "Dress." The year's work of yet another included meetings on "Waterways," "Animals," "Our Nation," "Socialism," and "The Simple Life." The program of another club offered in an exceptionally long season of twenty-one meetings, five of which were purely social, a program providing study of "the Bible, history, music, art, and literature." [25]

One factor in the great variety of subjects covered by the programs of many clubs may be the device of passing on to them through the Federation suggestions of various standing state committees for their work:

"I suppose," said the president of the Federation on one such occasion, "that the members of the state committees will want to put in special pleas for the subjects of their particular committees. I'm on the history committee and I'll say my say first. Now, I

[24] Cf. Ch. XVII for discussion of the music and art clubs which are members of the Federation. No other clubs have as clearly defined fields of study.

[25] While Biblical themes are still common, they do not bulk as large as formerly. Thirty-five years ago it was not uncommon for a club to follow some study of the Bible concurrently with other work. Such a mingling of the two strains appears in the following course of study for the twenty-one meetings in the year's program of one club: Meeting No. 1: Abraham; Egyptian women. 2: Isaac; modern Egypt. 3: Jacob; Greek women. 4: Joseph; Greek religion. 5: Moses; Roman women. 6: Banquet. 7: The twelve tribes; *Quo Vadis.* 8: Tabernacle; French women. 9: Jewish feasts; French palaces. 10: Idolatry; English women. 11: David. 12: Solomon; American women. 13: Temple; Yellowstone National Park. 14: Children's Day. 15: Queen of Sheba; [State] writers. 16: Joseph and Mary; Queen Wilhelmina. 17: Christ; Czarina of Russia. 18: Jerusalem; lives of Patti, Schumann-Heink, Sembrich, Melba. 19: Mary Magdalene; prophecy of the ——- Club ten years hence. 20: Modern Jews; *Harold, Last of the Saxons.* 21: Passion Play; modern painters.

think it would be really very fine if each club would include at least one meeting this next year on early state history. I don't see any reason why you shouldn't do it. We should all know more about the early history of our own state. The history committee would like you to have just one meeting on that. Now the other ladies may speak for their committees." Members of the Industrial Relations Committee and of the Committee on the Exchange of Bulbs and Seeds, and so on, followed with their pleas.

In addition to this parceling out of programs by energetic committees, there are certain standard subjects which recur; year after year appear meetings on "Origins of Thanksgiving," "Old Christmas Stories," "How the Flag Originated," "Flag Day and Its Meaning," "The Home Life of George Washington," "St. Patrick and the Shamrock," "Thoughts on Easter," "Friendships of the Bible," "Bible Types of Modern Women," "Birds and Flowers of the Bible," "Old Testament Heroes," "Beauty Spots of the State," "Middle Western Writers," "Riley and His Poems," "Readings from Eddie Guest," varied with "History of our Club" and "How We May Improve Our Club."

Amid these programs—winding along through "Ruskin, As Man, Author, and Critic," "The Xantippes of History" (with the comment, "If all had husbands like Socrates they would be found more numerous in our city"), "Etchings and Engravings," "Immigration," "The Power of Music," "The Effect of Friendship upon Character," "Intelligence Testing," "Modern Novels"—certain trends appear. A slow shift is taking place away from the almost exclusive preoccupation with "literature" as the heart of the things worth studying toward more active interest in the life of Middletown. The earlier aims of "the promotion of literary and social tastes" are somewhat giving way to "the social and intellectual advancement of women and united effort to further improvement in the community in which we live." "In the early days we were more interested in ancient Greece," said one club woman, "but now we are interested in what is happening in Middletown."

The oldest and largest of the women's clubs, which now has 168 members, began in the late nineties to form clubs within the club, or "departments." The Literature Department, which included art, music, and ancient history, comprehended the entire work of the club in 1890, but the Literature and Art Department today is in the minority both in number of meetings and of members, owing to the rise of popular Departments of Sociology and Civics (originally Philanthropy and Civics) and History and Current Events (originally Education and Home).[26] These new departments, made up of women having the ballot and aware of many currents in the life eddying about them, zealous to improve Middletown, find their way hard. One business class woman who says that she is "never able to get reading done," joined the History and Current Events Department whose program announced the study of Wells' *Outline of History* and the *Review of Reviews* News Letter. "But," she protested, "at the meetings the women, instead of discussing, read aloud little sections of the book and I get nothing from it at all. At one meeting they read from the *Outline* and at the next read an article on pioneer life by Dr. —— and at the next went back to Mr. Wells. I have stopped going, and I'm going to resign."

The Sociology and Civics Department, setting out in 1924-25 to study community problems with the help of Blackmar and Gillin's *Outlines of Sociology* and Miss Byington's Russell Sage pamphlet on *What Social Workers Should Know About Their Own Communities,* found itself bewildered in the attempt to cover in one meeting the political life of Middletown and in the next the religious life, trying to answer questions on the churches ranging from "Do their spheres overlap?" to "What

[26] The following changes in distribution of membership are illuminating:

1890: Entire membership literature, art, music, and ancient history.
1902-3: Lit. and Art 54 per cent., Philanthropy and Civics 27 per cent., Education and Home 19 per cent.
1909-10: Lit. and Art 45 per cent., Sociology and Civics 31 per cent., History and Current Events 24 per cent.
1919-20: Lit. and Art 45 per cent., Sociology and Civics 33 per cent., History and Current Events 22 per cent.
1923-24: Lit. and Art 32 per cent., Sociology and Civics 22 per cent., History and Current Events 34 per cent., Dramatic Art 12 per cent.

As late as 1899-1900, nine of the general club meetings were under the Department of Literature and Art and only five each under the Departments of Philanthropy and Civics and of Education and Home, whereas in 1919-20 Literature and Art had five meetings, Sociology and Civics six, and History and Current Events three, and in 1923-24, Literature and Art and History and Current Events had four each, and Sociology and Civics and the new Department of Dramatic Art, devoted to rehearsing and presenting plays, three each.

is the social welfare work of each? How is it financed?" Baffled by the difficulty of finding concrete answers to the complex current questions they are attempting to study in this and other clubs, Middletown women tend, not unnaturally, to fall back upon generalizations. Thus a paper dealt characteristically with the "problems" of the church in Middletown in such terms as:

"Whenever in poetry we hear exalted the beauty of night, it is always a night with stars. . . . No man with a soul in him can look at the stars and not see God. . . . Men would be hopelessly lost in the darkness of this world but for the light of the true Christian."

A paper on "Bolshevism in America" at another study club explained:

"There is room in this country for but one flag and that is the American flag. Put down the red flag. It stands for nothing which our Government stands for. It is against the integrity of the family, the State and Nation. It floats only where cowards are in power. . . . The whole Bolshevik movement in Russia was in the interest of and financed by Germany. In the United States German Gold also has been stimulating the Bolshevik campaign.

"Bolshevism has no root in America. . . . The I.W.W. represents organized Bolshevism in America. We have certainly shown that we know how to handle the I.W.W. . . . It would be foolish to magnify such disorders as Bolshevism. . . . The American people believe in America. They believe in doing things the right way, not the wrong way.

"We must take this movement of Bolshevism seriously. . . . We must not for one moment relax our vigilance. . . ."

"Part of it is contradictory," said the writer afterwards, "but I read different things in different places."

Still another club attempted the discussion of the "problem" of the movies in Middletown, "a subject of vital interest from a moral as well as commercial aspect"; a paper presented "Tendencies of Movies and Their Possibilities," and the program thereafter shifted into "a chatty round-table discussion of favorite screen stars, best plays, and why certain ones were chosen by club members."

Discussion of child-rearing, although it would appear to be a dominant interest of many of these federated club members, is confined almost entirely to the struggling Mothers' Council. This organization, replacing the moribund Parent-Teachers Association, is kept alive by the larger churches, each of which has its own group; at the monthly union meetings attendance ranges only from twelve to twenty. Programs consist, in addition to music, Bible reading, and prayer, of talks or the reading aloud of papers sent out by the national organization, followed by scattered discussion.[27] According to the emphatic testimony of one of the active members:

"The Mothers' Council is dying. Actually it is powerless to act in matters concerning the problems of home and school because it is connected with the churches. The only good thing it has done is to bring Catholic and Protestant women together and help break down religious prejudice—though even this was almost spoiled when a minister's wife bitterly attacked the Catholic attitude towards Bible study in the schools. The meetings really do next to nothing to help mothers to deal with their own problems in bringing up their children."

Only occasionally in the other federated clubs do papers appear on such subjects as "The Family as a Coöperative Institution" and "What Public Health Is Doing for the Child." "House Planning," "Interior Decoration," and other questions of homemaking are only less infrequent. And even such discussions tend to fall into the familiar generalizations, as, for example, in a 1924 paper on care of health:

[27] The concepts under which the "problems" involved in child-rearing are conceived by these groups appear from the 1923-24 program: "The Power of Organized Motherhood to Benefit Humanity"; "New Application of Music and Song in Education"; "Choosing Children's Books." followed by discussion led by the local librarian; "Coöperation between the Home and School"; "The Duty of Parents in Training Children for Citizenship"; "Thoughts on Religious Education," and "Christian Spirit in the Home"; "Courtesy—Respecting the Rights of Others." The 1924-25 program included: "What Constitutes a Modern Good Father?" (by a minister); "What Can We Do Toward Meeting Community Needs?" (by the head of the local hospital, whose principal interest in her talk was aligning the local women behind the hospital); "The Woman of the Home"; "A Real American"; debate on "Resolved, That a College Education is Necessary to Success"; a "Teen Age Meeting," including talks on the Boy Scouts, Camp Fire Girls, the Bible in the public schools, and review of the novel, *The Plastic Age;* "The Misunderstood Child." It is only fair to remark that these imported program subjects glorify out of all proportion the actual proceedings at the meetings.

"The physical laws of health, however simple and concise they may be, are not to be separated from the mental and spiritual condition of the patient, and no one will deny that their influence is of great importance. 'Know ye not that your body is the temple of the Holy Ghost which is in you, which ye have of God, and ye are not your own? For ye are bought with a price; therefore glorify God in your bodies, and in your spirit which is God's.'

"The subject of foods and food values is a broad one. . . . If people would but fully realize that by proper living it is possible for every one to be well, very few persons would be ill. The great hope of modern medicine is the prevention of disease."

It continued with a sketch of the work of Jenner and the Mayos, with no reference to the concrete problems involved in "proper living" in Middletown.

Press reports of club papers, while they frequently include summaries of the facts covered, tend to single out for eulogy "the charming and gracious manner" in which the paper was presented, or the "well-chosen and eloquent" words of the speaker; one of the most popular lectures to which Middletown club women listened in 1924 was described as "a poem in prose." As school education appears to be more valued in Middletown as a symbol for things hoped for than for its specific content, so this activity of "improving one's mind" through club papers may be regarded as serving in part as a focus of sentiments rather than as something to be put into practice and used. And, just as the city paid more attention to elaborate externals of its houses in the eighties, when there was little to be done in the way of adding material improvements inside, so these papers show a tendency toward verbal ornamentation not only in the familiar regions of literature and art, but also in the baffling new programs on "civic problems." In other words, here as elsewhere, stagnation or mild bewilderment tends to result in the proliferation of superficial external aspects. If, however, the pressure of maladjustment in local life becomes acute enough to force increasingly concrete discussion and action upon the emergent "Sociology and Civics" concerns of these women, we may expect to see a recession of this ornate verbiage before more definite action addressed to specific ends.

As the clubs have become less exclusively literary they have increasingly made sporadic forays into practical civic affairs.

No one of the fifteen federated clubs lacks an annual contribution to at least one of the social service organizations: the Free Kindergarten, Humane Society, Anti-Tuberculosis Association, Social Service Bureau, Day Nursery, Y.W.C.A., etc. Federated Club members point with pride, also, to the fact that they initiated the first Associated Charities in the city, the Visiting Nurses' Association, the Juvenile Court, the teaching of manual training in the schools; to their sponsoring of art exhibits in the schools, recognition dinners for honor students, a rest room for women in the Court House, anti-tuberculosis shacks, investigation of the working conditions of women in Middletown; to their agitation for regulation of dance halls, better motion pictures, and dry law enforcement; to their changing the name of one of the streets to Pershing Avenue and planting trees along it as a war memorial.[28] Discussion as to which woman's club originated certain of these projects is active, and in some cases a men's club disputes the claim, but the Chamber of Commerce is wont to turn to the Federated Clubs for help in a community program, and the men frequently admit that "it's usually the women who carry things through." This concrete satisfaction of something definite accomplished in Middletown appears to be an increasingly prominent feature of the work of the women's clubs.

And yet the essential function of these groups is probably no more to be found in their civic work than in their study programs. Of the twenty meetings yearly of one representative study club, three are luncheons with no programs, another is a banquet for husbands, a fourth an annual banquet for members, while an annual guest day, an annual picnic, and an annual business meeting are also without programs; eleven program meetings remain, at all of which refreshments are served and a "social hour" is enjoyed. In the early days of the largest women's club a banquet was held once every three years, whereas today there is an annual banquet, the year opens with a dinner or tea, there is a New Year's "Open House," a

[28] The Mothers' Council has fostered various movements in the community: religious education in the schools; an effort to remove obscene literature from the local news-stands; getting the sororities to recommend that their members do not wear such low-cut gowns to dances; urging probation work in the juvenile court; and persuading the high school authorities to set aside the first period in the morning for committee meetings so that the children can come home earlier from school.

"Daughters' Night," and a final garden party. The Record Book of the Federation for 1899-1900 records discussion of whether "the social phase of club life should be encouraged or not" and of whether a club "should have an object beyond self-culture." This is no longer an open question, and the attitude of the woman who said, "I have had several invitations to join women's clubs, but I haven't joined any yet. I am waiting until an invitation from the *right* one comes along," apparently reflects an increasingly prevalent tendency to utilize club membership instrumentally as a social stepping-stone. In the nineties, when books, magazines, and other opportunities for "self-culture" were relatively less available and informal contacts with friends more so, the social aspect was apparently less important; today, with this situation reversed, it must be served first.

As part of a democratic counter-trend traceable in part to the war, the oldest woman's club has made the important shift from an exclusive to an inclusive membership. From a carefully selected membership of thirty-nine in 1890, growing more slowly than the population of the city during the next thirty years, this club increased 125 per cent. between 1920 and 1923, its present membership being 168. Many members regret the change, as "it has brought in many women who want the social prestige of club membership but are not willing to work for culture." In such a club as this or the Matinee Musical, described above, the ranks tend to be swelled out at the bottom by the socially hungry and depleted at the top by the women to whom this is "just one more club." All other clubs are more exclusive, keeping their membership to between twenty and thirty and maintaining the established social levels of the city.

One club, which exerts considerable effort to keep "mental and social culture" as its sole aim by requiring attendance at all meetings and serving no refreshments, is nevertheless the most socially exclusive of all clubs. It has a strictly limited membership, meets during only seven months of the year at the conspicuously leisurely hours of nine to eleven in the morning, and allows no written papers, its purpose being "to revive the lost art of conversation." More than any of the other clubs it has held to the consecutive study of "literary" subjects, but of late years its programs have been interspersed with talks on travel, and one recent year it yielded to the current "civic" tendency to the extent of basing its program on "The Larger Citizenship." But its social exclusiveness remains, and it holds itself rather aloof from the other clubs of the Federation.

Appeal to social prestige was used by a publishing house to draw women into the club which most nearly resembles the Chautauqua Reading Circles of the nineties. The agents for this "clever book-selling scheme," as it is called even by some of its supporters, secured the adherence of some social leaders and used their names as drawing-cards for others. "But, Mrs. ——, you don't realize that this isn't just a study group," said the agent to one woman who had said that she did not care to join. "It will give you a chance to know the best people in town!" With this inducement, plus the fact that they would be receiving "the equivalent of a college course except for economics," 125 women invested $66.00 in a set of books to be used as a basis for a six-year course of study. And, in defiance of the customary social stratifications, the nucleus of the plan has persisted; women who, as one member explained of her particular group, "would never have come together in an invitation club," continue to meet. During the three years since the Chapter was organized, one morning group largely composed of society women has practically fallen to pieces; an evening group of business women limps on feebly, and two other morning groups of more mixed membership survive; the last two, like the first, display in their 9 A.M. meeting hour the same conspicuous leisure which marks the morning club described above. The study centers around the books which one member described as "outlines, about what you would find in an ordinary school history." The first year covered "Epochs of Human Progress" in eighteen meetings, the second "History of the Drama," and the groups in 1924-25 were studying the "History of Art." "I work harder on this than on any of my other club work," said one of the most active leaders of club life in the city. "I usually spend an afternoon at the library on each paper," said another, "and two evenings writing it." As many as six papers at a single meeting may cover, for example, "Egyptian Architecture," "Mesopotamian Architecture," "Solomon's Temple," "Greek Architecture," "Greek Sculpture," and "Greek Painting." At such a meeting one sees the persistence of the traditional emphasis upon "improving one's mind" by means of all sorts of knowledge. Following the reading of papers come the "questions for this time":

Ques. "Describe the plan of an Egyptian temple."
Ans. "They were very massive."
Leader. "Yes, and, of course, we know they had decorations and all."
Ques. "Contrast the cella of a Greek temple with the inner rooms of Egyptian and Mesopotamian temples."
Leader. "Of course, they were just about the same."
Ques. "Point out the differences between the Egyptian, Mesopotamian, and Greek religions."
Leader. "Of course, we know they were all pretty much alike. They all worshiped gods."

"I learned three things from this meeting," said one of the group as the dozen women were leaving, "there were three kinds of Greek columns, Roman architecture wasn't as good as Greek, and Alexander the Great lived before Christ."

Somewhat set apart from the other women's study clubs and including a few men among their numbers are the groups centering their work in "practical psychology." As eagerly as Middletown in the nineties thronged to hear "Dr. C——'s free lectures to men and women only on Solar Biology or our relation to the Zodiac—the mysteries of yourself," about 250 people, the majority of them women, came in 1924 to find out from two women lecturers "How We Reach Our Sub-conscious Minds." Following the free advertising talks, thirty women, some of them from the working class, paid $25.00 for a course of ten lectures giving "definite psychological instruction for gaining and maintaining bodily fitness and mental poise and for building personality," setting forth the *"Infallible Formula"* for "reaching, re-directing, and enriching the operations of Your Greater Self," and assuring that "we all come into the world with the same equipment of brain cells. You can become any sort of success you choose." Another group of two dozen women and two men comes together weekly in a meeting in the parlors of the Chamber of Commerce, beginning always with the prescribed ritual:

Leader. "How are you?"
Group. "Fine and dandy, why shouldn't I be?"

Leader. "Now we have started vibrations which have lifted us already to a higher plane."
Group, in unison. "I am relaxing, relaxing, relaxed. . . . I am wholly passive—Universal Mind has taken possession of me. . . . Knowledge is Power. I desire Knowledge. I have Knowledge—"

and proceeds to the study of concentration, attention, the subconscious mind, and so on from their textbook, *The Master Key.*

Despite the uneven and somewhat scattered nature of the work of some of these women's study clubs, it is chiefly in these groups that the intellectual traditions of the nineties live on.

If self-cultivation in women's clubs has become somewhat more diffuse as the population of the city has become so differentiated that the future of a woman's children and the business success of her husband are not remotely connected with the social level of her clubs, the educational work of the men's clubs, except for strictly specialized professional clubs and listening to talks in other clubs, is not simply diffuse—it is engulfed.

From 1878 to the chaotic gas boom days a group of substantial citizens met regularly in a Literary and Scientific Association, "one of the fixed institutions of learning in the city." Bankers, lawyers, doctors, merchants, wrote and discussed papers on "What is Mind?" "The Physiology of Life and Death," "Obvious Reasons for Evolution," "The Ultimate Destiny of the Earth," "Monopolies and Taxation," "The Legal Effects of Marriage," "The Religion of Asia," "Freedom of Speech," "The Evils of Our School System," "What Snall We Teach Our Boys?" "Patriotism vs. Dishonesty," "The Relativity of Knowledge," and "The Relation of Science to Morality." The intellectual give and take of this group was evidenced by the opposition it aroused,[29] and, after the first

[29] "This association, like all the others of a similar character," observed a local historian in 1880, "has failed to secure the good will and friendship of all professed promoters of progress. To some, it partakes too much of religion; to others, too largely of science; to some, not enough of either. Indeed, some whose cherished opinions and beliefs have suffered by coming into contact with the relentless argument of others, have irreverently and uncharitably charged infidelity as the prevailing sentiment. Occasions of its adjournment have been seized upon to write and publish its obituary by those whose wish has been father to the thought; but while numerous examples of the instability of human affairs are constantly passing our notice, the Literary and Scientific Association of [Middletown], now upon the threshold of the fourth year of its existence, never gave better promise of

breath-taking rush of the gas boom, it was resumed in the Ethical Society, including in its membership "any gentleman above the age of sixteen years, who enjoys a reputation for temperance, virtue, and is liberty-loving, truth-telling, and debt-paying," who received a unanimous vote of the members.

"The utmost freedom of speech will be allowed," continues the press announcement, "there are no fees or fines. . . . The exercises are to consist of essays, orations, recitations, and discussions. Any plan the purpose of which is to benefit our city morally or humanity collectively may be placed before the society. It is hoped that public sentiment may be aroused so that any and all palpable evils with which we are afflicted may be eradicated because the best element in society demands it. . . ."

The society started out with a membership of twenty-nine "of all denominations and political complexions" and shortly increased to fifty, including some women; attendance of members and guests frequently reached seventy-five or one hundred. It led a wandering existence, meeting at first in the "Blue Ribbon" Rooms and later in the Universalist Church, being requested to leave the latter place, because, it is reported, its Sunday afternoon meetings drew so many more people than the church services. Week after week in the nineties baker and nailmaker sat side by side with banker and doctor,[30] discussing such questions as the "Ethical Life of Man" ("The well-being of man and not the glory of God should be the subject of our efforts. Intellectual, moral, and physical culture and not piety is the prime condition of man's well-being. . . ."), "Physical Culture for Children in Our Schools," "Free Silver," or "The Meaning of Evolution." Nor was this the only discussion group; the Carpenters' Union conducted for a while a series of Sunday afternoon discussions, and lively discussion went on

continuing to edify all who will attend its meetings, and of wielding an influence for good in the community where it seems to have a permanent lodgment."

The important thing to note here, in the light of the current quietistic men's civic clubs, is that there could be a club of leading business and professional men in Middletown a generation ago that stirred up fierce indignation on philosophical and social questions. Cf. the discussion of the leveling effect of credit in Ch. VIII, and also Ch. XXVIII.

[30] A baker, describing the way he spent Sunday in the nineties, said, "In the morning a bunch of us bakers would get together with a keg and bicker over wages, flour, and what not. Then after dinner I'd go to the Ethical Society."

at the Workingmen's Library, where smoking and talking were allowed. A number of church and neighborhood groups met to argue the merits of favorite authors or disputed political questions.

Today lawyers, doctors, bankers, ministers, and now and then other professional groups, meet in their several associations to listen to papers on the details of their work, and, in the case of the Medical Association, at least, the papers by outside speakers frequently furnish stimulating additions to the professional knowledge of the members. Except for these specialized getting-a-living clubs, however, the discussion groups of the nineties are as far removed from the leisure-time pursuits of Middletown men today as are trolley rides, taffy pullings, and church socials.[31] The new civic clubs, which with golf, bridge, and motoring are the leisure interests toward which Middletown business men gravitate, are in no sense an outgrowth of the early discussion clubs; they are bred of a different stock and nourished at the breast of the local business life.[32] These largely non-overlapping groups, carefully selected for prowess in business, highly competitive, and constituting a hierarchy in the prestige their membership bestows, exemplify more than do churches or lodges the prepotent values of the dominant group of business men of the city.

And just as a small group of Middletown's business class exhibits with peculiar clearness trends operative in a much larger section of the population, so Rotary, oldest and most coveted of all the civic clubs, represents the aims of the others.[33] The

[31] In the twenty of the twenty-four adult study clubs in Middletown, exclusive of the professional associations, for whom figures were obtained, there are 710 members; of these, twelve (1.7 per cent.) are men, individuals scattered through three groups of women, two of them classes in "mind-power," and the third a district Parent-Teachers Association. In the high school clubs of the type built around art, literature, and other courses, from which data were obtained, 133 of a total of 385 members were boys.

[32] The present Chamber of Commerce is, however, a direct descendant of the earlier Citizens' Enterprise Association, both being concerned with promoting the business and industrial interests of the city.

[33] The following five characteristics of Rotary, from *The Rotarian* for September, 1924, are typical of the pronouncements of all of these clubs: (1) The unique basis of membership—singleness of classification; (2) the compulsory attendance rule; (3) the intensively developed friendships; (4) the activities for the betterment of the individual member and his business; (5) the requirement that members strive for the betterment of the craft corresponding to their classification in Rotary, particularly stressing higher standards of business practice.

eighty men who gather each Tuesday for lunch at Rotary come together under the rules of the national organization decreeing that a single outstanding man in each business and profession in the city may be a Rotarian; but since strict adherence to this rule would omit some of the business leaders, special "associate" and "honorary" memberships have been created so that one sees among Rotary members four lawyers, three bankers, and four millionaire manufacturers all engaged in the same plant. These chosen head men, meeting in the best hotel or at the Country Club, stand about chatting, observing the ritual of calling each other by first names, until the president shouts, "Let's go!" whereupon all crowd into the dining room. No "blessing" precedes this meal as in the other civic clubs, as the classification of ministers is unrepresented in Rotary.[34] Eating proceeds vigorously at the long tables for about half an hour. Ten minutes of lusty song follows—the latest Broadway hits and Rotary songs, chief among them:

> R-O-T-A-R-Y,
> *That* spells RO-TAR-EEE.
> R-O-T-A-R-Y,
> It's known o'er land and sea;
> From North to South,
> From East to West,
> He pro-fits most
> Who serves the best.
> R-O-T-A-R-Y,
> *That* spells RO-TAR·EEE.

When "every one is feeling good," the scrolls bearing the printed words of the songs are rolled up on the wall and the introduction of guests takes place. "I have with me as my guest Bill Smith, visiting Rotarian from Jacksonville," says a member. Bill stands up, and from all the tables rises a brisk volley of, " 'Lo, Bill!" "Hi, Bill!"

As at most meetings in Middletown, speeches form the *pièce de résistance* of the programs, these being of three kinds: (1) speeches from a Middletown member on his "classification"— "Being a motion picture exhibitor," "Making and selling high-tension insulators," merchandizing, advertising, the law; (2) speeches by the head of a local charity, the librarian, the direc-

[34] See Ch. XXII for the reasons for omitting a minister from Rotary.

tor of the vocational work in the schools, or by the head of organizations such as the state bankers' association; (3) speeches by "outside speakers" routed to Middletown through the International Rotary headquarters and speaking on "Sound Economics," "R.O.T.C. in the Colleges," "Tax Revision," "The United States and World Leadership," and similar topics. Nowhere is Middletown's predilection for a "real good speaker" or its ready acceptance of the views of a person who pleases it more apparent.

The civic work of these clubs with their slogans of "service" and "the under-privileged boy" [35] is likewise of three kinds: (1) certain annual affairs such as inviting the honor students of the high school to one of their luncheons, holding special club chapels in the high school, attending the county poor farm in a body at Christmas time and making speeches and distributing small gifts, conducting an annual Christmas party for one hundred or so needy boys, or an annual Easter egg hunt in a local park, at which hundreds of children search among the leaves for the lucky eggs drawing prizes of merchandise and money; each of the clubs does four or five of these civic things a year; (2) the sporadic good turn to meet some local need—giving a radio to the Orphans' Home, or agreeing to take turns week after week in bringing a crippled boy to high school in their automobiles; (3) considerably less common activities of the more ambitious sort, e.g., one club secured a summer camp for the local Y.W.C.A. The Dynamo Club of young business men, affiliated with the Chamber of Commerce and differing from the other civic clubs in that its membership is open to any member of the Chamber, secured a municipal golf course for the city and organized a local drive to utilize school yards as playgrounds.

The lack of coherence in the subjects of the speeches to which each of these clubs listens week after week and in the noncontroversial occasional charity which constitutes their civic work suggests that, as in the women's study clubs, the reason

[35] Four major subjects before the national Kiwanis Convention in 1924 were: "1. The emphasis and intensification of Service on behalf of under-privileged children. 2. The development of better relations between the farmer and the city man. 3. The aggressive development of a coöperative spirit towards the chambers of commerce and the coördination of activities. 4. The fostering of a fuller realization of the responsibilities of patriotic citizenship."

for their dominance lies in neither of these activities but in the instrumental and symbolic character of their organization. Not only are they a business asset, but by their use of first names, sending of flowers on birthdays, and similar devices, they tend to re-create in part an informal social intercourse becoming increasingly rare in this wary urban civilization. "It isn't Edward T. Smith, President of the So and So Corporation, you're addressing," said a speaker at Rotary, "but the human being, the eternal boy in Ed." "It makes you realize the other fellow hasn't got horns on and ain't out to get you," as one man put it. "You can't sit down at a table with a man and talk things over without getting to understand him better," said another. "There were a couple of fellows here I used to look at and think, 'What's *he* done?' and then I got to know them at Rotary and found they were doing a lot of fine things without waving their arms about them the way some folks do."

These genial, bantering masters of the local group find here some freedom from isolation and competition, even from responsibility, in the sense of solidarity which Rotary bestows. For some members the civic clubs have displaced lodges and churches as centers of loyalty and personal and class morale.

"Rotary and its big ideal of Service is my religion," said one veteran church and Sunday School worker of Middletown. "I have gotten more out of it than I ever got out of the church. I have gotten closer to men in Rotary than anywhere else, except sometimes in their homes."

All the clubs take pride in an accumulated sense of service. Said a speaker at a Middletown Rotary luncheon:

"The lowly Nazarene who walked by the Sea of Galilee was the first Rotarian, and the second great Rotarian was that other man who probably did more for mankind than any other man that ever lived, Abraham Lincoln. Could we spread the Rotary spirit to the coal mines, this ideal of service would end all strikes. Could it be spread to the Governments of Europe, France would not have entered the Ruhr, Germany would have paid. You remember what President Harding said at the Rotary convention in St. Louis: 'If I could plant a Rotary club in every city and hamlet in this country I would then rest assured that our ideals of freedom would be safe and civilization would progress.'"

Challenging him and his world at no point, often proving of actual cash value in his business, membership in these clubs may serve a Middletown business man as the symbolic repository of his ideals, assuring him that by virtue of carrying on his business and being a member of this club his daily life in the group is its own justification and has dignity and importance. This combination of utilitarianism and idealism, linked with social prestige and informal friendliness, is almost irresistible.

And yet the men's civic clubs are not without their local critics. Some feel that the civic club mountain groans weekly and brings forth—a slogan:

"When you come right down to it, what's the justification for the existence of this club?" said a loyal Rotarian in a burst of private candor. "Want to know what I think? We're just a bunch of Pharisees and hypocrites!"

"It's an awful job week after week getting up a three-ring circus to entertain these clubs," said an officer in another club, "and I sometimes wonder if they're worth the trouble."

"The whole town is over-organized," declared the wife of one Rotarian vehemently. "I don't think the men's clubs amount to much. They get together and some one talks about something for a few minutes and then they go off to business again and forget all about it. If all the men who meet in these separate civic clubs would get together, say once a month at the Chamber of Commerce, and discuss one or two things and act upon them, they might get something done!"

Certainly it is true that a wide gap exists between the activities of the civic clubs and the major maladjustments of which Middletown complains. In general, civic club members, like others, habitually regard these friction spots as inevitable accompaniments of life, and the city pursues its accustomed course with more or less creaking of the machinery in much the same manner as before the existence of the civic clubs. This situation presents few anomalies when it is realized that the clubs exist primarily as an adjunct to the business interests of their members and as a pleasant way of spending leisure; chiefly as a supplement to these interests and in regions where no enemies will be made or no ructions raised do the clubs become "civic."

And within Rotary itself a cleavage is beginning to appear. Some members complain that a certain group always sit together and play together and that this "cliquishness" will "spoil Rotary."

"Do you know what's behind all these civic clubs?" asked a member of another club. "Snobbishness. Each Rotarian goes home and spreads the Rotary talk about the Rotarian being the best man in his line in town, then all the wives tell all their neighbors, and then the wives begin forming their exclusive sets."

This tendency may later on eventuate in the splitting off of another, more exclusive group within Rotary. Just as forty years ago the lodges, by straining off a more exclusive group, cut into the churches as a center of social life and loyalty for Middletown men, so, as the lodges have become inclusive, civic clubs led by Rotary have cut into the lodges. The signs of fission beginning to appear in Rotary may point toward a repetition of the cycle in the future.

The great days of lodges as important leisure-time institutions in Middletown have vanished. At present, despite the heavy building programs of leading lodges, business men are "too busy" to find the time for lodge meetings that they did formerly; the man who goes weekly to Rotary will confess that he gets around to the Masons "only two or three times a year." Working men admit, "The lodge is a thing of the past to what it was eight or ten years ago. The movies and autos have killed it." "He belongs to a lodge but never goes," said more than one of the working class wives interviewed.

"Twenty years ago," said a business class lodge officer, "when we had 186 members, we used to get out 125 regularly for meetings. Even ten years ago with 300 members we got out 200. Then we had a president who wanted the lodge to grow, and we ran the membership up to 912. Now we have to have a supper before meetings to get any one, and at that we get out only about forty, or on initiation night sixty. The heart of the lodge movement went out of it when we all began letting anybody in, regardless of whether he lived the ritual. The lodges here are on the rocks."

The attendance problem among the working class lodges is also acute. The Eagles, e.g., give a dollar at each meeting to the fifth name drawn from the membership list if the man is present; if he is not present the dollar is added to the dollar to be offered the following week. On one recent occasion the lodge went for thirty-seven weeks without the fifth man being present; the next week the owner of the fifth name was present and received thirty-eight dollars, a five-dollar hat, a free shampoo, and a free cleaning and pressing of a suit of clothes.

In the main, business men join lodges today for business reasons—a gentile business man of any local standing can hardly afford to stay out of the Masons at least; and workers join chiefly for the sickness and death benefits, though even here the Workmen's Compensation system, group life insurance by employers, and the spread of the habit of independent insurance, are cutting into the lodges. The ritual is said on every hand to mean little today, apparently far less than even a generation ago. "No man or woman can follow the teachings inculcated here," said the press report of the founding of a new local lodge in 1890, "without being purer, nobler, more charitable, and more willing to speak kind and loving words to those whom misfortune has overtaken." And yet, "the other night," remarked a high degree lodgeman in 1924, "one of the Templars gave the rest of the order the devil for not making their practice and their professions square better. All of us know it is to laugh when a man is elevated to the Commandery or to some chair 'because of his diligence in performing the duties and learning the rites of Masonry,' when he knows and we know that he only learned enough to skin by, thanks to coaching, and really isn't interested in the rites themselves, but joined for business reasons."

In the race for large memberships to support ever larger competitive lodge buildings—the Masons lead the field with a new million-dollar "temple"—the old personal note has apparently dropped largely from membership. According to a man closely identified with local lodge life:

"It used to *mean* something when you belonged to a lodge—the lodge meant something and you meant something, and when you met a fellow member on the street or in his place of business one of the first things you'd think was that he was a fellow member. Now, lodges are so large you often don't even know a man's in 'em, and if you do you don't care."

The extent to which the lodges have lost this close brotherhood, in the eyes of some members, appears in the remark of a business man in speaking of the decline in fellowship in his church, "Why, you go there and it's as cold as any lodge."

As lodge interest has declined, interest in other clubs has tended to take its place among the business class, but not among the workers. The lodge and the declining labor union are almost the only clubs of the working men. The club life of the working class and business class groups interviewed is suggested by the following data on 123 and thirty-nine families respectively of the working class and of this group of the business class: [36]

	Business class wives	Working class wives	Business class husbands	Working class husbands
Total number of individuals answering	39	123	39	123
Belonging to 1 or more clubs, lodges, etc., of some kind	36	44	38	70
Belonging to 1 or more lodges	0	20	34	60
Belonging to 1 or more church clubs	23	17	7	0
Belonging to 1 or more labor unions	0	2	0	17
Belonging to other clubs	35	18	30	1

This disparity between club affiliations of the two groups becomes more apparent from the fact that seventy-seven of the seventy-eight total affiliations of this representative group of 123 working class husbands are either in trade unions or lodges, clubs of decreasing local importance, which many of them apparently seldom attend.[37] As pointed out in Chapter VIII, only eleven of the 100 working class families for whom income distribution was secured contributed anything to labor unions. Forty-eight paid lodge dues, amounts ranging from $3.00 to $43.00, averaging $15.20. In so far as Middletown may be "clubbed to death" this is largely a business class phenomenon. More working class wives belong to clubs than a generation

[36] See Table XIX. The Ku Klux Klan is here regarded as a lodge.

[37] Cf. the discussion in Ch. VIII of the much greater social significance of the Middletown labor unions in their heyday during the nineties.

Of interest in this connection are the percentages of the local social news in the two leading papers during the first week of March, 1890 and 1923, that were devoted to lodges, to clubs (including other organized groups such as labor unions), and to miscellaneous social affairs such as parties:

		Lodges	Clubs, etc.	Misc.	Total
1890	Paper A	32%	51%	17%	100%
	Paper B	29	67	4	100
1923	Paper C	20	77	3	100
	Paper D	12	76	12	100

ago, in all likelihood, but the pressure upon the working class as a whole appears to be not from the multiplication of their loyalties but from their social isolation.[38]

With greater organization has come increasing standardization of leisure-time pursuits; men and women dance, play cards, and motor as the crowd does; business men play golf with their business associates; some men in both groups tinker with their cars and tune in their radios; a decreasing number of men are interested in gardening, a few turn to books, one or two surreptitiously write a little; a few women "keep up music" and two or three paint or write; among the wealthy are a few who collect paintings and prints, two who collect rare books, and one who collects rugs. Interest in drama, as in music, art, and poetry, centers mainly in the high school. In 1877 there was even a Mechanics' Dramatic Club, "a local group of amateurs"; but today an occasional lodge revue is put on with much labor to raise needed funds, and now and then a sorority gives a revue, but the giving of plays is confined to the high school and to a few women in the Dramatic Department of the Woman's Club. For those who look wistfully beyond the horizon a hobby tends to be like an heretical opinion, something to be kept concealed from the eyes of the world. One family, unusually rich in personal resources, has recently built a home a little way out of town, set back from the road almost hidden in trees. So incomprehensible is such a departure that rumors are afloat as to what secret motive can have prompted such unprecedented action. Hobbies appear to be somewhat more prevalent among high school pupils than among their elders. Of 275 boys and 341 girls in the last three years of the high school answering the question, "In what thing that you are doing at home this fall are you most interested?" one boy was publishing a small magazine, one studying aviation, one practicing mental telepathy, fourteen doing scientific experiments, one girl was collecting books, one studying photography, one collecting linen handkerchiefs, two doing botanical experiments, and three girls and one boy writing.[39] But most of their answers show that standardized pursuits are the rule; with little in their environment to stimulate originality and competitive social life to

[38] Cf. Ch. VIII for other factors involved.
[39] Cf. Ch. XVII.

discourage it, being "different" is rare even among the young.

Men have adopted more rapidly than their wives the activities growing out of new leisure-time inventions: it is largely they who drive and tinker about the car, who build the radio set and "get San Francisco," who play golf, who first use such new play devices as gymnasium and swimming pool. Meanwhile, such new leisure as Middletown women have acquired tends to go largely into doing more of the same kinds of things as before. The answers of the two groups of women interviewed to the question, "What use would you make of an extra hour in your day?" bear witness to the narrowness of the range of leisure-time choices which present themselves to Middletown women. Both groups spoke of wanting time for reading more often than anything else, but as noted in Chapter XVII, this desire to read was both more marked and more specific among the business class. Only one of the thirty-two business class wives answering would use the time to rest, while approximately one in seven of the ninety-six working class wives gave such answers as "Rest," "Go to bed," "Lie down and rest, something I hardly ever do." More than a third of the working class group answered blankly, "I don't know." One in sixteen of each group answered, "Fancy work or crocheting," but, in the case of both of the two business class women so answering, in a tone of apology. In both groups a number mentioned getting out more with people, but the answers indicate different kinds of pressure. Among the working class it was frequently: "I'd go anywhere to get away from the house. I went to the store last night. I've been out of the house only twice in the three months since we moved here, both times to the store." "I have two daughters. One lives only a block away and I've been over to see her only twice in the last two months. The other lives ten miles out on the interurban and I never see her. If I had an hour I'd use it to see them." The pressure upon the group of business class women is apparently much less at this point; some of them say vaguely, "I'd like an hour in the afternoon for bridge or the movies," or "I'd like more time for reading, calling, visiting, and social life." Not one of these business class women answering referred to church work or Bible reading as a possible way of spending an extra hour, although to seven of the working class women answering, such work was their chief desire; two business class women, however, mentioned civic work among other things. No worker's wife spoke of more time with her children, but four of the other group felt this as their chief desire. No woman of either group spoke of wanting to spend more time with her husband. One woman perhaps summed up the situation of the business class mother whose children are not below school age: "I am busy most of the time, but I can always get out when I want to." Another expressed herself as actually having time to spare: "I am not pressed for time. I really have time for more civic activity than the community wants me to do." For a large proportion of the working class wives, on the other hand, each day is a race with time to compass the essentials.[40]

Much may be learned regarding a culture by scrutiny of the things people do when they do not have to engage in prescribed activities, as these leisure pursuits are frequently either extensions of customary occupations to which they contribute or contrasts to the more habitual pursuits. In Middletown both aspects of leisure appear. The reading Middletown people most enjoy, the spectacles of romance and adventure they witness on the screen, the ever-speedier and more extended auto trips, many —perhaps even today the majority—of the women's club papers, would seem to be valued in large part because of their contrast to the humdrum routine of everyday life. This seems to be particularly true of the working class. On the other hand, the whole system of business men's clubs is apparently valued in part for its instrumental character, its usefulness to the main business of getting a living, and even such an apparently spontaneous activity as golf is utilized increasingly as a business asset; this use of leisure-time groups as an extension of the main activities of life is appearing to a minor but seemingly increasing extent in the women's study clubs.

Finally, the greater organization of leisure is not alto-

[40] To summarize: Of the ninety-six working class wives answering the question, twenty-seven answered, "I don't know." Sixteen would use it for housework or sewing, fourteen for rest, eighteen for reading, seven for getting away from home and seeing people, seven for church work, two to write letters, one to earn money, and four said that they are not pressed for time and might therefore use the time in various ways.

Of the thirty-two out of this group of forty business class wives answering this question, fourteen want time for reading, three for housework, two for fancy work, four for their children, three for social activities, one for rest, and five stated that they are not pressed for time and might use it in various ways.

gether a substitute for the informal contacts of a generation ago; opportunities to touch elbows with people are multiplied in the mobile and organized group life of today, but these contacts appear to be more casual and to leave the individual somewhat more isolated from the close friends of earlier days. In view of the tightening of social and economic lines in the growing city, it is not surprising that the type of leisure-time organization which dominates today tends in the main to erect barriers to keep others out.

V. ENGAGING IN RELIGIOUS PRACTICES

Chapter XX

DOMINANT RELIGIOUS BELIEFS

In addition to their various pursuits concerned with the immediacies of life about them, Middletown people engage in another activity, religious observance, whereby they seek to understand and to cope with the too-bigness of life.

The more obvious manifestations of religion are the setting aside of one day in seven and of forty-odd buildings throughout the city for religious ceremonies. On the seventh day, and occasionally during the week, these buildings are opened and services are conducted in each by the priest or minister who gets his living as the leader of the group constituting each "church." Membership in one of the religious groups is generally taken for granted, particularly among the business class, and a newcomer is commonly greeted with the question, "What church do you go to?"

Middletown exhibits a wide variety of religious beliefs, but almost without exception, the beliefs of all groups center in the collected writings handed down in the *Bible*. With few exceptions, notably the Jewish temple, churches in Middletown represent some form of Christianity. When one asked various individuals in Middletown, "What does one believe in if one is a Christian?" they inclined at first to think the questioner was joking, a condition reflecting the general tendency to accept "being a Christian" as synonymous with being "civilized" or "an honest man" or "a reputable citizen." In general the answers were on this order, although, even within the Christian tradition, there are many shades of belief: "Why, God made Heaven and earth and sent Jesus Christ, His son, to save the world from sin. If you believe in Christ you will be saved." Perhaps the outstanding divergence in these statements appeared in the tendency of some either to add to or to substitute for "believe in Christ" the words, "do His will." A somewhat more philosophical variant is the statement of a Middletown minister that "the world is good, God is good, and His spirit wherein men are to live is love's spirit." Around this core has grown up an elaborate system of beliefs, prohibitions, and group-sanctioned conduct.

Prominent among these beliefs is a confidence in the all-sufficiency for all mankind of this religion. Of 241 boys and 315 girls in the two upper years of the high school marking the statement, "Christianity is the one true religion and all peoples should be converted to it," 83 per cent. of the boys and 92 per cent. of the girls marked it "true" and 9 and 3 per cent. respectively "false"; the remainder were "uncertain" or did not answer.[1] According to the minister of the largest Protestant church in the city, "Christianity can never be supplanted by any other system, for it gives the only perfect statement of right." Said the minister of the second largest Protestant church, addressing a Sunday audience at the Middletown Chautauqua:

"Christ's teachings are plain. . . . Twenty centuries have added to them *nothing* out of human experience, nor can any other centuries add to them, for they are the teachings of God walking in human flesh. . . . All human philosophy, reasoning, is alike; it is no more than the newspaper—just for the hour; but the Bible, read as a child would read it, is for the ages. Prophecy, knowledge, philosophy, creed, catechism, . . . all fail; the Word abides."

The same confidence is reflected in a paper on "The Women in Japan" read by a member before one of the federated clubs:

"Nowhere, except in Christianity, can the awakened mind of the women of Japan find the guidance which it needs in this day of new horizons."

The slight touch of disreputability commonly clinging to the doubter is reflected in the remark of the minister of a leading church in announcing to his morning congregation:

"Dr. W—— will address us at the Chautauqua tent this evening. I will be perfectly frank with you and tell you that he represents the liberal school of Christian thought; but I do not think he will give us anything of that sort this evening, as he is a very gracious gentleman and thinker."

[1] This, like the other verbal reactions to this "true-false" questionnaire, is presented as suggesting a tendency in the local life rather than as offering absolute proof.

Questioning the truth or adequacy of their religious beliefs is not within the conversational area of most people in the city.[2] The openly free-thinking group, represented in part in the Ethical Society of the nineties,[3] has disappeared and in its stead is an outwardly conforming indifference among a certain minority of the business class. "I've joined the church," said one prominent man, "because who am I to buck an institution as big as the church, and anyway it seems necessary to conform, but I guess my pastor knows all right how little my heart is in it." [4] For the most part, however, Middletown does not question the adequacy of the "great fundamentals" of its religion.

A second commonly held belief is that in the sacredness of the Bible. According to the minister of the largest church in the city:

> "The Bible is the only great book that even attempts a sane, reasonable interpretation of the origin, orderly conduct, and destiny of the world and of man. . . . The study of the Bible is the surest path to great ability which can be found. If you would develop character and be mighty for good among men, first be mighty in communion with God and commit to memory His thoughts."

Or, according to a prominent Bible class teacher, addressing a class of 200 adults, "A man who would change one word of the St. James [sic] version must be a consummate fool." A leading club woman assured the Literature and Art Depart-

---

[2] The emotional basis of this prohibition upon religious doubt is reflected in the statement by a teacher of teachers in the local normal college in an address before the Federated Club of Clubs—"Why can we trust what Browning says about God? Because he never doubted."

[3] Cf. Ch. XIX for an account of the Ethical Society.

[4] Religious discussions in the course of leisure-time intercourse are rare and largely taboo among the business class. Such reticences are not so rigorously observed among the workers, and one occasionally hears frank questioning of these "fundamentals" among them. Such a conversation as this, e.g., which took place in a group of men and women talking informally before a meeting of a labor group, would scarcely be possible among the more socially wary business class:

A painter. "I was paintin' for a woman the other day and she says to me, 'Do you believe in Hell?' and I didn't know what was comin' but I says, 'Sure I do—Hell on earth.' And she says, 'That's the way with me, too, and I believe in a Supreme Being, too—but you can't tell me He came down on earth and walked round as "three persons." Not but that I think Jesus was a good man. I *do.* But I don't believe in this incarnation business.'"

Woman member. "That's what I say, too. Of course, we all believe in a Supreme Being."

ment of the Woman's Club, "The Bible is . . . above all else
the inspired book of our religion and on its teachings is based
the hope of the whole world." Said a local lawyer to one of the
largest Sunday Schools in town, "I'd rather have a person graduate from Sunday School with a knowledge of the Bible than
from Harvard or Yale without it." Fifty-eight per cent. of the
241 boys and 68 per cent. of the 315 girls answering the junior-
senior questionnaire marked "true" the statement, "The Bible
is a sufficient guide to all the problems of modern life"; 26 and
20 per cent. respectively marked it "false," and the remainder
were "uncertain" or did not answer.[5] Phrases from the Bible
are cited on all manner of occasions, in or out of their context,
as carrying special truth; so a popular minister in an appeal for
funds for a new church building addressed the following paragraph to his members:

"God is spoken of as the creator of the heavens and the earth
(Gen. 1:1), as the possessor of the heavens and the earth (Deut.
10:14), as the one whose is the earth and the fulness thereof
(Ps. 24:1), as the one who owns every beast of the field and the
cattle upon a thousand hills (Ps. 50:10), as the one to whom
belongs the silver and the gold (Hag. 2:8). God has never quit-
claimed or in any way relinquished His right, title, or interest in
the earth and the fulness thereof." [6]

[5] Cf. Ch. XIV for the answers by the same group to the statement, "The
theory of evolution offers a more accurate account of the origin and history
of mankind than that offered by a literal interpretation of the first chapters of the Bible."
There is a general feeling in Middletown that the book is "perfect" and
free from inconsistencies "if our finite minds could only understand." One
of the most intelligent women in the city was chagrined because she
could not explain "that awful lesson about the Gadarene swine" to her
class of some twenty-five girls of high school age: "I didn't know what
to say about the thing, so I spent the thirty-five minutes making the girls
talk. I got about three-fourths of the way around before the hour ended
and told them I might tell them what I thought Sunday after next. Meanwhile, I am looking it up in all the commentaries I can get my hands on."
Over against this mood should be noted, however, the remark of the
minister of the one doctrinally liberal church in Middletown to his men's
Bible Class touching one of the "miracles": "You could drop that miracle
out of the Bible and it wouldn't worry me. I don't get any sense out of it."
Another minister to one of the most thriving working class groups is
wont to say whenever baffled by a passage in the Bible, "Well, that's one
of the things we'll have to wait till we get to Heaven to find out." Such a
statement elicits a general nodding of heads in assent from his flock.
[6] This emphasis upon ability to repeat verses from the Bible is a marked
feature of the religious life in this culture. A summary in the daily press
of the work of the "daily vacation school"—conducted by the largest Prot-

A third outstanding belief held with varying shades of emphasis by most of Middletown is that in God as completely revealed in Jesus Christ, "His divine son." Seventy-six per cent. of the 241 high school boys and 81 per cent. of the 315 girls marked "true" the statement, "Jesus Christ was different from every other man who ever lived in being entirely perfect"; 17 and 15 per cent. respectively marked it "false," and the remainder were "uncertain" or did not answer. This same habit of thought is reflected in the statement by the minister of a leading church that "Jesus lived the only perfectly sane human life the world has known. . . . Wherefore he is the world's only competent leader; other leaders are worthy in the measure and the degree they help their followers to get close to Christ." It is noteworthy that the minister of the one religious group in Middletown who, though Christians, are generally regarded as questioning the "divinity" of Christ, is not invited with the other ministers to address the high school pupils at their chapels. The minister of the most fashionable church disposed of the Christian Scientists as "not Christians because they do not believe in a personal God, the deity of Christ, and the reality of sin." This emphasis upon the unique importance of Jesus and the Bible is reflected in the statement in a sermon by the minister in Middletown most interested in progressive religious education:

"The Bible is the greatest book in the world; we might lose all the other books in the world and not be much worse off. Jesus is the greatest man that ever lived; we might lose all the other men that ever lived and not miss them much, but Jesus is indispensable."

A fourth generally held belief is that in a life after death in "Heaven" and "Hell." An extreme form of this belief was tested in the true-false questionnaire to high school juniors and seniors: 48 per cent. of the 241 boys and 57 per cent. of the 315 girls marked "true" the statement, "The purpose of religion is to prepare people for the hereafter"; 36 per cent. and 35 per cent. respectively answered "false"; 12 and 7 per cent.

estant church to "promote the idea that giving happiness is one of life's fundamentals, and to inculcate early a knowledge and love of the Bible by methods which cannot fail to intrigue the youthful mind"—says with evident satisfaction, "and it was amazing to see the children, called upon at random, stand and name Bible books or recite Bible passages or chapters."

respectively were "uncertain"; while 4 per cent. of the boys and 1 per cent. of the girls did not answer.[7] The literalness of this dominant belief in a "future life" is typified by the following statement from a sermon by one of the most popular preachers in Middletown:

"Put Jesus Christ first, not for your three-score years here but for eternity. . . . He is the Messiah and you dare not give Him second place in your life. . . . My God, brethren, when you look in His face, a whole eternity of looking in His face, what will you say, what will you do, if you haven't put Him first here!"

Another expression of this belief appears in the following from the "calendar" of a leading church:

"God does not want to deprive you of your hard-earned money. He wants you to take it to Heaven with you and enjoy it forever, but the only way you can do this is to convert your cash into character, into Christian manhood and womanhood. Then you can get into heaven and look upon it and enjoy it through all eternity. God advises you to put your money where it will be safe, and where you will receive some benefit from it in the next life. God asks that you entrust your money to Him for your own benefit. 'Lay up for yourselves treasures in Heaven.' Rich in good works, generous, laying for yourself a good foundation against the time to come."

To many citizens it is unthinkable that one should seriously doubt a hereafter, for which, many believe, "we are simply in training here on earth." "Do you mean to say," retorted a prominent lawyer incredulously to one who questioned this commonly held belief, "that you think it doesn't make any difference in the hereafter how a man lived here?" In some of the working class churches this belief is held in extreme form. In an unpainted little building on the South Side where a Sabbath morning service was in progress, a man of forty-five was stamping about the platform, shaking his fist and shouting:

"I'm through with this dark, cold old world and I thank my blessed Jesus that there are a few that are saved and that we can

[7] Along with the sex difference here suggested, there appears to be a difference between children of the working class and of the business class: 155 (three-fifths) of 254 workers' children and twenty-eight (less than a third) of seventy-four children of the more prosperous business class parents marked this statement "true."

know we are saved already!" He sat down and the congregation one by one trooped up to shake his hand.

Another leader arose and denounced "this vile world," saying, "I know not the day or the hour, but my back is turned on this life of corruptibility and I am ready to go home where Jesus is."

Belief in Hell is apparently dying out somewhat, however, even as compared with so short a time ago as 1890, while belief in Heaven seems also to be diminishing in intensity, especially among the business class. Eight of the eighty-three Middletown women from the combined working and business groups answering the question appearing below concerning thought of Heaven, volunteered the unsolicited information, "Most people think of Heaven less than they used to." And yet the belief appears to remain with most of the population. Middletown, one gathers, thinks less upon the hereafter than it did in 1890, and at the same time actively questions it little. On the one hand, people do not go so often to the cemetery to visit the graves of their dead. On the other, the subject is not urgent enough now to evoke the protest it did from the free-thinking objectors of 1890. No one today stands before his peers to assert as did a leading local medical man before the Ethical Society in 1890, "The intelligent moral man knows that no human being has any knowledge of a future state of personal existence." No man today has his relatives announce over his open grave, as did more than one member of the Ethical Society, that the deceased believed death to be the end. Today, so much is "the hereafter" taken for granted that a business class woman, though she rarely goes to church herself and doesn't "think of that sort of thing much," remarked in an awed tone at the death of one of the last survivors of the Ethical Society that she was "shocked to learn that a fine character with such a beautiful home life didn't believe in God or immortality."

With these four beliefs in the supremacy of Christianity, of the Bible, of God and Jesus Christ, and in existence after death, all touching what is called "doctrine," combined with at least a nominal belief in practical ethics, "loving one's neighbor as oneself," goes commonly a belief in the institution through which these religious habits of thought are taught. This reverence for the church appears in such statements as:

"The greatest sin a man can commit is not to become a member of a church." (*From a sermon in a church including both workers and business class folk among its members.*)

"And above all you must not criticize the church, for that is the cause of all dissension in it." (*Instructions to children being baptized in the largest Protestant church in Middletown.*)

"The church is an absolute necessity. For us to attend church services is part of our Christian duty." (*Sermon in a business class church.*)

"There cannot possibly be any Christians outside the church, because it was for the church that Christ died." (*Sermon in a working class church.*)

"You cannot worship God unless you, yourself, in His House, bow your head and in person glorify Him who holds you in the hollow of His hand." [8] (*Statement by the minister of a working class church prior to "go-to-church Sunday."*)

"I do not need to lecture you people who are here, but the people who call themselves Christian who are today motoring and playing golf, who will not see the inside of a church today, they are traitors to the Kingdom of God." (*Sermon in a leading business class church.*)

It appears that there is a strong disposition to identify the church with religion and church-going with being religious: "Even if going to church doesn't give us anything else, it at least gives us the habit of going to church," said a rising young business man speaking at a Sunday evening service in a leading church. A working class girl announced proudly, "I haven't missed Sunday School one single Sunday in seven years." [9] It

[8] This emphasis upon the church building as "God's house" is responsible for a tendency to narrow still further the identification of religion and the church to an identification of religion and the physical church building. Thus, one of the women in a growing church that has had to expand its Sunday School to the residence next door complained that "to me Sunday School just isn't Sunday School when it's held in a house and not in the church."

[9] The emphasis upon regularity of church attendance in Middletown's religious life is the more understandable when we realize that, according to a study of 1,974 Sunday School teachers in Middletown and other parts of the state, who have a considerable share in shaping the religious habits of the oncoming generation, 96 per cent. of these teachers attend church services regularly, and "the typical Sunday School teacher regularly supports two church activities in addition to the church school." In other words, it is an ultra-regular group that carries on church work and helps pass religious habits on from generation to generation. Walter S. Athearn and

is possibly suggestive of the importance placed upon the church and its support that 369 high school boys and 423 girls rated "being an active church member" third among a list of ten qualities "most desirable in a father" and third and fourth respectively in a somewhat similar list of qualities most desirable in a mother. It is possibly significant of a varying rate of secularization that whereas 144 (nearly a third) of this group of 451 working class children selected this quality as one of the two "most desirable" traits in a father, only fourteen (less than a fifth) of a group of eighty-three children of leading business class people so marked it; similarly, 132 (more than a fourth) of this group of working class children and only nine (one in nine) of this business class group [10] gave this degree of prominence to the same quality in a mother. Both business and working class mothers, however, appear to be laying less emphasis in bringing up their children upon "loyalty to the church" than did their mothers in training them.[11]

By way of further insight into these dominant religious beliefs may be noted the answers of the sample housewives interviewed to the following three questions:

(a) What are the thoughts and plans that give you courage to go on when thoroughly discouraged?

(b) How often have you thought of Heaven during the past month in this connection?

(c) What difference would it make in your daily life if you

associates, *The Religious Education of Protestants in an American Commonwealth* (New York; Doran, 1923), p. 380.

[10] Among the girls eighty-three (a third) of the 242 working class girls and two of the forty business class girls included this among the two qualities most desirable in a father; seventy of the working class girls and none of the business class included it among the two qualities most desirable in a mother. This was the only one of the ten traits which none of the business class girls included among the two most desirable in a mother. See Table XV.

[11] Cf. Ch. XI on the training of children and Table XIV. The difference between the present generation and their mothers is probably considerably greater than would appear from the figures given in this table, as, in the case of loyalty to the church more than in that of any other item, women frequently said, "I suppose I *ought* to mark that 'A,'" and often marked it so, regardless apparently of whether they themselves went regularly to church or gave the church much thought. The proportionately greater number of workingmen's wives who emphasized it, however, undoubtedly reflects a real difference.

became convinced that there is no loving God caring for you? [12]

Of the seventy-three working class women who answered question (a), the largest single group—twenty—said that it is their religious faith or devotions which give them courage. Representative of this group are the following:

"Just thinkin' of the Heavenly home that's waitin' me."

"I used to cry when I was discouraged, but that didn't help any. Now I just git down on my knees and pray and that gives me strength."

"When I'm discouraged I just read the Bible and think of the coming of the Kingdom."

"I pray and think of God. I've got lots of faith in prayer. My baby was awfully sick and almost died last year; a lady came and prayed and folks in church went up and knelt round the altar, and the baby got well and is better than ever. The doctor says it was prayer and nothing else did it."

"I pray and try to think things'll be better soon."

Other variants on this same theme were:

"I get to thinkin' of God's plans and goodness and the need of His other children for me."

"I just keep tellin' myself it was meant to be that way anyway. The Good Man up yonder knows what's best."

"Even though things are so unequal in this world and people that work hardest have the least, I know God will care for me somehow."

The next largest group—seventeen—turn to thoughts of their family and home:

"Usually I can get over it by thinking and planning about the children."

[12] Answers were secured to these three questions from seventy-three, sixty-eight, and fifty-five working class wives respectively, and from five, fifteen, and nineteen business class wives. While mere verbal responses in this region of behavior are notoriously unreliable, the interviewers felt that these answers in the main reflected the spontaneous reactions of the women in informal conversation. The numbers answering these questions are small, but they are believed by the interviewers to be representative. Cf. Appendix on Method, regarding the method of gathering these answers.

"Nothing cheers me up except sometimes the children. Sometimes the blues last a week."

"I just think, 'When evening comes we'll all be together.' "

Four of the other women answered that they "almost never get discouraged." Four more "just look around and see there's lots of others worse off," while a fifth varies this by "just singing and forgetting my blues." Four others "just go to work and work off" their depression. Another said, "I just sit down and look facts in the face and try to see what can be done about them." Two sleep off the blues, and four mentioned "going somewhere and getting out among people." Eleven said in substance, "Nothing helps me. I just try to grin and bear it," while two more added that although "nothing helps," they "sometimes think of God and Heaven," and an additional three "just go and cry it out." [18]

Of the sixty-eight working class women answering question (b), "How often have you thought of Heaven during the past month in this connection?" the largest single group—thirty-one —answered "often" or "every day":

"Land sakes! I don't see how people live at all who don't cheer themselves up by thinkin' of God and Heaven."

"My first husband died, leaving me with five children. I'm trying to bring them up to be good Christians, for that's the only way they will ever get to see their father. I just *know* he's up in Heaven now waiting for us!"

"Often. Seems like it's the only place we can ever rest."

"The trouble is most people just live for today and don't think enough about Heaven. I *know* there's a Heaven and I know there's lots of these people expectin' to be there that'll be fooled."

"I often wonder about the future. I study and study over the passages in the Bible that tell about the judgment day."

Sixteen said in substance, "I ought to think of it but don't much, though I reckon I ought to go there as I've always tried to live right." Twenty-one said they "almost never" or "never" give Heaven a thought:

[18] The answers of the business class women responding to this first question were so vague and scattered that they are not given here.

"Just today and tomorrow are the things I think of."

"There's too many other things to think about today. I hardly ever think of Heaven like I used to—only when a relative or friend dies."

"Almost never. I used to believe in Hell. Now I don't know what to think, but, anyway, I don't spend any time worrying about it."

"I don't never think of it. What I believe is you get what's coming to you in this life."

Among the fifteen business class wives answering this question, all but two said they never or almost never think of Heaven, and of these two, the answer of one was too vague to be classifiable, while the other said, "Never, I guess, but I have an uneasy feeling I ought to." The following are characteristic of the answers of the thirteen:

"I try to think of God when I'm discouraged, but prayer doesn't mean a great deal to me any more. I just never think of Heaven and I don't think people think of Heaven as much as they used to. More practical things in religion are emphasized now. We don't know anything about it anyway, so why should we think about it?"

"The modern trend in religion is away from this emphasis. Ideas of conversion and salvation were too much emphasized in my childhood and I've reacted against them."

Finally, with regard to question (c), "What difference would it make in your daily life if you became convinced that there is no loving God caring for you?" seven of the fifty-five working class wives answering rejected the question outright as unthinkable:

"I couldn't ever become so bitter over things as to doubt God's love."

"It's a sin not to believe in God and His care, and any one who doesn't is just stubborn."

"Nobody could make me believe there isn't a God."

Thirty-two more were so emphatic as to say that life would be intolerable or utterly changed:

"I simply can't understand how people live when they lose faith in God."

"I just couldn't go on. I know a man whose family were Methodists, but he didn't believe in God and he went blind; and he sat all day and said, 'My God! My God!' and when he'd done it one day for half a day suddenly he could see again. You bet I believe in God!"

"I couldn't keep on at all if I didn't believe God cares for me! It's so wrong to lose faith; it isn't God's fault that there's trouble in the world—it's men's sin."

"When I was left a widow with a baby on my hands I thought there couldn't be a kind God, but I've made myself come to see that we must believe what the Bible tells us."

"If there wasn't a God there'd be no point in life. You wouldn't care how you lived or what you done. Why, there wouldn't be no meaning to anything!"

"Life would just go black."

"It would take all the courage out of me."

"How on earth would we get along without the thought of God? I could never have stood the last election without the thought of God helping the right side."

"Life wouldn't hold out no hope for me or all my children. I just trust Him and *know* He won't disappoint me."

"It would make a lot of difference—even though I never go to church."

Nine of the fifty-five said, "It wouldn't make much difference in what I do every day," and only six said it would make no difference.

"I don't know whether there's a God or not. People ought to be good for the sake of being good and not for rewards."

"I haven't nothing against the church or God—I just don't bother about it all."

"There are things we can't understand and I don't try to."

One final answer voices a troubled doubt which there was reason to believe is secretly entertained by many in Middletown:

"I don't think He's very loving! It isn't so bad that people have to die, but it's the way they die. As long as God *can* do something about it, I don't see why He *doesn't.*"

Yet, whereas thirty-nine of the working class wives were emphatic that it would affect their lives seriously to lose belief in a loving God, and only sixteen said that it would make little or no difference, the nineteen business class women answering appeared to be less sure in their faith; only eight felt the belief very important, while eleven said its loss would affect them little or none at all. Here, too, some rejected the idea of even questioning this belief:

"I never thought of such a thing. The fundamentals don't change."

"My ancestors were all from New England. We have never questioned such things."

"You have to believe the fundamentals! The church means a lot to me, but of course it can't help me with my children."

"I couldn't imagine such a thing. I believe just as my mother did."

"It would make a good deal of difference, I think. I haven't a very definite personal sense of God, but guess my whole outlook on life is pretty much built around the idea of God."

For others, however:

"My belief in a personal God and in prayer are pretty vague, but I have a feeling that I may be losing something important, and I don't want this to happen for the sake of my little girl."

"The loss of the belief would make very little difference to me, I'm afraid. I have frequently felt that the feeling of connection with a personal God was too little with me—but then I think that Christian theology emphasized too much God's care for the individual in a material sense."

"The sense of a loving God is not as vivid to me as it used to be, nor is prayer, but I somehow feel that life loses meaning without God and that a sense of God is essential for character building. So I'm trying to give it to my children and to recover it myself. I am teaching the children to pray."

"I think, of course, that we must believe in some kind of loving Power or Father above us. But what really makes the difference is if when we go to bed at night and check over our lives we find we have paid our debts and been decent to other people and provided a good home and good nourishing food for our children—that is what counts. My husband never prays in public or makes civic speeches like a lot of men. He isn't that kind. But he pays all his debts and is fine and honest. He never tries to persuade any one to take more than they can afford in his business."

"I haven't any belief in a personal God. It's what we do ourselves that counts."

"I don't think of that sort of thing much and haven't a very vivid faith in God. Really I just don't know what I think about all these things. I pray every night to be a good woman, a good wife, and a good mother—and if you're all those, it's all you can do."

"I try to think of God—but prayer doesn't mean a great deal to me any more."

Although these groups are small, the points which emerge from the preceding answers confirm the observations based upon the multitude of staff contacts throughout the year and a half of the field work. First, members of the working class show a disposition to believe their religion more ardently and to accumulate more emotionally charged values around their beliefs. Religion appears to operate more prominently as an active agency of support and encouragement among this section of the city. A second point is the shift in the status of certain religious beliefs during the past generation—notably the decline, particularly among the business class, in the emphasis upon Heaven, and still more upon Hell.[14] A third point is the persistence of a vague belief in God even among many who are lax or skeptical in other beliefs. And finally the range of belief is impressive. Just as this city performs identical services in

[14] In this connection should be noted the seemingly greater self-consciousness of religious belief today, particularly among the business class. No graduating class in the Middletown high school today would print in its annual a "Class Ode" ending as does the following from one of the class books of the early nineties:

> "The flower of life now bursts in bloom,
>   Its fragrance perfuming the air;
> But its delicate freshness soon will be gone,
>   Its priceless beauty so rare
> Will be scorched by the blaze of passion and sin,
>   Will fade in the dearth of despair;
> Unless it be fed by beauties within
>   And drink of the fountain of prayer."

its homes by processes ranging all the way from the use of primitive hand implements like the broom and wash-board to the use of electrical labor-saving devices, so in this region of its religion one observes many shades of belief addressed to this common purpose of coping with the too-bigness of life. Even the dominant Christian doctrine varies from such unquestioning affirmations as that of the working class wife who said at a Mothers' Council discussion of religious education, "We don't need religious education in our church. In our church a little child only so high can tell right off the bat who God is and what He does and all about Him," to the troubled questionings of some of the business group who, retaining few settled convictions themselves, are unsure what to teach their children. One such business class mother said uneasily:

"People are questioning and wondering now about religious matters. There are just a lot of things you have to try not to let yourself think about."

Another thoughtful mother, one of the most enterprising young matrons of the city, said, "We don't go to church much, but we are sending our son to Sunday School and in addition giving him a book to read similar to the first part of Wells' *Outline of History,* written, however, for children; it explains scientifically the geological and anthropological beginnings of the earth and man. I am hoping that from this he will draw his own conclusions about religion. I have taught him nothing about religion but simply try to answer all the questions he asks—but I have a terrific time doing *that!* In order to instruct children in religion you have to have a pretty definite belief. It's hard to teach them without lying to yourself."

These questioners of established beliefs have, in the main, to work out their adjustments in isolation in Middletown. One conscientious mother who marked "loyalty to the church" zero as a quality to be emphasized in training the young, remarked, "I can't see that loyalty to the church accomplishes a thing— but I'd hate to have any one in town know it!" Even those best equipped to effect a synthesis of the new and the old often flounder along alone; thus a science teacher in the high school, a college graduate with some post-graduate work, walking home from a sermon on evolution and religion, said in a bewildered tone:

"I wish he had said something about Jonah. A whale's throat is *so* small it simply *couldn't* swallow a man. Of course it might have kept Jonah in its mouth, but I don't see how—all *that* time! And then he never said himself what he thought about evolution. He just said that you *could* believe in it and still be a Christian. I don't see that that helps much."

Questioning of the dominant Christian beliefs in public appears to have declined since the nineties, but one infers that doubts and uneasiness among individuals may be greater than a generation ago.

Chapter XXI

WHERE AND WHEN RELIGIOUS RITES ARE CARRIED ON

Middletown carries on its religious observances founded upon these major beliefs chiefly in the special houses of worship noted at the outset. There were in Middletown in 1924 forty-two of these church buildings associated with as many different religious groups, which ranged in size from 2,000 persons to a couple of dozen.[1]

This division into forty-two religious groups, almost all of them representing some branch of Christian faith but each centering its worship in a separate building, arises first from traditional or "denominational" differences in doctrinal beliefs or forms of worship diffused from without, and secondly from convenience of location within the city, with various economic and social considerations operative here as elsewhere in Middletown's life. Thus there are twenty-eight different denominations represented in Middletown, divided into five Methodist groups, five Christian, four United Brethren, two Baptist, two Lutheran, and so on. A major cleft among these twenty-eight groups marks off Catholics and Protestants, there being roughly fifteen Protestants for every Catholic in the city.[2] Most children are born into at least nominal allegiance to one of these

[1] Negro churches are not included here. There is, in the main, a rigorous taboo upon the mingling of Negroes and whites in religious observances, including the Y.M.C.A. and Y.W.C.A.

It was difficult to discover the exact number of religious groups and buildings in Middletown in 1924, owing to the intermittent nature of the services of some of the very small groups. Making a similar count for 1890 proved impossible.

[2] In addition to this major division into Catholics and Protestants there are a few smaller groups—Jews, Christian Scientists, Spiritualists, and so on.

In view of the fact that the community is so overwhelmingly Christian and Protestant, the discussion in this section will be based chiefly upon these Protestant churches as representing the dominant characteristics of the religious life, although services of all kinds in the one Catholic church were visited by the staff.

two major groups and almost never change from one group to the other; likewise one is born into one of the twenty-eight denominational groups and tends to remain in it throughout life. Traditional denominational differences, for the most part, mean relatively little to Middletown today; the Quaker church in this Methodist community, for instance, has so far forgotten its historic traditions as to use a popular book of revival songs. People cling from force of habit to the church into which they were born and become acutely aware of differences between sects chiefly when such a movement as the Ku Klux Klan temporarily sets one group to exalting itself at the expense of another, or when a "revivalist," called in "to stir up the spiritual life" of a church, galvanizes fading differences into new life—as one well-known revivalist did in 1924 when he thundered at the city, "You might just as well expect standing out in the rain to save you as to expect 'sprinkling' to do you any good." In general, preoccupation with the daily necessities of life and with such new emotional outlets as the automobile, together with the pressure for civic solidarity, tends soon to dull such occasional acute doctrinal self-consciousness.

And yet, according to the testimony of people who have lived in Middletown during the entire period under study, there is more subtle church rivalry today than formerly, as financial and social competition, particularly among the business class, have tended to replace earlier doctrinal differences as lines of cleavage.[3] The interdenominational mingling of an earlier day has apparently declined somewhat with the growth and differentiation of the community, competitive building programs, and national denominational financial burdens. As one housewife remarked, "When I was a girl here in Middletown people used to go about to the Christmas entertainments and other special services at the Methodist, Presbyterian, and other churches. People rarely do that sort of thing today." In the diary of a Middletown merchant one reads for the year 1890, "The new

[3] The tendency for the church affiliation of a newcomer to be determined in part by its instrumental character has been pointed out; the remark of one young business man, an officer in a leading church, reflects a widespread attitude: "The —— church has done a lot for me; when I first came here it got me to know —— and —— and ——, and there aren't keener business men in town." And yet, having made these contacts and become solidly established locally, this man was, in 1925, trying to resign from his church office.

Christian Church dedicated today. No service at the Methodist or Baptist churches on account of it." And later, "No service at our church [Presbyterian] on account of the dedication of the new Baptist church." And later still in 1890, "No service at our church—dedication of the Lutheran church." The competitive spirit made itself felt in the church of the writer of this diary under the stimulus of these new churches being built by other denominations, and the local press in 1890 records: "Rev. G. A. Little preached a boom sermon at the Presbyterian church yesterday, presenting strong grounds in favor of building a new church. He predicted that in five years the Presbyterian church would occupy the position the Presbyterian church in A—— does now and the size of our city would compare with [this same neighboring city]." And sure enough, in 1894 the Presbyterians outdid all the other new buildings by building the first cut stone church in the city. Another diary in the early nineties, that of a Catholic workman, records, "Last Thursday eve. went to church to hear Father W—— lecture on 'The American Citizen.' Lots of Protestants there." Today Protestants are practically never seen within the doors of the one Catholic church, the closing of one Protestant church for the services of another is almost unheard of save in the case of the union evening services in summer continued since 1890, and it is regarded as disloyal when a member of one church goes to a service at another.[4]

This apparent tightening of denominational lines, despite much talk of inter-church unity, is possibly not unrelated to national denominational organization. In 1925 the national council of one of the leading religious groups in Middletown passed a resolution, reprinted in the local press, that: "We regard with thanksgiving the fact that Presbyterians are responding liberally in support of a great variety of religious, philanthropic, and educational causes. We urge upon Presbyterian churches, however, . . . that full measure of loyalty which inspires the fulfillment of obligations contracted by denominational agencies, before Presbyterian beneficence is poured upon

[4] A characteristic manifestation of this trend appeared in 1925 when a prosperous "uptown" church debarred a smaller outlying church of the same denomination from the use of its kitchen, which the smaller church had formerly used for money-raising church suppers; the uptown church was at the time bending all its own resources to a campaign for a new building.

causes outside the Church." The second largest Protestant church in Middletown, which gives $25.00 a year to the local Social Service Bureau and nothing to the Visiting Nurses' Association—although the head of the latter remarked jocularly that she might as well move her offices down to this church since so many of her cases come from among its members—has three missionaries in the foreign field and contributed through the church proper, independently of money raised by its missionary societies, $4,486.57 during the fiscal year 1923-24 to all missions and benevolences, of which only $500.00 was spent locally.[5] Speaking to the Mothers' Council on week-day religious education, a local minister warned the women, "It is no use for us to expect to get backing for week-day religious schools from the churches, as they all have their own financial interests to look out for." Indignation was expressed by certain business class women present at "the idea that our churches can't support such a Middletown movement. They give heavily to missions and can't help here when there's so much to be done. Why can't the Protestant churches do as much for religious education as the Catholics?" An influential business man, himself a church officer, said emphatically:

"Every church is working for its own interests and is afraid of everybody else. You go to the meeting of the official board and the men there aren't interested in civic affairs. They're all just interested in their own narrow little church concerns and seeing that the other churches don't get ahead of them. I resigned all my church offices awhile ago, and they insisted on reëlecting me, but I haven't told them I'd accept. What I'm interested in is Middletown and civic affairs right here."

While this vehement attitude probably does not express a majority view, it is representative of an undercurrent at work even among church members which is reported to be stronger than a generation ago.

[5] The strategy of this draining of local "works" away from the "message" and organized program of Middletown churches is possibly open to serious question: "The thing that gave lodges a chance to get started," said a Middletown business man, "was the fact that they did more systematically the thing the churches did in only a hit-and-miss fashion, that is, look out for the poor and sick." As noted elsewhere, it is the local civic emphasis, rendering vivid its religious message of "service," that is making Rotary a successful rival of the churches with some of its Middletown members.

The economic and social considerations which appear to be becoming more potent in marking off one religious group from another are, as suggested above, reflected in the houses of worship. Some of these buildings are imposing structures of stone and brick, while others, particularly in the outlying sections inhabited by the poorer workers, are weatherbeaten wooden shelters little larger or better built than the poorer sorts of dwellings. Generally, however, a church building is larger than a dwelling, and it commonly consists of at least four rooms: the main auditorium with rows of seats for from 200 to 800, in which the principal services are held; a second room where the young are trained in the religious habits of the group; usually a room below fitted with cooking facilities where church suppers are held from time to time; and a small pastor's study. With the exception of the Catholic and one Protestant church, the Episcopal, the main auditorium comes to a focus in the platform or pulpit where the preacher stands.   As pointed out above, pictures and other means of decoration demanded in the homes of Middletown are conspicuously absent in the Protestant churches, although when members of the city's women's clubs write papers on art they are chiefly concerned with religious art. The buildings housing Middletown's five largest religious groups were built between 1888 and 1895. While the high school buildings, used to diffuse the secular habits of the group, have taken on radically new forms and provide for new activities since 1890, these buildings through which chiefly the religious habits of thought are diffused have remained relatively unchanged.[6]

But although church buildings themselves have virtually stood still throughout the last generation, other buildings under religious auspices have been built, second to none in being "up-to-date." These buildings of the Young Men's and Young Women's Christian Associations not only provide places for

[6] The plans of new buildings which two leading churches hope to erect at some time in the future include wider facilities than are contained in the present edifices.

Six leading Protestant ministers were asked whether their churches were open daily for those wishing to drop in alone for devotions. Two reported their churches not open; one open but "none to use"; one open and used by "possibly one person a week"; one always open but not heated in winter; one "always open, heated all winter, and used by possibly thirty-five to fifty a week." The Catholic church is used far more frequently during the week.

carrying on limited religious training and services outside the churches but also for sharing largely in the secular life of the community, offering swimming, basket-ball, weekly movies, Boy Scout teams, social and club life to their members. The school buildings of the city, also, serve as centers for an extensive program of religious training carried on under these organizations.[7] Six to ten factories in the city have weekly shop meetings of fifteen or twenty minutes conducted each year from November to April under the Y.M.C.A.

Each home of the city constitutes also traditionally a scene of religious activities. But it is generally recognized in Middletown that the "family altar," the carrying on of daily prayer and Bible reading by the assembled family group, is disappearing. Ministers of six leading churches, representing widely varying shades of Protestant opinion, were asked how many families in their respective churches have daily family prayers:

I. (Church of 922 members.) "Three or 4 per cent. probably."

II. (1,600 members.) "Five per cent., if that many."

III. (1,250 members.) "Not over 10 per cent." When the interviewer commented on this being a larger figure than that given by some of the other pastors the minister responded, "When you get down in these shacks and poor homes on the South Side you find an amazing number with that sort of faith—homes I'd never think of leaving without reading from the Bible and prayer."

IV. (184 members.) "One—and possibly another. There's only one woman who asks me to pray when I call."

V. (804 members.) "Eighty-eight families out of 350."

VI. (247 members.) "Possibly 10 per cent.—though I fancy that is high."

Still another hard-working pastor of a church frequented mainly by the working class said, "Not 10 per cent. of my people have family prayers. The fathers and children get up at different hours, and it is almost impossible in many homes to assemble the family regularly."[8] The same group of six ministers were asked how many members of their respective churches read the Bible daily or regularly:

---

[7] See Ch. XXIII for an account of these Bible classes.

[8] Many families used to hold family prayers before retiring at night. The heavy increase in occasions for various members of the family being "out" evenings now (cf. the discussion of clubs and of children in Chs. XIX and XI respectively) makes this increasingly difficult.

I. (922 members.) "Possibly 20 per cent. read it some time during the week."

II. (1,600 members.) "Twenty-five to 30 per cent.—maybe more."

III. (1,250 members.) "Twenty-five per cent. at least."

IV. (184 members.) "We sold fifty calendars of daily Bible reading at five cents each at the church door, but I've no way of knowing how many read their Bibles."

V. (804 members.) "Eighty-eight families out of 350."

VI. (247 members.) "Ten per cent. would be an outside figure."

The religious ceremony of verbally blessing food before eating is common, although according to the testimony of both ministers and church members it appears to be less frequent and less spontaneous than a generation ago.[9]

The early custom of using the home for meetings of "prayer bands" and "cottage prayer meetings" survives today chiefly in outlying working class neighborhoods. A Middletown minister, addressing the Federated Club of Clubs, summarized the shifting status of the home as a place where religious rites are carried on:

"The home is failing to instruct the children religiously as it should. The family altar, set times for devotional reading and discussion, are not as common as they were formerly. People do not seem to be ready to acknowledge that they value such things any less, but excuse themselves on the ground that they are too busy. It has been crowded out of the family program."

Club meetings traditionally open with "devotions," but only six of the nineteen women's study clubs, as noted above, do so today.

[9] This habit of "saying grace" before eating is generally observed at important public dinners, a minister being specially invited to be present to perform the office. This habit is showing signs of attenuation: thus the Chamber of Commerce groups do not have "grace" at weekly luncheon meetings but only at large evening dinners such as the annual banquet; Rotary does not say grace except at the ceremonial Christmas luncheon; Kiwanis and Exchange, on the other hand, each have a clergyman member say grace at the regular meetings.

It is possibly symptomatic of differential degrees of secularization in the habits of different sections of the group that while all but one of the eleven township "farmers' institutes" were opened with an "invocation" by a clergyman, the county farmers' institute held in the Chamber of Commerce in Middletown opened instead with a professional funny man.

As in the place of religious observances, so also in the time set aside for them, conflicting trends are at work. On the one hand, other activities of the community as they become more secularized are forcing upon religion a narrowing place and time specialization: as religious rites tend to concentrate in church and allied buildings rather than pervade home, club, and civic and social groups, so, too, there is a tendency to center them more exclusively in the one day in seven that traditionally "belongs to the Lord." Despite the efforts of the Ministerial Association to keep Wednesday evening free from other community activities for prayer meeting, the community no longer makes any effort, even outwardly, to observe this former custom; instead, the prayer meeting occasionally alters the hour or even the day of its meeting to give place to other events. Even special services must make way; the press reported in 1925, "The [union] prayer service, a part of the Week of Prayer program, to have been held tonight at the —— church, has been postponed because of the dedication of the —— gymnasium." The increasingly non-religious character of any church social meetings which may be held during the week is part of this same trend.

On the other hand, just as organized religion, being forced back into the church buildings, is reaching out into the schools and shops through the Christian Associations, church clubs,[10] and allied organizations, so through these same agencies it is reëstablishing itself during the six days of the week at the same time that the more traditional religious observances are being increasingly confined to the seventh.

Both processes, the tendencies for other community activities to invade the traditional preserves of the church and for the church in turn to adapt itself by taking over extra-religious activities, appear in the battle raging for the possession of the stronghold of religious tradition—the Christian Sabbath. The Sabbath of 1890 is vividly described by one housewife:

"We went to Sunday School, morning church, young people's society, and evening church—*every week*. When we girls in our family missed Sunday School we weren't allowed to go out in the evening all the following week. We weren't allowed to play games

[10] For the attempt of the churches to meet community activities on their own ground through an elaborate network of clubs, see Ch. XXIII.

or even to crack nuts or make candy on Sunday. Father would never take pictures on Sunday. We couldn't read a newspaper or any weekday reading, but only things like our Sunday School papers."

Such rigorous Sunday observance was far from general in 1890, but this reflects the sort of Sabbath that many God-fearing families sought to maintain. There was even a law upon the state statute books that:

"Whoever, being over fourteen years of age, is found on the first day of the week, commonly called Sunday, at common labor, or engaged in his usual avocation (works of charity and necessity only excepted), shall be fined in any sum not more than ten nor less than one dollar."

Even as late as 1892 an ordinance was passed in Middletown that:

"Any person convicted of having, on Sunday, within said city, pitched quoits, or coins, or of having played at cricket, bandy, cat, townball, or any other game of public amusement, or of having discharged any gun, pistol, or other firearms, shall be fined therefor in any sum not less than one dollar nor more than five dollars."

But cracks were beginning to appear even then. In July, 1890, the local press notes:

"The universal opinion among our patrons of ball is that we do not want and will not have any Sunday games. . . . It is to be hoped that the manager and board of directors of the Middletown team will see matters in this light and strictly observe the wishes of the people."

In 1892 a compromise was reached and a sacred concert was regularly combined with the ball game, "the band playing at intervals."

The press noted in 1890: "Sunday is becoming more a day of recreation than of rest every week." In 1893 the Trades Council raised money for the Homestead Defense Fund by staging *Enforcing the Sunday Law in Middletown, A Farce.* In a Sunday afternoon labor lecture in 1892 the speaker said, "The day has gone by when Sunday gatherings of this nature can longer be decried by the church. The purpose of these gatherings is as laudable and elevating as is to be found within the walls of a church." In the industrial city of 1900, surging with its new population, the press gives half a column Monday morning to reporting, with no unfavorable comment, a Sunday morning cock fight between cocks put forward by the iron workers and by the glass workers. Sunday morning trade union meetings were reported regularly in 1900.

Today ministers still insist that "the Christian Sabbath belongs to the Lord," and "if it is His day, then He has a right to say how we shall spend it"; a Lord's Day Alliance speaker assures Middletown that "it is the European Sunday that has brought about the downfall of Germany and Russia"; and in announcing the coming of Branch Rickey, manager of the St. Louis Cardinals, to lecture under the Lyceum of the Ministerial Association, a church calender states, "Mr. Rickey never plays on Sunday." But, meanwhile, advertising pages continue to urge Sunday motoring; games are played throughout the city parks on Sunday, and more men play golf at the Country Club Sunday morning than any other half day of the week, though organized Sunday baseball is still condemned by churches and by a good share of the population. In 1924 the Gun Club shifted its weekly shoots to Sunday tentatively, apprehensive of public objection, but no objection was made and the experiment became permanent.

Just as the habits of daily prayer by small children and of sending children to Sunday School, as pointed out below, persist more strongly than the praying and church-going habits of the adults, so it is in the Sunday leisure-time outlets for small children that Middletown is most conservative: the heavy week-day programs of school, Y.M.C.A., and extracurricular life stop utterly on Sunday. A number of churches held Sunday School in the long Sunday afternoons of 1890, but today the afternoon is usually swept clear of services. Ministers of six leading churches were asked, "Does your church attempt to offer any Sunday afternoon or evening program other than the six-thirty Young People's meeting to interest young people who might otherwise go to the movies?" Three answered, "Nothing"; one answered, "Only the Intermediate's Christian Endeavor at four"; one, "The Christian Endeavorers give a late tea one Sunday a month that extends on into the Christian Endeavor service, but that's all"; and one said, "Yes, three things: we conduct the Sunday School at the Orphans' Home; our young people visit and sing at the hospital and other places —often as many as four autos full of them; and prayer bands meet here and in the homes. I keep pushing this work and urge our young people to form prayer bands and report their numbers to me." The Y.W.C.A. holds vesper services on four or five Sundays immediately preceding Christmas and Easter, and a monthly Sunday afternoon social hour.

Many mothers spoke of Sunday as a "difficult" day for the children:

"All the other days of the week they are busy and active and have plenty to do, but on Sunday they have nothing to do and nearly tear the roof off. I let them go to the movies because, while I don't like the idea and was never brought up that way myself, I'm glad to get them out of the house!"

"Sunday is such a tedious day for the children. My daughter sews, which I was never allowed to do, but I don't let them play cards or dance or go to the movies on Sunday yet. I give way at some new place each year."

All along the line parents may be seen fighting a rear-guard fight, each father and mother trying to decide "what's right": one family lets the children swim on Sunday at their summer place but not play golf; another does not approve of Sunday card playing and attendance at football games, but the parents have begun both in order to keep close to their son; others make popcorn on Sunday but not candy; others object to Sunday evening bridge but allow Mah Jong. Meanwhile, although there were in 1890 no Sunday theatrical performances, Sunday is today the biggest day of the week with the movies, and these Sunday evening audiences are the oncoming generation. The Sunday night house of a leading exhibitor is almost solidly young people, and unless he can attract boys and girls and young men and women on that evening, he states, his house is empty. One popular high school boy whose family forbade his going on Sunday night complained to his parents that he was "out of it at school on Monday because everybody talks about the new shows they saw with their Sunday evening dates and I can't ever talk about any of the new shows until Tuesday." It is possibly indicative of a differential rate in the breaking down of the taboo on "Sabbath breaking" that eighty-three (a third) of the 254 working class children and only twelve (a sixth) of the seventy-four children of the more prosperous business class parents in the three upper years of the high school wrote "true" after the statement, "It is wrong to go to the movies on Sunday"; suggestive, too, is the fact that only 23 per cent. of the 241 boys marked this statement "true" as against 40 per cent. of the 315 girls. With this generation the "Sabbath" of 1890 is being increasingly secularized into the "Sunday holiday," and the males appear to be throwing off the older tradition, as noted elsewhere in respect to other habits, more rapidly than the females.

Like art and music, religious observances appear to be a less spontaneous and pervasive part of the life of the city today, while at the same time this condition is being met by more organized, directed effort to foster and diffuse these values.

Chapter XXII

LEADERS AND PARTICIPANTS IN RELIGIOUS RITES

At the head of each of these forty-two religious groups is the man who gets his living by "ministering to" that particular group. His duties are varied. As the use of verbal symbols holds a prominent place in these religious activities, he is first and foremost expected to be a good talker; this capacity is widely exploited;[1] he is called upon to talk on short notice upon "President McKinley," "Thrift," or "Six Years After the War," at the D.A.R.'s annual dinner, at the dedication of a public building, or at a meeting in behalf of anti-tuberculosis work. He is also expected to be a "sympathetic pastor," a "good fellow among the men," and to "draw the young people."

Pastoral activities are changing markedly among the business class; there is apparently considerably less overt religious behavior during pastoral calls—kneeling and praying and reading from the Bible—than was customary even as recently as a generation ago, and the pastoral call is tending to attenuate among those who are not "in trouble" to a brief social call. Among the working class the earlier pastoral call is still common, but other agencies, such as the secularized organized charity bureau and visiting nursing service, are tending to diminish the rôle of the minister even when families are in trouble. The shrinkage of this pastoral activity eventuates in at least two things: increased emphasis upon preaching as the principal means of drawing young and old, and the scattering of the preacher's energies among various kinds of civic work as he mingles with the men. In pioneer Middletown, as noted in Chapter III, a preacher was one of the major sources of infor-

[1] In an editorial characterization of one of the leading ministers in 1924 a local paper begins: "Dr. —— is a brilliant pulpit man, a sound thinker, and a splendid speaker. His sermons always are of the highest type, and he has been greatly in demand among the neighboring churches for pulpit talks." Similarly, the diary of a prominent citizen notes in November, 1890: "Rev. —— preached for us morning and evening. Think we would be satisfied with him—good plain speaker—clear voice. Hear every word he says— so with even those who are hard of hearing."

mation in the community. The increase in popular education, the heavy diffusion of new channels of information, and the general sophistication of Middletown,[2] together with the decline in interest in the general discourse noted in connection with changing lecture habits, tend to dwarf the relative significance of the sermon in Middletown's life at the same time that it is becoming relatively more important among the things done by ministers with declining pastoral work. "When are you going to carry out your threat to come hear me preach?" asked a minister in a bantering tone of the man across from him at a civic club luncheon. "Well, I read your fiction in the Monday morning papers [3] and that's all I can stand," the other rallied. There was a chorus of laughing comment over the word "fiction," and one prominent church member said under his breath, "Stick to it, Ed—you're an honest man!"

These men must usually prepare two speeches for each Sunday and one for the mid-week prayer meeting. Interviews with six leading Protestant ministers revealed all of them hard-pressed to talk so much.

"The trouble is," said one of them wearily, "that these people don't realize that a constantly changing presentation of the Gospel is needed in a world changing as fast as ours is. A man was in here just this week saying to me, 'We don't want any of this modernist business. What we want is the plain old Gospel.' The older generation in this church are about fifth graders, their children are high school graduates, and *their* children are beginning to go to college; it may be another generation or so before the general level will be high enough so that people will expect a minister to have time and energy for study if he is to be worth shucks as a minister."

The reading of these six men is all done "on the run." "My only chance is late at night before I drop off asleep," said one of the most energetic of the six. "It's not really reading, I'm too tired, and then, too, I read *for* sermon material rather than giving myself up to what the author has to say." [4]

[2] Cf. in Ch. XVIII the Middletown man who prefers to listen to radio sermons because there is no preacher in town as good.

[3] Verbatim reports of sermons as sent in by the ministers appear in the Monday morning paper.

[4] These six leading ministers read:

1. "One book a month and dip into others that my wife often finishes. I'd be snowed under if it weren't for my wife's reading. I read four or five

All the ministers in the larger Middletown churches feel the constant strain of their scattered and often irrelevant activities at a time when the church and Sunday sermons are meeting increasing competition in the fuller life of the city today. One of them summarized the outstanding features of his year's work in 1924 as follows: "Nine hundred and fifty-nine pastoral calls of ten minutes or more, 133 sermons, 101 talks (high school, civic clubs, prayer meeting), thirty-five addresses (commencement, Memorial Day, etc.), twenty-nine funerals, thirty-four weddings, innumerable 'phone calls and letters, the latter all longhand as I have no secretary." Another stated as the chief tangible aspects of his four years' work in Middletown: "Three hundred and eighty-six sermons, about 100 special talks or addresses, about 2,700 pastoral calls reaching about 6,000 persons, eighty-nine weddings, 111 funerals, 403 additions to the church."

These ministers of the six leading Protestant churches, with two or three others, including the Catholic priest, are practically the only ministers the Middletown business class ever goes to hear. None is a local man; the fathers of all but one

books during my summer vacation." Asked for the five outstanding books on any subject read during the past year, he mentioned Jefferson's *Five Great Controversial Subjects*, "Fosdick's new book," "Studderd Kennedy on *Lies*—a dandy," Hutton's *Victory Over Victory*. "I didn't finish Robinson's *Mind in the Making*—couldn't make much out of it."

2. "Two or three a month." The only book he could recall was *Seeing Straight in Sunday School*.

3. "About one a week the year through—though I have to depend on my vacation to bring it up to that. The outstanding books this year have been *Mind in the Making*—that's fine!—and *What Is Man?* by J. Arthur Thompson—good!" He could not recall others.

4. "Not five a month. It bothers me a lot not to read more. Among the outstanding things are Studderd Kennedy's *The Wicket Gate* and Sheila Kaye-Smith's *End of the House of Alard*. He had difficulty in remembering others.

5. "I aim to read one a week and just about do it." The only outstanding book mentioned was George Muller's *Life of Trust*—"a great biography!"

6. "Not more than one a month, though that's chiefly because my eyes bother me. It's hard to pick; the list includes Corbin's *Decline of the Middle Class*, Wallace's *Sociological Studies in the Bible*, Robinson's *Mind in the Making*, the volume on the Reformation in the Cambridge Modern History and also the volume on the Puritan era in England, a book by a Columbia professor on social forces, Ross' *Social Control*, Starr's *Talks on Geology*, and I re-read James' *Pragmatism* and Darwin's *Descent of Man*." It is worth noting that this man is the pastor of the small, unpopular liberal church.

It should be borne in mind that these leading ministers stand almost alone at the top of the local profession.

were of the working class; two of the six worked as working men before entering the ministry, two others were teachers, and the other two have always been ministers; two are graduates of large Eastern universities, three of small Middle Western colleges, and one is not a college man; one holds a master's degree and all have studied at divinity schools, three of them graduating; three of the six have traveled out of the United States; all are Masons; and all but one has been preaching in Middletown from four to ten years. The living of these six leading men ranges from about $3,000 (the salary of the mayor, twice that of a policeman, half again as much as that of the city librarian or a high school teacher, and two or three times that of an elementary teacher) to $5,200.[5] It compares favorably with that of perhaps 40 per cent. of the eighty members of Rotary and a considerably higher percentage of the other civic clubs.[6] The amounts earned by the remaining thirty-odd Middletown ministers, again with only two or three exceptions, fall off sharply from this high plateau; in many of the outlying churches attended by the working class it often becomes, particularly when "times are bad," a nip and tuck matter to raise the minister's meager salary. Conditions have not changed markedly among some of the smaller of these struggling congregations since 1890, when an announcement in the press extended "thanks to the kind donors who contributed $32.50 to the —— church for the purpose of making a present to their minister, to whom the ladies presented a fine suit of clothes. The reverend gentleman was delighted with the useful present."

Emphasis in selecting a minister upon his success with young people and with men indicates points of strain in the church. In general Middletown ministers are more highly esteemed by the women than by the men. The men usually accept the religious beliefs of the civilization in which they live, but as one hears their conversation over a period of many months one gets a distinct impression that religion wears a film of unreality to many of them and that its advocates in the pulpit constitute a group apart. Absolute proof in this region is impossible; the

[5] Where "parsonage" is "thrown in" with a minister's salary it is here counted as approximately the equivalent of $1,000 additional salary, on the basis of rents in Middletown.

[6] Cf. Ch. VIII for the distribution of those getting Middletown's living by income tax returns and also the incomes of the sample workers.

behavior and folk talk of the people must be the chief guide. Such talk for the most part reflects much the same attitude as the poem read at the anniversary of the Masonic lodge in the nineties which referred to that past time when "preachers 'rode circuit' on horseback like men":

In regard to the censorship of books and movies a Middletown editorial asks, "Who is qualified to be censor?" and continues by way of answer, "Not police officers, not prosecuting attorneys, not single-minded professional critics who see bad in most of the things others pronounce good and good in the things that others pronounce bad, not ministers of the gospel, not extremists of good or evil—not anybody who by instinct and training may be unable to look at literature and art except through the eyes either of prejudice, ignorance, or inexperience."

Speaking of a local man who, after breaking one golf club, topped his next stroke, the local press said in 1924, "Fortunately no ministers or women were present."

Another press article was headed, "They Preach on Sunday But They All Have Their Recreations," and continued, "Any one who has an idea that just because various local citizens happen to be ministers and preachers of the gospel they aren't able to enjoy themselves, has another thought coming. . . . Middletown's preachers do the same things that other persons do. Baseball games and other athletic exhibitions, the theater, when the program of the day is a worthy one, the civic clubs and the like help the local preachers to remember that there's plenty of joy in life."

A group of half-a-dozen influential business men, all college graduates, were discussing the ministry: "I'd never advise a boy to go into it," said one emphatically. "I never heard a preacher yet," added another, "who didn't make me mad by standing up there where he knows you can't talk back and saying things he doesn't really believe and he knows he can't prove." The general opinion of the group was that the ministry is "played out."

"What's a minister without his salary, anyway?" asked a labor union man of a group sitting about talking before a labor meeting. "You take away his salary and he's nothin'! When he gets a chance to make more money at another place he calls it a call from God!" (*Laughter.*)

(*Another man.*) "Yes, and you take these churches—what do they work for? Take the Presbyterian church—what does *it* work for? For the Presbyterians! Even the Baptist church is the same."

(*First speaker.*) "Yes, it is, but I like Rev. [the Baptist minister]—*there's* a preacher not afraid to soil his hands! He's as common as you or me. He'd touch a corpse or anything. Do you remember when that man shot his wife and children and then hisself? Rev. —— went there and even helped lay 'em out."

Among the business class there is some tendency to "help out" a minister with a free membership in the Country Club or little salary gifts "on the side." This slightly patronizing attitude in matters involving money is shared to some extent by both husbands and wives. Referring to a former minister's wife who had had some money of her own, one business class wife remarked, "It was so nice to have a minister's wife one did not pity."

The conspicuous omission of ministers from Rotary is possibly indicative of the attitude of Middletown business men: some explain the omission by saying that if any minister were included it would have to be the pastor of the prominent millionaire manufacturers, and since they are Universalists, the more orthodox would object; others say simply that inter-church rivalry is too keen to allow the selection of any one minister for this important recognition; the upshot of the situation is that no minister is included among these leaders of the city.

Such insights gleaned here and there, although they reflect accurately the more frequently expressed opinion, must be set over against a widespread attitude of respect and in many cases warm affection and esteem for these religious leaders, especially among the women of Middletown:

"I like Mr. ——," said one woman. "It always makes me feel better just to meet him on the street."

A 1924 editorial commending a minister called to another city says, "The fact that he is a minister of the gospel has not kept Mr. —— from mixing with his fellow men in the many and varied avenues of life outside the church. He often has been referred to as a regular 'he man.'"

A prominent member of Kiwanis said of his new minister, "My boy and girl used to go to church under protest and sit and read their Sunday School papers through the sermon, but now my boy says, 'Dad, I *want* to stay to church. A big, regular man like Dr. —— is worth hearing.'"

"I've known just two preachers that were real fellows," remarked a young business man, "—— and ——, and of the two

—— could drive his car a little faster, play a better game of golf, and was not averse to using a little judicious profanity on occasion. He always wound up his sermons at quarter to twelve, no matter where he was, and he always wound them up with a bang, too. We were the first church in town to let out.[7] He had the actor's instinct to quit while they still want more. He was a real business man."

When a minister formerly in Middletown returned to the city for a single service, the church was packed to welcome him and many spoke warmly of what he had meant to them.

Several mothers spoke with appreciation of the work of the pastor of a small business class church among the boys.

That the somewhat condescending tone of many of the above remarks represents a substantial sentiment in Middletown, however, may be inferred from the fact that at least four leading ministers expressed bitterness over their work:

"I've been wanting to get away for a long time," said one, referring to a source of difficulty. "Of course I can stay and fight them, but I don't want to leave here with an unfinished fight on my hands, and rather than run the risk of having to stay here to see a fight through, I prefer not to start it."

Another man in speaking of his difficulties said again and again by way of explanation of his "not throwing the whole thing up and leaving," "But I won't be a quitter! I won't be a quitter!"

From both ministers and their congregations an outsider gets an impression of the ministers as eagerly lingering about the fringes of things trying to get a chance to talk to the men of the city who in turn are diffident about talking frankly to them. On the one hand more than one man explained, as did a member

---

[7] This qualification of winding up his sermon so as to dismiss early has become an acute criterion of preachers in this day of the automobile: "I don't like our new preacher," said a worker's wife. "He talks too long— keeps you there till 12:05 or 12:10, and when you get home and have dinner and do the dishes your afternoon's gone. Now look at Rev. ——; in summer he's through at 11:30—just an hour. All the churches ought to do that." A business class woman, walking home from a socially prominent church, said in a tone of resignation: "We've been spoiled by our former pastor. He always let us out at a quarter to twelve and we were home early. It is hard to have the service hold till twelve or after."

In at least two churches the minister was heard to announce at the close of the Sunday School that he wanted all present to stay to church and would promise to let them out "at a quarter to twelve sharp!"

of Rotary, "I'm afraid to talk to a preacher and tell him what I really think because he'd think me an atheist or something like that," and on the other hand one of the keenest ministers lamented to a staff member regarding the business men of the town, whom he was desperately anxious to get closer to, "They'll talk to *you*, but they won't talk to me because I'm a minister—and it's the same with all the other ministers in town."

The week-by-week minutes of the Middletown Ministerial Association over a period of fourteen years from September, 1910, through December, 1924,[8] offer perhaps the most concrete evidence on the activities and general outlook of Middletown ministers in this day when many of their traditional activities are being drained off by secular agencies. The meetings consist of "devotional exercises" (prayer, Bible reading, and occasionally devotional talks by members); passing of resolutions of sympathy, thanks, approval, or endorsement on various occasions; listening to talks by members of the Association, local citizens, or outside speakers; and discussing various questions of general interest to the members. The work done by the Association during these fourteen years divides itself into activities initiated elsewhere but endorsed publicly by the Association; activities endorsed privately by the Association; activities inaugurated by the Association involving no sustained work; and problems or movements actively worked at by the Association more or less steadily during the period.

Among the activities which the Association did not initiate but on which it passed a vote of approval are: *Certain campaigns:* for Y.M.C.A., Y.W.C.A., Boy Movement, Near East Relief, Anti-Tuberculosis Association, various evangelistic meetings, Red Cross, Community Chest; celebration of certain organized "days" and "weeks";[9] *the work of certain organi-*

---

[8] The meetings were held weekly or bi-weekly during this entire period except for two months each summer. The minutes for these fourteen years, the only time for which they are available, were read in detail. No similar records could be obtained for the 1890 period.

All ministers in Middletown, with the exception of one leading minister who resigned because "the Association wasn't doing anything" and some of the ministers in the very small outlying working class churches, are members of the Association.

[9] Father and Son Week, Children's Week, Life Insurance Week, Health Week, Week of Prayer, Poppy Week, Good Will Sunday, Anti-Tuberculosis Day, Anti-Saloon Field Day, Go-to-Church Sunday, Bundle Day (for

zations: Woman's Franchise League, Family Altar League, Lord's Day Alliance, Swear Not League, the Veterans of Foreign Wars in "keeping history correct in the school books," American Bible Society, Humane Society; and *certain other diverse movements:* work of city officials in suppressing vice, 6 P.M. Saturday closing movement, appointing delegate to Trades Council, action of Cemetery Board against Sunday funerals, saner Saturday night, promotion of the Middletown Chautauqua, Church federation, Passion Play for local auditorium, Lyceum lectures, inviting W.C.T.U. convention to Middletown, visit of Field Secretary of World's Sunday School Union to Middletown, selection of Middletown as location for a state basket-ball tournament.

Outstanding among the activities which it endorsed privately but on which it did not make its endorsement public are the League of Nations and a movement to secure a reform mayor to clean up the city.

The list of activities inaugurated by the Association during these fourteen years is: closing of post office on Sunday, opening of all churches at 11 A.M., evangelistic campaigns, securing speakers at County Infirmary and Children's home, Armistice Day service, petition for a special session of the legislature to correct the law holding it not a crime to possess liquor, singing of church choirs at the hospital, community school of religion, daily vacation Bible School, parent meetings in the churches to discuss the problem of vice, investigation as to whether Middletown laboring men get one day's rest in seven, publication of statement to inform the Negro vote of the true meaning of a constitutional convention, resolution that there should be more care in selection of books for public library, protests against certain motion pictures.

The interests at which the Association worked openly and persistently during this period are: doing away with Sunday motion pictures, prohibition, keeping Wednesday evening free from social engagements for prayer meeting, religious education in the schools, opposing prize fights, and discouraging marriages by justices of the peace.

The following excerpts are characteristic of the Association at work:

the Near East), Decision Day, Prison Day, Lafayette Day, Civic Righteousness Day, Mercy Sunday, Patriotic Sunday, Sabbath Observance Day.

"It was decided that the Chairman appoint a committee of three whose duty would be not to raise money but to take such action as would show the interest and sympathy of this Association in the starving millions in China."

"A motion was carried that the Ministerial Association send a telegram to the miners' and operatives' conference at Washington asking them to invoke the Divine Wisdom in deciding the questions before them."

"A statement was made that the —— Auditorium [owned by the Middletown Y.M.C.A.] had been used by the High School pupils on the previous Sabbath for playing basket-ball, thus desecrating the sanctity of the Sabbath Day. This matter was referred to the Social Betterment Committee." [Next meeting.] "The Social Betterment Committee reported that the auditorium would not again be used in violating the Sabbath. . . . [At the same meeting.] "It was voted that we as an Association endorse the Boy Movement Campaign for our city and pledge our coöperation through a committee empowered to represent this Association."

"The vice conditions in our city were pronounced deplorable . . . and a vote was taken by common consent that we as an association keep our ears and eyes open for an opportunity to move versus this deplorable state of affairs."

"Mr. —— mentioned the fact that no real demonstration had been made in the interests of the Sunday Schools of Middletown and that a general plan that would involve a parade and all-day gathering of the Sunday Schools would be good."

A vigorous discussion on local political corruption was followed by the vote that "Bro. —— be instructed to present to Mr. —— an urgent request from this Association to become a candidate for Mayor and this action be withheld from the newspapers." At the next meeting a rival candidate, getting wind of this request, asked leave to address the Association, and "Mr. ——'s statement having been heard, the Association went on record that any discussion of any candidates that was indulged in last Monday morning was altogether impersonal and without view to action now or in the future."

"The Social Betterment Committee, reported by Dr. ——, told about the psychology of defeat and discussed our attitude toward Mayor ——, spoke on the subject to the end that as individuals we receive and encourage every good word and work of the

Mayor but that official expression of the Association be post-poned until some fruits of repentance are manifest."

Individually the ministers make further efforts to share in the life of the city. Thrust into an undesired aloofness as earthly representatives of an ideal popularly regarded as impossible of achievement by "poor, weak human nature," charged with persuading the community to adopt a way of life at variance with its current concerns, they seize upon every opportunity which seems to offer a means of sharing in the life about them without yielding too much of the principles to which they are committed. Two attend Ad. Club luncheons, one works in the dominant political organization, one testified at the state capital in the trial to impeach the judge of the local Circuit Court, and one is the head of the State National Guard. One tries to get the high school to change the last line of its cheer from "the whole darned team" to "our whole team"; another, hesitant to start a forum for fear it might cause dissension, tries to organize the collection and distribution of the old periodicals of his church to the poor; and yet another, eager to share in the noon shop talks in the factories of the city under the Y.M.C.A., a man who "just puts his foot up on a piece of machinery and talks along as common as an old pair of shoes," is not asked to talk again, because, "while he talks good sense, the men don't like it because he doesn't appeal to their emotions and isn't religious enough for them and they dislike his unorthodoxy." As one energetic minister remarked, "I've never been so 'stopped' and perplexed in my life. I just don't know where to take hold of the situation here." [10]

Leaders of the Y.M.C.A. and the Y.W.C.A. are not chosen like the ministers mainly for their ability to talk well, but to draw the young people, chiefly through athletics, to organize clubs, and to carry on executive work. The directors and assistant directors of both organizations in Middletown are college graduates and have had experience in similar work outside of Middletown.[11] They coöperate to some extent with the work of the churches and are becoming increasingly prominent as religious leaders in Middletown.

[10] This man considered the renting of an office in a down-town business building in an effort to get closer to the dominant interests of the city.

[11] See Ch. XXIII for discussion of the volunteer teachers of Sunday Schools and Y.M. and Y.W.C.A. Bible classes.

Associated with and supporting each minister are the members of his church. Children "join the church" for the most part between the ages of nine and seventeen. Slightly over half of all members join before they are sixteen.[12] A cross-section of church membership indicates a preponderance of females; in the largest Protestant church in 1924, 62 per cent. of the members were females; in another prominent church 60 per cent. were females. That this ratio has held fairly steadily for the past thirty years is indicated by the fact that in the first of these churches in 1893, 64 per cent. of the members were females.[13]

An attenuation of church membership vows similar to that observed in lodge membership is apparently in progress. Virtually anybody can join a church simply by signifying his intention to do so,[14] and once in, it is practically unknown for any

[12] Based upon Athearn's distribution of 2,302 Sunday School teachers and officers in this state according to age of joining the church and also of a sample of 6,194 members of five Protestant denominations in forty-three states; in the former group 56 per cent. joined before they were sixteen and in the latter 59 per cent. (*Op. cit.,* pp. 372-4.) According to Athearn the "median age of joining the church is 14.9 years. The predominant group, however, joined church at twelve, thirteen, and fourteen years." (*Op. cit.,* p. 64.) It is noteworthy as possibly having a bearing upon the fact commonly remarked upon in Middletown that "you can't tell the difference, from the way they act, between church members and those that don't belong," that although legally in this culture a minor may not make binding contracts and a girl under sixteen is regarded as below the "age of consent," it is nevertheless considered proper for a child of ten or twelve to make a presumably lifelong contract by joining the church. In fact, early "decisions" are encouraged; sermons frequently urge them upon children; says a Middletown church calendar regarding the allied decision to support the church financially: "The question sometimes comes, At what age should children be taught the practice of tithing? Very early in life, when the brain is plastic, memory retentive, and impressions permanent. The tithing child puts God first. Later he realizes he has been obeying Christ's command, 'Seek ye first the Kingdom of God.' "

The point here is not the rightness or wrongness of joining the church at one age or another but simply the disparity in point of view regarding the age at which one becomes competent to enter contractual relations in different segments of Middletown's life.

[13] This was the only church for which membership records for the early period were available.

[14] Only one case indicating any selection came to the attention of the research staff. In this case a group of the active men in a prominent church met for supper with their pastor at a downtown hotel preliminary to selecting likely men to "win" for the church. A list of names of men, chiefly newcomers to Middletown not affiliated with any church, was read, and some one present engaged to interview each one. When one name was read, one of the church elders present mentioned the fact that "he doesn't pay his bills," whereupon the minister rejoined, "That so? Well, give me that card back. We don't want that sort of man in our church." Another of the

one to be ejected. A few churches, notably the Catholic and Episcopal, give some instruction regarding the duties of membership to children about to join; such instruction to adults is far more rare. The nature of the instruction given to children is suggested by the statement of the pastor of a leading church: "Oh, I just instruct them in the meaning of the church, the sacraments, and about God and Jesus and the Bible."

Three things may bind a member to his church: money contribution, attendance at services, and church work. Of these the last is, according to the ministers, the least common. In five of the six leading Protestant churches the minister set the percentage of active members, i.e., those who do something in connection with the church other than attend services, at 25 per cent.; in the sixth church, with less than 200 members, the figure was placed at 50 per cent. "Lack of leadership and complacency" were the answers given over and over again by ministers to the question, "What do you consider your greatest problem here?" [15]

The meaning of church membership in terms of money contribution is indicated by the portion of the family income given to the church. Traditionally every Christian "returns a tenth of his substance to the Lord." A few families in Middletown continue this practice of tithing, but, according to the statements of ministers and church treasurers, the great majority contribute far less than a tenth. Of the sample of 100 working class families for whom income distribution was secured eighty gave less than 2.5 per cent. of their year's earnings to the church, fifty-eight gave less than 1 per cent., thirty-nine less

names suggested was that of a man who "is getting a divorce" and again the reply was, "Well, we don't want him, then." It is extremely unlikely, however, that either of these men would have been refused membership if he had applied.

[15] The ministers of the six leading Protestant churches were asked what percentage of the leading men in each religious group, "elders," "deacons," "vestrymen," and so on, read their Bibles regularly. They replied:

I. "Possibly 25 per cent. of the twelve members of the Session and possibly 20 per cent. of the twenty-two trustees and deacons."

II. "Less than 50 per cent. All the eleven elders do but this isn't true by any means of the twenty-eight others."

III. "I'm pretty sure that seven out of twelve of our deacons do, but only two out of eight trustees."

IV. "Two or three—certainly not over four at most—out of ten."

V. "Every one, I think."

VI. "I suspect four of the ten do."

than 0.5 per cent., and nine gave nothing. Of the twenty giving 2.5 per cent. or more of their earnings, only one gave as much as a tenth to the church.[16]

Income distribution, including church contributions, could not be secured for a sample of the business class. Some idea of the money nexus between the business class and the church can, however, be gained from the contributions of members in certain churches. The members of the most fashionable church in Middletown, almost exclusively business class people, include roughly, according to the church secretary, one person who contributes more than $1,000 a year, ten persons who contribute between $500 and $999, fifteen between $250 and $499, seventy between $100 and $249, and 425 less than $100.[17] In another prominent church with a scattering of well-to-do business people and many working class members, only one person gives as much as $260 a year, one $200, and the great mass of the members from $14 to $52 a year.

That church membership does not necessarily mean even a minimum money contribution appears from a distribution of 918 members, a random sample of approximately 50 per cent. of the total, of the largest Protestant church in the city on April 1, 1924. Their financial nexus with their church was as follows:

566 of the 918 (62 per cent.) had pledged themselves to contribute money to the church.[18]

177 of these 566 were in arrears at the close of the year for payments due in the first six months.

273 of the 918 (30 per cent.) made no pledge to contribute, and only seventy-five of the 273 subsequently paid anything other than possible incidental contributions placed in the collection plate.

Three of the 918 were listed in the published membership roll as "Exempt" and information was not given as to the contributions of seventy-six more members.[19]

[16] No check was made of church membership. The significant point is that although ninety-one of these 100 families contribute to the church, only one continues the traditional tithing. See Table VI.
[17] Cf. Ch. VIII for estimate of the incomes of Middletown's business class.
[18] Four hundred and sixty-seven of the 566 also pledged something to "benevolences," i.e., denominational Christian activities other than the support of current expenses of the local church.
[19] Forty-four of the seventy-six names left blank were adults, fourteen more were wives who were not listed as contributing though their husbands were, and eighteen were children, presumably children of the forty-four adults.

Financial support of the church appears in Middletown at all levels, from the spontaneous giving of the traditional tithe on the part of a few ardent church members, through the stage of giving "because I wouldn't like to live in a land without churches" or because it is "expected," to not giving at all. In the face of increasing financial demands of the individual churches and their denominational boards, "giving of one's substance to the Lord" is meeting with greater competition as community-wide charity is becoming secularized and the outward dominance of the church in the community declining. Meanwhile, according to statements by the ministers and by the treasurers of two business class churches, the increasing money needs of the churches and their denominations and the declining dominance of the "message" of the church with many Middletown people are increasing the relative prominence of the money tie between a church and its members as compared with a generation ago.

No check of church attendance in terms of membership was made. Although the tradition is that "every one goes to church," a check of attendance at the services of the forty-two religious bodies during the four-week period from October 26 to November 22, 1924, shows that an average of only eleven males in each 100 of the white male population and eighteen females in each 100 of the white female population attended Sunday morning church services on each of the four Sundays.[20] Ten males and fourteen females in each 100 attended the Sunday evening service, and, according to the estimates of the ministers, about half of these were people who had not attended in the morning. About eighteen people of both sexes in each 100 went to Sunday School [21] and five to the mid-week religious service. The equivalent of eight in each 100 of the male population aged five to twenty-one and eleven in each 100 of the females of that age attended the Sunday evening "young people's" services, though actually few under ten years and, in

[20] See for these and the following figures Table XX. Population was here taken as 38,000 minus the same percentage of Negroes as in 1920. The percentage of males and females was also assumed to be the same as that shown by the 1920 Census. In 1920 there were 103 males for each 100 females in the white population of Middletown.

[21] Athearn found that 49 per cent. of a sample of the urban population of this state under twenty-one years of age was enrolled in Sunday Schools. (*Op. cit.,* p. 62.)

the working class churches, a number of persons over twenty-one were present at these meetings and are included in the totals. Over the four-week period an average of 20,632 persons per week was counted at all five types of religious services, though this would mean far less than that number of different individuals.[22]

Careful estimates by the research staff of the ratio of the sexes showed 68 per cent. to be females in a total of 2,030 persons attending sixteen Sunday morning services in churches of all types. This follows closely the ratio of female to male membership in the largest Protestant church.

The habit of attending religious services appears to be declining as compared with 1890, according to the almost universal testimony of ministers and church members and according to the study of the church-going habits of the forty business class and 123 working class families interviewed. In fifteen of the former the entire family goes to morning church either three or four Sundays a month, as compared with twenty-four of the families of the parents of the present housewives in the early nineties. In seventeen (a seventh) of the 123 workers' families the entire family goes to Sunday morning church three or four times a month as compared with fifty-three (nearly half) of the wives' parents' families. Attendance by the entire family at the Sunday evening service has disappeared among this group of business class families, though it is still continued by eighteen of the working class families three or four times a month.[23] Sunday School attendance, while remaining roughly

[22] It is perhaps worth noting in this connection that some 31,000 admissions are paid to Middletown theaters each week, although this total includes people from the surrounding county.

[23] The Sunday evening service is a "problem" for ministers only surpassed by that of the prayer meeting in Middletown's business class churches. Of the half-dozen large churches in the center of the city it was commonly said in 1924 that only one man could get a crowd Sunday evening. The largest Protestant church in the city, with nearly 2,000 members, was experimenting with moving picture services, devoting, e.g., four Sunday evenings to pictures of the life of Lincoln, in an effort to find and hold a crowd. The one uptown church that consistently gets a crowd is that conducted by a brigadier-general who led an infantry brigade overseas during the war. The congregation is made up of working class people and less prosperous members of the business class. The minister is a Rooseveltian person who asserts, "Any man who dares stand up as the spokesman of Christianity and say it stands for pacifism is a liar. . . . If you find any one whom you can't talk religion into, by the good Lord you can knock the devil out of him!" Confident of his message, unperturbed by modern-

constant with the children of both classes as compared with a generation ago, exhibits some decline in both cases among adults. The increase in non-attendance at any religious services is marked among the workers.[24]

Reasons given by the women interviewed for going or not going to church, while they cannot be taken at their face value, throw some light on the changing status of this long-established habit: [25]

WORKING CLASS

To thirteen of the thirty-three working class women who usually attend, church-going would appear to be an unquestioned custom which needs no explanation:

"We're Catholics and go regularly."

"We believe in holiness and sanctification."

"I and the children are Catholics and go regularly. He's a Baptist, but he works seven days a week and don't feel like church."

ism, preaching on "Shall We Monkey with Evolution or Accept the Challenge of Evangelism?"—telling his crowded audience with a boyish grin that he wants to hurry the service along "because I've got a whale of a sermon for you!"—using the catchy old Sunday School hymns that everybody knows, always introducing innovations that bring the congregation into the service, it is not surprising that his church begins to fill up at seven and that frequently every seat is taken when the service starts at seven-thirty.

[24] See Table XXI. These data should be used cautiously; samples are small, and the factor of memory is involved, though in an activity such as church-going, performed over and over for years, relative accuracy of recollection is probably more nearly attainable than in less often repeated and less ceremonial activities. Every effort was made to check estimates, and the interviewers felt that answers approximated closely actual conditions in both periods. In cases where there was difference in attendance habits between winter and summer months, attendance was based upon the winter months of more regular church-going.

[25] Thirty-three of the 107 working class wives answering this question and twenty-five of the thirty-six business class wives go to church frequently enough to give reasons for *going* to church rather than for *not* going, although a larger number, particularly of the working class, go occasionally. Where the church-going habits of the wife differ from those of other members of the family, the reasons given represent the wife's view. The reasons quoted here are selected as most characteristic. The classification of answers is necessarily loose, and some comments could be classified in more than one group.

Thirteen answer in terms of definitely prizing the service:

"We like it. It does us good for the whole week, especially Mr. ——'s sermons."

"I feel better all week for going and looking forward to it."

"I get a lot of good out of it. Real Christians want to hear others' opinions in regard to the future."

"Folks can be better when they go to church. It keeps you in touch with the Bible teachings. People are influenced by things around them like church."

Three stress their children in this connection:

"After I was married, we didn't bother to go to church or Sunday School but when the boys started in he and I began going again too."

"I like missions better than regular churches because they do more for the children. My father used to be a mission preacher. We usually attend mission churches when we go."

"We were all raised to go to church. It helps you in bringing up a family to do right—and that's hard enough anyway in these days!"

Two emphasize a poverty of other alternatives:

"It's the only place I go, and I get just everything out of it."

"I like to go to church. It's the only place, about, I ever do go."

Two more go under the influence of special revivalist teaching:

"In April my man got converted and he's since been ordained a preacher in the Christian New Light Church. I used to go to church some but he never went. Now we both go all the time—wouldn't miss it! We read the Bible every day. We used to go to the movies once or twice a week, but since February we ain't gone at all, 'cause our church says it's wrong—and it saves money, too."

"Sin is the chief cause of unhappiness and evil in the world, and salvation is the one cure. The Scripture is being fulfilled in the evils that are abroad in the world now and I look for the second coming soon. Only a few will be saved. My man and I try not to let the physical interfere with our faith, but, as he says, it has to be provided for. I don't approve of no socials and doings like that—they're all of the world."

Reasons among the non-church-goers are more varied. Twenty of the seventy-four who do not attend regularly either have to work on Sunday or, after their week's work, prefer a day of rest and freedom to the bother of church-going:

"I could pick up and go to church but I just don't—I'm too tired mostly. My husband's a member of the Methodist church and he don't like to work Sundays but he just has to to keep his job."

"By the time I get the children ready it's too late to start and I'm too tired. He works hard all week and rests Sunday. I approve of church all right but just can't scrape up time and energy to go."

"Too many kids. The mister never went to church. I like to go but it's too much work to get the kids off. The boys used to go to Sunday School but they acted up so bad that I just took 'em out."

"Even when the DeKoven Chorus sang at the church my husband wouldn't go. He says he's at the factory so much he wants to stay home when he can."

"He don't like to go to church Sunday night. We've been away from church this summer more'n ever since we got our car."

"In the summer we'd rather get out in the car, because we've only the one free day during the week. In the winter I don't like to have dinner so late, so I don't stay for church. It's like my husband says: sermons are too long anyhow."

Twenty others say that they simply have not the habit of going:

"I just never got in the way of going. My oldest boy, he goes to the Epworth League. I ain't got nothing against church but I just don't bother to go."

"I used to go to church some in the country and used to miss not going when I first came here to town, but I don't even think of it now any more."

"We haven't got started in Middletown, though we used to go some in ——. Guess none of us are really interested. Some of our friends here criticize us pretty much for not going to church. I always try to act right, though. I feel awfully bad if I hurt anybody, and I try not to, but I don't find church helps me any in that."

"We used to go to the —— church, but it's too far away to go since we moved. The children go to the Brethren Sunday School. When I was working I didn't have no time for church anyway."

One of these gives a variety of reasons for having lost the habit:

"It's just awful and I'm ashamed, but we never go. The reason people don't go to church as much as they used to is lack of discipline in the home and lack of interest. Then there's other things: unless a person is prominent in the church or socially, people in the church have no interest in her. My mother used to be a Baptist, but she got broader in her views as she grew older—really a Universalist. Then she was in the amusement business, and the church condemned amusements—'specially Sunday amusements—so *that* shut her out of the church. I'm not the praying kind. Churches used to teach that God would do everything, but science is teaching that the early Bible legends are just like fairy tales, the Bible written by a man, and that we must rely more on our own five senses."

Thirteen speak of not finding enough of interest in the service to draw them:

"I don't find it worth while enough to make up for the sacrifice it means for me to go."

"I don't see anything to it. Just like going to a funeral."

"Just never got interested in going to church, though I go to revivals and things like that. My mother always takes the children to Sunday School."

"We don't like the minister at our church. He's a good man, but he can't seem to preach good sermons. My husband says he can get more from reading at home than he can from hearing him and I don't like to go to Sunday School and then not stay for church, so I just don't go at all."

"Church somehow isn't interesting, though I like the preacher at the Holy Roller church. I used to be Catholic and wish I'd brought my children up Catholics since all this Klan disturbance. The Catholics are better Christians than most Protestants and they certainly are better church-goers."

"I like to go to church when I can understand the sermon, but you know how it is, sometimes they are just too deep for you—you can't understand them."

"I used to go but sorta got out of the habit because I didn't know nobody and it's too hard to keep the family dressed. When one's shoed, the others are out. I don't miss going anyway. The church is just a mess of people that's went for years and years. There ain't never nothing new, and you get tired of the same old thing."

Eight appear to regard keeping the family together on Sunday as more important than church attendance:

"I used to go to evening service pretty regular, but the rest of the family don't want to go, so now I stay home. I like the family to be together. Just as good people stay outside the church as in it anyway. My children don't get anything out of Sunday School."

"Both he and I are tied down all week and need to get out Sundays. I feel it's better to keep the family all together than for me to go to church Sunday morning and my husband to be at home alone."

"My man has gone to ——'s church with me sometimes, but he says he just feels out of place there and I think it's better for the whole family to be together at the movies than separated and part of them at church. My two youngest go to Sunday School and never miss. Funny, isn't it, when we are children we love to go to Sunday School and when we grow up we can't bear to go to church!"

For six, financial and allied social considerations predominate:

"It cost too much and we just didn't get much out of it. People talk about tithing, but what's a tithe to some ain't a tithe to others. It's all right always to give to others, but when you see other people shiftless and you're saving and scrimping why ought you to give to them when they get in trouble through not having saved."

"I believe in bringing up a child in going to church; it gets him in with a good class of people. We don't have a close church connection any longer, though, because once when we couldn't pay our regular contribution they dropped us from the roll."

"It takes too much money to buy nice clothes for church, and if you aren't dressed up nice it's no place for you. Then, too, after the children get off to Sunday School I'm too tired to go. I won't go off and leave my house in a mess."

"We've scarcely gone to church at all since we came to live here eighteen years ago. There's too much dressing up for church these days; he used to go to church in his overalls just as good as anybody. But now it won't do for people to go without being all dressed up and we just haven't got the clothes for that."

"I like Sunday School more'n church. They talk more direct to you than in church. My man he don't like to go to church—says he doesn't know many people and feels queer."

"I don't care much about going. If you aren't dressed as well as other people they don't pay any attention to you. Oh, I get something when I go, but I don't go much. I like the Ladies' Aid pretty well."

Five say that they are kept away by the character of the people who attend:

"There's too many hypocrites in the churches. They're just all the time asking for money, trying to squeeze it out of you. This church over here is just full of Klan members, and I just ain't got any use of the Klan."

"After I married I drifted away from church. Then I began to decide I didn't *want* to attend anyway. There are too many people in the church that worship the almighty dollar—too many mean people that oughtn't to be there. People oughta live right, and then it won't matter about church. My girl sometimes wants to go to Sunday School and I won't keep her away—because she *might* learn some good. But I won't force any child of mine to go to Sunday School or church, either."

"I got tired of hearing hypocrites get up and testify so I just stopped going to church, but he goes to the other U.B. churches about town when there ain't a service at his own church."

"Sunday's the only morning we have to sleep. He says church is only a money-getting scheme anyway, and people are mostly hypocrites. If they really believed the things they say in church they would do something about all the things that are wrong. If all the people who have called themselves Christians for 2,000 years had really believed in Jesus and his way of living, they'd 've made the world a mighty different place from what it is now."

For one, "sin" is the obstacle:

"My husband hasn't salvation. There's no real reason for people's lack of interest in church. They just don't want it, because if they did, they'd have it. It's free for all."

And one is kept from the Catholic Church by fear of the Klan.

BUSINESS GROUP

Nine of the twenty-five business class women who make a practice of going to church say that they go because they find distinct value in the service:

"I think the church is the finest thing in the world. It means everything to me."

"I love the Episcopal service and always get something from ——'s sermons."

"My ideal is to be at church every Sunday. I always get something from church which helps me all through the week just as I do from reading *Daily Strength for Daily Needs.*"

"I don't go for the sake of seeing people as they are usually cold and formal. But I get something from the service and the sermon."

"You can usually hear a good sermon when you go to church, and I really have more respect for myself when I do attend church."

To seven, church attendance has never presented itself as a problem or they go as a matter of principle:

"Why, it just never occurred to me to question church-going!"

"I was brought up to go to church and I just feel uneasy as if something is wrong with the day if I don't go. I often get something from church that helps me through the week."

Seven others, if they have raised the question, regard regular church attendance as something to be done without debate, whether they enjoy it or not:

"I would hate to live in a community where there was no church. I must confess I am not as interested in church or in church work as a good many. Lots of the time I am just plain bored by church, but I feel I ought to go."

"Church going is one thing I insist on with my family. Loyalty to religion is the foundation of all the rest of life and the church is the one institution that represents it."

"Church doesn't fill the need with a great many people. I get something from it but lots of times I wish I didn't feel I ought to go."

"Mr. ——— and I go as a matter of principle. People ought to support churches. The children go to Sunday School because they enjoy it. The church used to be the social center in the community but it isn't now."

Two attend church primarily for the sake of their children:

"I haven't any very definite beliefs, but feel about going to church pretty much the way I do about bringing up children: I'm afraid not to go for fear it will be unwise and particularly for fear my child will be missing something valuable if she is not established in habits of going to church and Sunday School."

"I go chiefly because I think it sets a good example for the boys—though I really do get something out of it. When my older boy went to college I told him that of course I did not expect him to go to church and Sunday School every Sunday like he always did at home, but I wanted him to go occasionally and to have faith in something. If a boy has faith in his Creator, I think he will be all right whether he goes to church and Sunday School or not."

Six of the eleven who do not attend say that they are not interested:

"I just can't get to church. I have to spend so much time on the children that there's no time or energy left for me to go to church. Really, though, I don't miss going to church because I get so awfully little out of it when I do go. It's not that I think [my minister] insincere, but the service is so formal and just doesn't get at real problems. Then, too, our minister doesn't look at things theologically the way I do. I'm an admirer of Dr. Fosdick and he isn't."

"We don't go. I believe the church is a good thing for the community, and the beliefs it stands for are probably a good thing, but I've just pretty well lost any belief in church and God and immortality I ever had. If you do the very best you can, that's all any one can do anyway. Now our children go to Sunday School, but, when I come to think of it, I don't know why we send them. I never gave it much thought—guess we just do it like many people go to church, from habit. It's all pretty much a matter of habit—and, too, being afraid of what the neighbors will say if you don't."

"I went to church all the time as a girl and just got utterly sick of it. The first Sunday after I was married I said, 'I'm not going to church!' and I've never been much since. My husband just can't listen to sermons; he has tried over and over to go, but just can't swallow them. He often says if he expressed any of his real views people would call him an atheist. I don't miss going to church in the least and think Sunday a much finer day now than it was when I was a girl. When the children were little they wanted to go to Sunday School as the other children did, and I let them go. They went for a while but did not get much out of it and soon gave it up. My boy still says his prayers every night and he says I taught him to, but I can't ever remember doing it."

"I just don't see any particular reason for going to church, though I believe in a general sort of way the church is a good thing for the community."

Two say that they need to rest on Sunday or that too many other things fill the day:

"Of course I think going to church is the right thing to do, and I guess I usually get something from church when I go, but in summer we mostly go out in the car Sundays."

"I like to go, but it is just too much of a rush to get off in the morning. I like to go to prayer meeting better than to church because it isn't such a rush to get there."

Three have lost the habit of going:

"We don't ever go to Sunday night service now. It used to be that people didn't have anything else to do and went as a matter of course. People are just as religious as they used to be, but they don't always go to church to express it."

"I always went to all church services from the time I was a little girl. After my first baby came I decided to give up regular church-going and I never miss it now. I guess if that's wicked-ness, I am wicked [laughing]. Lots of people just go to church from habit and as a matter of course."

"We outgrew the Lutheran church we were brought up in and then went to the Methodist Sunday School for a while to hear —— and then just drifted away and now don't go."

Over against these statements as to why people do or do not go to church in Middletown may be set the reasons for going set forth by the churches in their advertisements and public statements by their ministers prior to Go-to-Church Week:

"It will make you feel better."

"Very helpful to every member of the family, from gray-haired Grandpa down to Wee Baby Bessie."

"The purpose of this church is to help humanity. We want your help in our effort to help others."

"The church is one of the fundamental institutions and should be supported."

"If tomorrow every church of every denomination in Middle-town should be filled to its capacity, that fact alone would give impetus to all good works indefinitely in this community. . . . Our physical presence in a church is a helpful thing spiritually and practically."

"Come and join us in our morning services next Sunday. . . . Note how much better you will feel the whole day long."

"No city has a more wonderful bunch of churches than has Middletown. Select one and make it your Church—the place where you can worship God, learn of His world program, be-come associated with the best people, and where you will find a glorious opportunity for service. . . . No man can be happy as a Christian unless he is active in some Church—working, giving, praying. Our ambition is to serve the largest possible number of people. We can help you in sickness, in sorrow, in trouble. But our greatest blessing for you is simply an opportunity to make your life count big for the uplift of people right here in Middle-town and unto the uttermost parts of the world."

Yet another reason for church-going, one that probably cuts deep into the roots of the habit, is voiced by the local press in an editorial, "Why Attend the Church?" "In the church of the right kind there is an atmosphere of soul peace and contentment that comes more nearly meeting human needs and longings for better things than anything the week days hold." In other words, as in the case of civic loyalty and patriotism, in church question marks straighten out into exclamation points, the baffling day-by-day complexity of things becomes simple, the stubborn world falls into step with man and his aspirations, his individual efforts become significant as part of a larger plan.

Chapter XXIII

RELIGIOUS OBSERVANCES

As giving money to the church and attending church meetings constitute the core of the formal religious activities of church members, so the Sunday morning church service is the core of these religious services. Around it the other activities of the church have been built up.[1] In most Middletown churches its chief feature is a sermon by the minister; accompanying the sermon are Bible reading, prayers, and the singing of religious songs. Fervor on the part of the congregation during the service appears chiefly in the small churches frequented by workers or in a few evening services during the singing of old, familiar hymns. The minister of a Protestant church of over 1,500 members said of his congregation: "My people seem to sit through the sermon in a kind of dazed, comatose state. They don't seem to be wrestling with my thought."[2] And yet,

[1] The discussion of religious observances in this chapter is based upon detailed reports by members of the staff of sixty-seven Sunday morning services, ninety-six Sunday evening services, thirty-one Sunday School services, eleven Young People's meetings, twelve prayer meetings, a total of 217, plus numerous reports of Young Men's and Young Women's Christian Association religious and social meetings and Bible classes, missionary societies, flower missions, men's club dinners, church annual meetings, and so on. These detailed staff reports are supplemented by long abstracts of sermons sent in by the ministers and appearing each Monday morning in the press, from one to six appearing each week, and by interviews with ministers, Y.M. and Y.W.C.A. secretaries, and other religious leaders.

All available comparable material for 1890, such as sermon reports in papers and announcements and reports of church meetings in the press and in record books, was utilized as a basis of comparison, but such early material was disappointingly meager.

[2] Evidence of this inertia is possibly afforded by the answers of leading ministers to the four questions:

(a) How many times during the past year has a boy between fourteen and twenty-one years come to you with an intelligent question growing out of one of your sermons?

(b) How many letters have you received during 1924 taking issue with what you said in the pulpit?

(c) Do you have a question box?

"People here are so soaked with the habit of listening to sermons," according to another leading minister, "that it's hard to get them to take any other interest in the church."

It is these sermons, forming the focus of the religious life of Protestant adults in Middletown, which are the chief means of diffusing habits of thought and action approved by the church to its members. A study of sermon titles shows them concerned with much the same themes today as a generation ago. All sermon subjects mentioned in the local press during 1890, a total of eighty, and all announced in the press during the two sample months of April and October, 1924, a total of 193, were classified by titles according to general emphasis—Biblical, theological, practical ethics, and so on—but the titles were, in most cases, so vague that the classification was felt to be of little value; no change since 1890 was apparent.

The prevalence of sermon subjects too general to afford an adequate clew to their content, "Not a Taste," "Three Men," "Pinch Hitters," "Law and Liberty," and so on, in a period when the general discourse is tending to disappear elsewhere, may be in itself a significant indication of their character.[3] And even the more definite sermon subjects suggest a type of

(d) How many questions has it averaged per month during 1924?

	Church 1	Church 2	Church 3	Church 4	Church 5	Church 6
Number of members	922	1,600	1,250	184	804	247
Question (a)	Very rarely	0	Not answered	0	3 or 4	0
Question (b)	1	0	0	0	2	6
Question (c)	No	No	No	No	Yes	No
Question (d)					⅙	

Answers to (c) and (d) were illuminating:

Church 1: "We had one once for three months and I took up the questions at the evening service, but I got only about one question a week, if that, and had to make up the others."

Church 2: "I got two or three questions a week when I tried it out three years ago—but they were all fool questions, like, 'Where did Cain get his wife?'"

Church 3: "We have no regular box, but I put one up from time to time when preaching a special series."

Church 4: "I didn't get one question a week—the whole thing was extremely negligible."

Church 5: "I get only one or two a year."

Church 6: (No comment.)

[3] It may be a commentary upon this institution of the sermon on a general theme that, when two high school teachers who had been to a leading church

diffuse treatment which, while still found in other public addresses today, is less dominant than in the nineties. Today, as in 1890, the subjects run week after week: "The Journey to Canaan," "The Sorrow of Jesus," "Sunday Observance," "Why Jesus Went to Church," "Play Square with God," "Jesus as Son of God," "Will Jesus Be at the Feast?" "The Importance of Faith," "The Call of the Universe," "The Seven Joys of a Christian," and so on. Indicative of what Middletown is wont to expect from its sermons is the comment of a veteran church worker asked to open a Sunday evening experiment in the form of a church forum on "What Kind of a Mayor Does Middletown Need?" He began, "There's something queer about this. This is a church all right, and it's church time—but this subject isn't the kind we're used to in church." An effort to make sermons gear in with the dominant values of the city appears in such titles as "Christ: From Manger to Throne" and "Business Success and Religion Go Together."

The determination of sermon subjects is suggested by the procedure of the minister of one of the largest Protestant churches: "Each year I follow a kind of 'Christian year' of my own—five sermons on the great doctrines of the church (the divinity of Christ, etc.); another series on heroes of the faith (Luther, etc.); a series on missionary heroes—these are pegs to hang our missionary appeals on; usually a group on the social implications of the gospel; four or five on denominational matters, and so on. My people eat alive the sort of sermons I deliver to children. My evening sermons are usually more popular—a series on the 'Music of Matrimony,' the youth in his teens being the solo, and so on through the quartet, drew especially well." Five of the six leading ministers interviewed said that they note no fluctuations in attendance due to an-

the night before were asked on Monday how they had liked the sermon, the following conversation ensued:
    First Teacher. "Let's see! I can't remember what it was about."
    Second Teacher. "Neither can I. I don't remember a thing he said."
    First Teacher. "It was awfully vague—he talked all over the place, and you couldn't put your finger on anything definite. Oh, I know! It was called 'A Country Without a Church'!"
    Second Teacher. "Oh, yes, he talked about a country without a child, a country without a Christ, and a country without a church. He said he couldn't imagine anything worse than a country without a church. I'm sure I can!"

nounced sermon subjects; the sixth said, "Moot doctrinal points draw the largest crowds and sermons on social questions least." The purpose of the sermon is suggested by the remark of a leading minister: "I put something at the close of each sermon that will exalt Jesus Christ." Another always ends his sermons "with a push for something to be done—something practical."

Adequate classification or summary of the content of these sermons to which Middletown listens is impossible, but their general trend may be gathered from the following excerpts selected as most representative of the sermons heard by the staff or sent in to the press by the ministers during the period under study. One should visualize the audiences as made up of about three women to every two men, with few children in the business class churches and many in the working class churches. The former audience is well dressed, many of the women being members of the exclusive clubs; the men, members of civic clubs, will be singing Monday noon to the tune of "Over There"—

> "Dy-na-mo, Dy-na-mo,
> We're the bunch, we're the bunch,
> Dy-na-mo!
> We're alive and coming,
> And we'll keep things humming;
> We're surely always on the go."

Tuesday noon they will be singing Rotary songs and perhaps being assured by their invited speaker of the general rightness of their world. The working class men and women come from cottages and bungalows; their world frequently presents more obvious discrepancies from their heart's desires. Sitting there, row after row, they listen to "the Word" expounded as follows:

"Why is Jesus Remembered?" (Press report; prominent business class church.)

Text: "The Son of man hath power on earth to forgive sins."

"This is the reason why we remember Jesus. It is because 'the Son of man hath power on earth to forgive sins.' The world's universal necessity is a remedy for sin. Your deepest need and mine is a cure for our sin. It is what every member of the human race must have if he is not to go lame and despairing out into the great beyond, for 'we have all sinned and come short of the glory of God.' . . .

"Oh, why, yes, some make light of sin. Some give themselves a hypodermic psychic injection of the suggestion that they are all right. Those that dispense such opiates are far more dangerous than those who dispense opium, morphine, or other narcotics. One affects only the body, the other deadens the soul. There is a vast difference between cauterization of the mind and regeneration of the heart. The auto-suggestion hypo may quiet the mind, but it never touches the fundamental necessity of the cure of man's sin and until man's sin is cured, his heart can never be right. That cure was made possible by the sacrifice of the Son of man upon the cross of Calvary."

(Staff report; business class church.)

"Soul winning is not a privilege of man but it is a *prerogative* of man! And there is a big difference. Why did not the angel tell Cornelius the way of life? Because it would have been treason against the eternal God had the angel told Cornelius. That was not the angel's business but Peter's. So, I say, soul winning is not only man's privilege but man's prerogative."

(Press report; business class church.)

"Oneness of purpose makes the cable that holds us true to the purpose whereunto we are called. . . . Oneness of spirit is the second strand of our cable. . . . Hope is the third strand of our cable of Christian unity. . . . One Lord and one faith are two more strands. . . . The sixth strand is one of baptism. . . . The seventh is one God and Father. . . . I pray that every one of these seven twisted strands that make our cable of Christian unity will be accepted by you and me so that when the stress and strain of life comes, it will be able to hold us up."

(Press report; business and working class church.)

"The Kingdom of God is the reign of truth, the assertion of righteousness, the acceptance of the divine mind by the human mind. It is divine thought, emotion, purpose saturating the human. It is love, sympathy, faith in every relationship. It is peace. It is power. It is victory. All nature waits the behest of the mind dominated by the divine. Seek the divine and the world, like a baying hound at the huntsman's call, will lie down at your feet. The Spirit's power is like the air, waiting for you to breathe it, but not forcing an entrance to your life."

(Press report; business class church.)

"There are certain great lines that can not be ignored if the republic is to endure. The first of these is a just respect for law.

. . . The second great essential is virtue upon the part of the individual and social purity upon the part of society. . . . The third essential is the acknowledgment of the absolute right of the law of God in human affairs. . . . If Christianity should fail, the republic would fail. And if America should fail, where is the hope of the world?"

"The Mighty Magnet." (Press report; working class church.)
Text: "And I, if I be lifted up from the earth, will draw all men unto me."
"We are greatly indebted to men of science all down the ages for their discoveries. . . . In the making of the earth, the All Wise Creator, placed within, and around, all the needed influences for the various working forces, whether seen or unseen by mortal vision. . . . Magnetism is an unseen force in the realm of nature, but a useful force.
"The All-Wise Heavenly Father planned for the uplift of humanity, and the attractive influence was His Only Begotten Son. Science tells us about magnetic ore, loadstone, magnetic oxide of iron. These forces have been wisely distributed, at no point overdone. . . . The greatest magnet known in heaven, or earth, is Jesus. At a point between the beginning and ending of time, when the human race needed an uplifting and saving power for man, the Heavenly Father gave His Son.
"We believe that Jesus is coming again; but the day or hour of His coming is not for us to know. The work of salvation and the kingdom forces will keep us so everlastingly busy we will not have time for speculative subjects. Evil influences are keen, and if it were possible the very elect would be deceived. The spiritual man must examine his chart frequently. The cross attracts. . . .
"The power of the cross has stirred the world more than any other influence."

(Press report; working class church.)
"The Treaty of Versailles is a joke and worse than a failure because it was framed and manipulated by greedy political tricksters rather than by Christian statesmen and unselfish patriots. It was a farce from beginning to end because the God at the center of the universe, whose throne is immovable, was absolutely ignored. His name was not mentioned. His blessing not asked. Infidelity wrote that document, and then we expect the good God to see it through. You will have to change every page of human history if you do not expect God to pass judgment on such infamy as that. . . .
"We are in the midst of a period of criticism in which certain so-called 'intellectuals' are trying to destroy the Bible and men's

faith in the fundamentals of Christianity, but the Bible still stands.
. . . The Bible comes from God and is the divine authority and
His revelation to men. No great piece of oratory or literature can
come from any other fountainhead. There could have been no
Patrick Henry, Abraham Lincoln, Wendell Phillips, or Daniel
Webster without it. . . .

"These may be dark days but there have been others in history.
. . . In the darkest hour of human history [Jesus Christ] is
coming back again. When He comes the world will have a King
that is above every other king and who swings the scepter of
peace and rules the nations of the earth."

(Staff report; working class church.)
"I have been asked, 'Is it right to pray for the Bearcats to win?'
by one of you who tells me he no longer believes in prayer because he prayed as hard as he could for the Bearcats and they
lost. I believe that prayer should be used only in cases where a
moral or spiritual issue is at stake. God *could* favor the weaker
team, but that would be unsportsmanlike of God."

(Staff report; Catholic church.)
"He who does not partake regularly of the Mass is saying to
Jesus that he will have nothing to do with him, and therefore
without the Mass no one can be saved. It is only through the
Mass that any one of us can escape eternal damnation. But I do
not want to threaten you. I want rather to make you feel the *joy*
of Holy Mass. Holy Mass is identical with Jesus' death on
Calvary. In it Jesus is praying for you."

(Staff report; working class church.)
"When a person gets religion he has (1) a new head to know
Christ and therefore a greater knowledge of Christ, (2) a new
tongue to tell of Christ, (3) a new heart to compass Christ.
Mohammedanism and the pagan religions are all hollow—there's
nothing in any of them. The only reason so many follow them
is because they are kept in ignorance—the same as the Catholics."

Sunday evening services are traditionally a somewhat more
informal and abbreviated edition of the morning services but
with the sermon still the central feature, and this type of evening service still predominates in Middletown, although it is
here rather than in the morning services that innovations are
beginning to be introduced.[4]

[4] For modifications of the Sunday evening service see below in this chapter.
It is characteristic of social change that it should come more slowly in
the morning service which is more heavily weighted with values as the
most solemn focus of the religious traditions of Middletown.

A special adaptation of the preaching service, a revival, occurs about once a year in many Middletown churches. Twenty-one of these revivals were reported in the press during 1924, six of them conducted by local ministers in their own churches and fifteen by imported evangelists. Revivals were frequent in 1890—one gathers, even more frequent than today—and in January, 1891, the press reported: "Revivals flourishing in all churches. The watchword is 'Middletown for Christ in January.'" A less constant feature of Middletown's religious observances than the Sunday morning and evening services, they are almost equally representative. Observation of one such revival conducted by the largest Protestant church in Middletown in 1924 exhibits many of the traditional characteristics of Middletown's religious customs as well as their modification under the influence of other changes in the local life.

"Why are you people having a revival?" a Rotarian and prominent member of this church was asked.

"To limber up the church. The church never has done anything without stirring up its emotions—same as the war spirit with its 'pep' meetings and community singing, or this 'Bearcat spirit' over here in the high school, or the Ad. Club's two competing teams, or the Klan, or a political convention—they're all due to emotions You come around to our meetings and see for yourself."

It was a revival conducted by the minister in person. A publicity man was engaged to keep the campaign in the press, a professional song leader was brought from the East, and the members were organized into a body of shock troops; weekly noon luncheons of the men of the church were inaugurated. The Friday before the revival started there was a " 'thuse session" for men at a church supper. After the professional song leader had taught a French song sung by the troops overseas, the *major domo* called on a man present to speak on "Organization."

This man, a jeweler, held up the framed parts of a watch and explained that, so disconnected, it could not run. Then he held up a map of one-quarter of the city with red dots, and blue dots marking off "fifteen groups of five to seven men each. By this organization Brother —— can in an hour's time transmit to the farthest corner of the parish any work he needs done. And not a man is refusing any task, however strenuous. But before

I close, men, there's just one thing: we've got to 'get right.'
I was touched the other night to hear a man I've always regarded
as a Christian say at Prayer Meeting that he was not 'right' and
was going to 'get right' before the meetings start—and, men, we
all ought to do it."

The chairman now called on an automobile salesman, who
said:

"Men, I said to a fellow in my shop, a good fellow, the other
day, 'Jim, are you a Christian?' 'No,' he said, 'but I am just as
good as Christians.' 'No,' I said, 'Jim, you're not. You may act
as well as I do, but if you have not accepted Jesus Christ as your
personal Savior you'll not be saved.' He went away and in two
days came back and said, 'Ben, you may be right. I want you to
show me how I can find out whether you are. And if you're right
you know I'll say so, and if I find you wrong I know you'll say
so!' . . . I got on my knees and asked God Almighty to show
him how to accept Jesus as his Savior. . . . Well, men, Jim came
back to me in a couple more days and said, 'I want to talk to
you.' And a tear ran down his cheek and he said, 'Ben, you're
right and I am accepting from now on Jesus Christ as my personal Savior.' Men, it is a wonderful feeling to lead a life down
the aisle to the altar and to Jesus Christ and to salvation. Men,
if you take the Bible and go out to try to show Jesus Christ to
other men you may not get very far, for there's lots of men
smarter than you. But, men, if you go out with Jesus Christ in
your heart through Him you can conquer all things. . . . Men,
we're set for a great meeting. I know we can have a rousing revival if we will!"

The minister was then called upon, the whole assemblage,
by now bubbling over with enthusiasm, rising and clapping:

"Men, I had another address prepared but a young man
came to me this afternoon and after we had talked he asked:
'Why is it that the church is so far behind other departments of
our life? Industry, science, invention are making strides rapidly,
but religion—!' I tell you, men, religion, the religion of Jesus
Christ, is the most up-to-date thing in the world. Philosophies,
sciences have all had flaws and have had to submit to change
but the gospel of Jesus Christ has never had to be changed in
the slightest detail. Behind the times? It is 2,000 years ahead
of the knowledge of the best men, and when we get up to it we'll
find it still unchanged, perfect. The reason why the church is
dead . . . is because the people outside so often can't tell the

difference between church members and others who are not members.

"I want to talk to you on 'Who is your boss?' Jesus Christ is the only one that has ever been able to bend men by his leadership and shoot them on and on to his great end. It is not we with our puny strength, our science and our philosophy, that will ever reach this goal. We are helpless without God Almighty. We are forever saying, 'I haven't time to do so and so for the church.' But I tell you, men, the only thing worth while is living the Life that God Almighty calls you to. Business is not the end of living—and I tell you the Christian business man gets more for himself and for his family than the other man. . . ."

What a revival in full swing looks like may be seen from the great meeting "For Men Only" on "The Follies of 1924"—promised as "A red-hot arraignment of present-day vices," and according to the advance newspaper notices of the campaign publicity manager, "the hottest meeting of the campaign." Tickets resembling theater tickets were distributed through all the downtown offices in advance of the meeting. On the appointed Sunday afternoon men packed the church auditorium, while 3–400 women crowded every inch of the basement to hear the minister's wife speak on "The Marys of Yesterday and Today." Upstairs the singing worked up in a crescendo to "The Old Rugged Cross," the Eastern song leader singing the verses and the full chorus choir of men the choruses. Meanwhile long wires were passed around on which those present impaled their greenback offerings. Then came the sermon:

"The follies of education and science have jazzed up the whole works, until it takes a man considerable time to find out what is true and what is false, but these are not the real follies of which I am to speak. . . . I want to introduce to you a fine crop of chorus girls who are positively indecent in their over-exposure. They are plump and well kept by the wealth of the modern society. I will now lead on the first group of three: Godlessness, Hypocrisy, Flippancy. Godlessness stands before us: tall, full-bosomed, long of limb, with all the make-up that allures a man. She is the mother of all the follies. Godlessness is at the root of the whole trouble with America today and with the whole world today. She is born of the devil and the whore of hell. . . ." [5]

[5] It is noteworthy that the appeal used both here and in the women's meeting to get out the crowd, by this church which condemns dancing, card-playing, and the theater, was identical with that used by the more sensational local movies. That the use of this appeal was not fortuitous is indi-

The sermon was followed by an appeal to come down to the altar to take "a forward step," and slips of paper bearing Bible texts and a friendly word from the pastor were distributed to the score of men who responded.

At other meetings enthusiasm ran equally high. At one a brass band was present, the church packed, and after closing a sermon on Daniel in the Lions' Den with "I tell you God will protect a man from anything, ANYTHING, if he will only pray to Him," the minister added, "We have promised a gift for every child here tonight and we've kept our word." Down the aisle streamed a score of men, each holding great bundles of toy balloons—red, green, yellow, purple—each inscribed with the name of the church and of the pastor. A gasp arose from the crowd. The minister climbed on a chair, put one foot up on the pulpit and smiled down on the excited business of distributing the balloons. Silence was restored long enough for a prayer, and audience and balloons surged out on to the streets.

After these revivals there frequently comes a let-down. The pastor of a large church exclaimed after a long campaign, "After years of importuning by my strongest men I yielded. But never again. My church reacted splendidly under the revival. It gave us a distinct quickening, but the curse of this church is the revival attitude toward religion that palsies a seven-day-week energy." Ministers of several leading churches oppose the revival type of service on similar grounds. But revivals continue to be held, and the tradition persists strongly; one leading business class church in 1924 made its revival preliminary to launching a heavy building campaign.

Although the sermon is traditionally the characteristic feature of most Middletown church services,[6] a few people complain, "The minister seems to think the sermon is the whole thing" and "The congregation ought to have more part in the service." Such comments usually come from the business group, but when one goes to the services in business class

cated by the press announcement of another great revival a year before that, "One of the greatest meetings of the campaign will undoubtedly be that for women and girls only at the tabernacle tomorrow afternoon, when Dr. —— speaks on 'Tricks of the White Slavers and God's Woman.'"

[6] The question may properly be raised whether verbal conditioning through the accustomed religious symbols utilized in preaching can expect to hold its own in the face of the massed conditioning of commercial advertising, the movies, and other agencies preaching to Middletown day in and day out through the eye.

churches particularly designed for participation by the people, the mid-week prayer meetings, one usually finds but a handful of the faithful present and their activity in prayer, discussion, or singing only forthcoming after considerable urging from the minister, who here, too, acts as leader. "Prayer meeting is the modern minister's nightmare," exclaimed one pastor. "He realizes its value, potentially, and the desirability of its quickening influence throughout the church, but it is desperately hard to make it go." "The preachers have talked prayer meeting to death," was the summary of a Rotarian, an active church worker, while the ministers insist that they talk only because nobody else will.

The usual prayer meeting in the largest Middletown churches consists of twenty-five to fifty persons, the large majority of them adults over forty, with women predominating as in the Sunday services. A new minister in the largest church in Middletown succeeded in raising the attendance temporarily to 150 by reducing the usual prayer meeting to half an hour and following it with a more informal discussion meeting. In another church, with the vigorous ex-brigadier general as pastor, over a hundred people are commonly to be found at the mid-week meeting, the majority of them working class people. In this church, as elsewhere, the subjects treated are much the same as in the Sunday services, and the central feature of the meeting is a talk by the minister differing from a Sunday sermon chiefly in being somewhat more informal and practical in its application. Here, too, some prodding is necessary to get the people to take an active part in the service, but one cannot listen to their earnest "testimonies" or see their eager response to the boyish enthusiasm of the pastor without realizing that it is something more than an habitual form, more even than the friendliness which peculiarly characterizes this church, which brings these people together, many of them coming out after a nine- to ten-hour day of factory labor. In most churches, however, attendance at this "family prayer service" of the church is no longer regarded by the majority of members as either a privilege or a duty; they simply omit it.

As children are encouraged to join the church at an early age, so attendance at church services is expected of children as well as of adults, but special types of services, Sunday School and Young People's Meetings, address themselves especially to the young. The former are in Middletown largely attended by adults as well, some even substituting Sunday School attendance for attendance at the preaching services.[7] Sunday School consists, in part, of singing, prayer, and occasional brief talks at the opening and close, but its central feature is the study of a "lesson" in small class groups modeled somewhat on those of the secular schools; opening and closing exercises usually occupy thirty to forty minutes and the lesson twenty to thirty minutes. In almost all Middletown Sunday School classes these lessons are taken exclusively from the Bible. The aims of the Sunday Schools were variously stated by the ministers: "To make intelligent, well-instructed Christians who have convictions." "To bring the pupils to Christ and teach them in Christ to be real working Christians." "To develop Christian character through Biblical instruction." "To teach the children righteous living."

Direction of the Sunday Schools and the actual teaching of the classes is entirely in the hands of volunteer workers; none of the six largest Protestant churches, which exhibit religious training in Middletown in its most developed form,[8] has a paid director of religious education or any paid teachers or administrators. One church has a committee on religious education, made up of both members and teachers of the Sunday School which meets monthly, two others have such a committee which according to the pastors "never functions at all," and the other three have no such committee. In each church the superintendent of the Sunday School is the person in charge of religious training; of the superintendents in these six churches in 1924 one

[7] The ratio of adults and children in Sunday School varies from church to church. Fewer adults attend in the more sophisticated churches. In one of the largest churches of the city, a favorite among the more up and coming workers but also having a considerable membership from the business class, the average Sunday School attendance in February, 1925, was 645, of which the adults averaged 165 a Sunday, the young people's department 135, the junior department 115, the intermediate 105, the primary fifty-nine, the home department sixty-six. In another prominent church, whose young married people's Sunday School class of over 200 constitutes a church within a church, a leader in the church remarked bitterly that the class was a "positive menace! The function of the Sunday School is to train the young so as to feed the church. Instead, these people turn out to Sunday School two or three hundred strong and hear a lecture and then don't stay to church, or stay to church and don't come to evening service."

[8] This applies to the Protestant churches only. The Catholic church has nuns of the church to instruct its children at certain hours during the week, and also conducts a private school for Catholic children who pay to attend.

was a banker, two were manufacturers, one a wholesale grocer eighty-four years old (superintendent of this school for forty-five years and succeeded in 1925 by a cattle breeder), one a department manager in a men's clothing store, and one a coal dealer. These men, with one exception, have full charge of the "service of worship" as well as the rest of the program; in the one exception the superintendents of the various departments with each class in rotation are in charge. Two are high school graduates, one "went to school when he could not farm," one had "probably only a grade school education" according to his pastor, one "went to college but may not have finished"; no information was secured regarding the sixth superintendent except that he certainly did not go to college. None of the six has had any training in pedagogy or in religious education; as one minister said, "Our Sunday School staff are just plain, well-meaning folks," and according to another, "I can't push my superintendent in these things, or he says he is not well trained enough for the work and wants to resign." None of the six had read Athearn's study of religious education in the state, including the city of Middletown, published two years before, though one had "seen reviews and mentioned the book" to his pastor. Two had tried during 1924 to introduce more beauty into the service, one by having each class in turn conduct the service of worship, the other by attempting "to break away from jig tunes, to break up the sameness of the service, and about twice a month to have various classes take charge." One Sunday School has a capable school teacher supervise the teaching; in another the minister "tries to supervise"; the other four make no attempt at supervision.

Asked as to the number of their students who prepare their lessons, the ministers replied:

"That's a sore spot—not 10 per cent., I suppose, though I believe 75 per cent. of the teachers do."

"My wife estimates that 20 per cent. read it over, but practically none really study it. And the girls have out their vanity cases and are powdering up before the lesson is even through!"

"Certainly not over 25 per cent."

"It varies by departments, but in all our secondary department work it may possibly run as high as 10 to 15 per cent."

"I never thought of that—doubt if there are 10 per cent., though. As to teachers—well, you know Sunday Schools!"

The sixth minister did not answer this question.

With only a small proportion of the pupils preparing the lesson, conduct of the classes falls doubly upon the teachers. Teaching of the lesson usually takes the form of the reading aloud of the Biblical passage of the day with comments and questions on its application to conduct by the teacher and occasionally some discussion on the part of the class; at least as much as in the public schools, the lecture and question and answer method predominates over discussion.

Abridged reports of a few classes, selected as most representative of those visited, set forth the nature of this activity: [9]

After the opening exercises in the auditorium of one of the largest churches, a class of twenty-two young men from about twenty-five to thirty-five, chiefly from the working class, retired for the lesson. The teacher, an earnest man of about forty-five, expounded the lesson on the text, "He came unto his own and his own received him not":

"Men, if we ask God, He will help us to get much good from this lesson." He then prayed that God would "touch the members of the class," and began explaining the passage for the day. In conclusion he said, "God doesn't use just prominent people in His work. He uses everybody. Your influence can work miracles for Him. If you go out on an interurban you'll see in the woods full of old stumps beautiful flowers. God placed those there just to show men He could place them there where no man planted them." At the close of the hour one member of the class congratulated him on "that fine talk"; members of the class took no part in the lesson but listened attentively.

Three hymns which seemed rather unfamiliar to most of those present formed the opening exercises in the largest Sunday School in Middletown; much chatting went on during the hymns. A class of girls of eighteen to twenty-five years of the less prosperous business group was visited; thirty were present out of a total enrollment of 111. The teacher, a woman of

[9] The thirty-one classes visited were a random sampling, covering classes and churches of all types, without reference to the lesson or program scheduled for the day.

about fifty, explained the lesson on "Nehemiah Restoring God's Ways in Jerusalem":

"This rebuilding of the wall can be applied to us. After we have sinned we must tear down all the débris and start anew again. When the wall is completed, we then have shut out from our lives everything of which God disapproves. For this reason, He desired that the people of Jerusalem be walled in, because they were His chosen few. . . . Today we probably have as many if not more temptations than these early people, but we have no excuse for sinning, because we all have access to the Bible. We should study it daily. . . . The peddlers tried hard to gain access to Jerusalem on the Sabbath, but the doors were ordered locked and the tradesmen kept out. Now today civilization has reached such a stage that some people have to work on Sundays, but they don't have to dance and spend all day at —— [a popular resort]. . . . We should all take a decided stand against this desecration of the day and should let people know what we think of the matter."

The entire lesson period was a lecture by the teacher. She said later that she had formerly tried to have class discussion, but the girls did not like it. She had been in the habit of handing out slips to the girls in rotation, assigning a different phase of the next week's lesson to each girl, but she found that those who had slips were inclined to stay away the next week, and the girls agreed that they would rather hear her talk. At the close of the lesson all classes, old and young, reassembled for the closing exercises and were taught by the pastor the following motto, which they repeated in unison after him: "Sunday School Scholars Saved from Sin and Satan."

A Bible class of some seventy-five men in the second largest Protestant church, most of them in their forties or older, and nearly all working men, studied for more than a year the book of Revelation. The teacher, a Middletown doctor, was set forth on the card of welcome handed each person as he entered as the state's "great Bible teacher." One Sunday morning the class was greeted by the words on the blackboard at the front of the room:

"'Blood unto the horses' bridles'—what does it mean? Rev. 15:7, last plagues," accompanied by a drawing of the plan of the Jewish temple. Much time was taken up discussing the meaning of the phrase and also the statement of the teacher that "none of us will enter the 'Holy of Holies' until all of the last seven plagues have been fulfilled and all men and all churches, each and every one, have been scourged." One or two men stood and read from the Bible corroborating passages from the Old Testament. Another man rose and asked, "Christ was in the grave three days. Now a day, we are told, is 'as a thousand years.' Doesn't that mean we'll have to wait three thousand years after Christ? Two thousand years have passed, and therefore we'll have only a thousand more to wait?" "You're exactly right," rejoined the teacher. "It won't be long now." Another member of the class rose towards the close and said good-naturedly, "Doctor, what's the use in our going home as you suggest and reading and thinking about this? We'd just simply get to arguing about it. I want to read another passage that's different from all we've been saying. It's here in this same Bible." He read a passage likening men to animals in their ultimate end. The teacher passed swiftly over the interruption.

In the Sunday School of another large Protestant church, after the singing of three hymns, the superintendent, a trembling white-haired man, announced, "We will follow a different order for devotional exercises today, repeating scripture verses and having sentence prayers. The time is very short, so lose no time in giving your verses and prayers." A dozen or fifteen people, all adults, repeated Biblical verses, and about an equal number, most of them apparently active workers, such as the pastor and his wife and Sunday School teachers, gave prayers. Then the classes marched to their various rooms to the accompaniment of lively music by the Sunday School orchestra:

A class of forty-eight, eleven boys and thirty-seven girls, all in their late teens, largely from the business class, was taught by the pastor's wife. The lesson period consisted of discussion by the teacher on the baptism of Jesus, punctuated by her requests for the reading of certain verses and such questions as "What does water symbolize?" "What does the dove symbolize?" "Does any one happen to know what the old idea of the coming of the Messiah was?" In the course of her talk she brought out the following points: "Jesus was thirty years old before he began to take up his life work. If any of you young men and women begin to be impatient when you get to be twenty-five or twenty-eight and have not yet accomplished anything or found your work, just think of this long training of Jesus. . . . God has a special place for each one of us where we can best do just exactly what He wants us to do. . . . Jesus was a perfect man, our ideal and our pattern, who had never committed any sin; why did He have to be baptized? . . . Satan is always going about trying to take advantage of every opportunity. How do we receive warnings from God against evildoing?" Leaflets containing the lessons for the "quarter" were passed out at the beginning of the period and collected at the close; apparently there was no expectation that any one would study the lesson during the week. The class closed with an earnest prayer that "each one of you young men and women may hear the voice of conscience."

The Mothers' Class, of twenty women between forty and sixty, in a business class church was exceptional in being taught by a school teacher, in not confining its study to the Bible, and in having a large amount of discussion. The teacher began by speaking of the possibility of employing a supervisor of religious education for the public schools, saying that it was a disgrace that Middletown had lost the one person trained in religious education it had ever had, because the church which temporarily employed her would not pay the $1,200 a year which she was now receiving in a neighboring city. Members of the class discussed the question freely:

"I think the Bible should be put back into the schools and read every morning to the children."

"I don't think so; I don't want just anybody interpreting the Bible to my child. One teacher *admitted* to me that she is an atheist."

"Oh, isn't that dreadful!"

"Yes, I know who you mean. She is a very intelligent person and very fine, but she certainly has one great lack."

"If parents would see to it that children were definitely affiliated with some church before they go to college they would have something to keep them steady and not get side-tracked."

"I've heard that lots of college professors are atheists and many teachers in schools in this country are atheists or Catholics."

The teacher brought the lesson back to the Bible by asking the class to name three of the great kings of the Bible; one said Hezekiah and was corrected. Several members spoke enthusiastically afterward of the class and the teacher, saying that she had introduced discussions which they all enjoyed.

The Sunday School classes described above, are in the larger, uptown Protestant churches. The small, South Side Sunday Schools are usually held in the church auditorium. A noisy earnestness characterizes the opening exercises, punctuated with much bustling to and fro by late-comers and secretaries. Hymns are sung lustily, especially the favorites with repeated choruses such as "I'm going to go to glory by and by." During the prayers of the superintendent adults may utter frequent "Amens," while restless children stare about or idly pinch each other. The whole school listens to the reading of the lesson, after which it breaks up into little groups scattered through the room, each with an earnest teacher leaning over the back of the pew in front, facing his class.

In one representative adult class hour the lesson for the day was "The Calling of the Twelve":

"Who was the most popular disciple?" queried the teacher. This led to a question from a member of the class as to whether Judas Iscariot was a popular disciple, and it was decided that since Judas was "a tool of Satan," he probably was popular, "because they are always popular among worldly people." Whereupon a woman in the class asked, "Well, isn't a Christian always unpopular?" The class decided that "a man or woman cannot abide by the teachings of the Bible and still be popular with the wicked men of the world today." Thereupon a man asked, "Who is a Christian?" and the question was capped by the answer, "He that doeth the will of the Father." The discussion swung back to Judas and some one asked whether it "was foreordained that Judas should betray Jesus." Opinions differed, some feeling that Jesus had been sent into the world to be betrayed by Judas, and others that the betrayal "just happened." At this point the class was interrupted by a bell from the platform, and the superintendent asked for "anybody wishing to leave a message"; there being no answers, the collection was announced, school papers distributed, and the school clattered out into the sunshine.

The Young Married People's Class of a thrifty working class Sunday School, with twenty-two women and seven men present, met in a bare little room off the auditorium with a stove in one corner; several had children in their arms, and one young mother was giving her baby its bottle. The teacher, the minister's wife, spent nearly half of the lesson period discussing ways by which the class could raise money toward the church debt:

"We raised $46 last year by giving a play. This year they want us to give $150 instead of $100 as we did last year. If we get an early start this year and give another play, perhaps it won't be so hard."

"Is this a personal pledge?" asked a man.

Volunteers were called for to pass handbills for a bazaar to be held the following week; the woman with the tiny baby and several others volunteered. Then, "What is the lesson about?" asked the teacher, and two or three persons, reading from their leaflets, answered, "Difficulties and Rewards of Confessing Christ." "Now read in unison the first verse." Verse by verse the lesson was read by the class, each verse followed by a few comments from the teacher, such as "I know it's hard to see His guiding love with so much unemployment, but we must just continue to trust and know that He will bring us through."

These bald statements, set down as accurate descriptions of what a person going to church and Sunday School in Middle-town sees and hears, cannot adequately represent the all-important consideration of what these services mean to the Middletown people themselves. When business class people send their children to Sunday School in the effort to conserve and pass on what they feel has been an essential foundation of their own lives or to recapture something lost and wistfully remembered, no mere piling up of verbal descriptions of the services can set forth the thing they seek. When an overwrought working class woman, ill-dressed and unkempt, rose in a noisy Pentecostal church and cried, "I'm tired of this ol' garlic and onions world! I'm going home to Jesus!" no simple description can convey the earnestness of her wailing words. The crowds which press forward with eager questions to the medium in the Spiritualist Temple every Wednesday and Sunday evening, asking, "Is my boy still living?" "Shall I go into the business deal I am considering?" "What is my next step?"—have all the earnestness of those who throng to touch the hem of the robe of a Savior. It is not the quality of the music which makes the singing of "Let the Lower Lights Be Burning" or "There Will Be Glory for Me" significant to the listening Sunday School. And when a woman, said to be one of the most conscientious mothers of the city, prays in the Mothers' Class of a leading Sunday School that "we may be forgiven for our many grievous sins of this past week," it is not what she says but the complex of tradition, human ties, and common difficulties and aspirations that makes the moment urgent for many of her hearers.

As is apparent from the above reports, some variations in conducting Sunday Schools exist, but conferences with six leading ministers and conversations with many Sunday School teachers and members of classes confirm the reports on the classes visited that, for the most part, going to Sunday School is much the same sort of activity in all classes and in all churches and has changed relatively little since the nineties.[10] Efforts to change the traditional methods of conducting Sunday Schools appear from time to time. In 1924 the Ministerial Association launched an ambitious twenty-hour night-school in religious education aiming to impart to Sunday School teachers some of the newer methods being adopted by the secular schools; 103 persons, largely Sunday School teachers, attended at some time, although only thirty completed the course.[11] Three of the six Sunday Schools studied use graded lessons for children up to high school age, aiming to base each year's work on selections from the Bible suitable to the needs of each age group. One other stresses particularly having pupils in each class of approximately the same age, although all classes study the same lessons. In several Sunday Schools kindergarten methods have been introduced for the youngest children, making the Sunday School hour for them a thing of games and handwork as well as the telling of the Bible story.

[10] See Athearn, *op. cit.*, for a study of Sunday Schools in this state, which included Middletown in its investigation.

[11] The courses taught were: The Organization and Conduct of the Sunday School, taught by a professional worker in religious education using Roll's *Organization and Administration of the Sunday School* as text; How to Teach Religion, taught by the same woman using Betts' book of that title; The Teachings of Jesus, taught by the professor in a small neighboring college, also using a text of that title; Bible Manners and Customs, taught by the same man using Mackie's *Bible Manners and Customs;* Biblical Geography, by a local minister who had lived some time in the Holy Land; How to Teach Intermediates, by a local minister, using Lewis' book, *The Intermediate Worker and His Work;* Lessons from the Old Testament, also by a local minister; Child Psychology, by the public school psychologist, using Myer and Chapell's *Life in the Making.* According to the director of the school, the course stressed the Bible as the center of Sunday School work, and the aims of the Sunday School were stressed, in the order listed: (1) To make better boys and girls; (2) to arouse interest in the Bible and teach the facts contained in the Bible; (3) to train up church workers; (4) to arouse interest in religion in a wider sense than exclusively the Christian religion.

When new departures crop up in adult classes they encounter more opposition. These innovations, like the others mentioned, are almost invariably confined to business class churches, and the difficulty they meet is illustrated by two such ventures in one of the most prominent of these churches:

In the class of older men, aged forty and over, which had "always been a dead one," according to one of the members, a teacher was secured who "was thoroughly alive; he explained the Ten Commandments in terms of Hammurabi's code and suggested that Moses cut them on stone himself on Sinai. We studied the Crusades with him—criticizing them freely, too! Each week he would assign collateral reading from really authoritative source books to two or three of the members and they'd report on them the next week. But some of the other members began to think him too liberal, and they'd just sit glumly twirling their thumbs all through the lessons until it got so uncomfortable for him that he quit."

There were glowing accounts about town from some of the most active young business men of a class of about forty men in their twenties conducted in this same Sunday School by an energetic teacher in the local normal college. Under his teaching the class grew and extended its work, in its heyday contributing $3.50 a week to help a girl through high school. A member of the staff attended this class on "rally day" in the fall of 1924. There were only six present and no teacher, and it was moved and seconded that the class disband and return to the older men's class. Here was the story as it came out that morning: "The trouble with —— [the once popular teacher now no longer with the class] was that he just used the Bible as a side issue. Why, he'd spend the hour discussing heredity and environment or reading aloud and discussing things like that book on the Syrian Christ. Or he taught the psychology of religion here Sunday after Sunday—good stuff, you know, but not the Bible, and the Bible *must* be central."

"Yes," interrupted a second speaker, "my experience is, the best classes are the ones that study the Bible and don't make it secondary. I think we ought to start at the beginning and go right through the Bible."

"A group of us members of the class," interposed a third man, "met over at the 'Y' two or three times and talked of tipping [the teacher] off how we felt."

"Yes, but he felt the opposition all right and began coming less and less till he finally stopped."

The Sunday School is the chief instrument of the church for training the young in religion; of secondary importance is the "young people's society" meeting in each Protestant church Sunday evening for an hour before the evening service. Like the prayer meetings, these discussion hours attract few people, attendance ranging from about fifteen to fifty, almost invariably with girls in the heavy majority. As in the case of the Sunday Schools, some adults attend. The purpose of these young people's groups as expressed by the ministers of the six churches in which they are most active is:

"To help the young people to grow in Christian character; to express themselves religiously and thereby deepen their religious character and so serve among the young people."

"Training for service."

"Expressional religion."

"My ideal is that this will be the expressional outlet for our religious educational program, but I haven't had much success. They don't carry out any projects and about all we actually accomplish is to keep the stream flowing and hope some day some leadership timber will come along and we can really get the society started. It's our old problem of leadership. I get an occasional real potential leader, but the complex social and high school program promptly gets these youngsters away from me."

"To link the spirit of the young with religion. The church everywhere is mainly middle-aged and snubs the young."

"Same as our Sunday School—to teach the children righteous living."

Five of these young people's meetings follow in the main set programs issued by their respective denominational headquarters, and the sixth centers its work in the Bible. The programs for each of the six for the four weeks preceding the interview illustrate their character:

Church I. Friendliness as Expressed Through Medical Missions.
How to Read the Bible Helpfully.
Installation of New Officers.
How Can We Develop the Spirit of Friendliness?

Church II.   Denominationalism.
              Missions.
              Friendliness and Good Will.
              "Mrs. —— told story of underground railway and
                  other high points in the history of the denomina-
                  tion."

Church III.   Here Am I—Send Somebody Else.
              Do We Spell Christmas with a Dollar Sign?
              What Kind of Home Life Does the World Need?
              Why Obey the Government? Why Obey the Church?

Church IV.   What My Bible Means to Me.
              What My Church Means to Me.
              What My Church Means to My Community.
              What My Church Means.

Church V.   Missions—Debate on Evangelical vs. Medical Mis-
                  sions.
              How Make our C. E. More Efficient.
              How to Develop a Spirit of Friendliness.
              How Can We Read the Bible Helpfully?

Church VI.   Immigration Restriction—Debate.
              Preparedness—Debate.
              Missions in Japan.
              "A man from a neighboring town talked on re-
                  ligion, ideals, etc."

These services in all churches follow much the same plan;
the abridged reports given appear to be widely representative: [12]

In one of the largest business class churches the meeting opened
with the singing of "Come to the Church in the Wildwood" by
four young girls. The leader, after announcing the subject, "Industrial Missions," and relating various stray facts about the denomination's missions in China, called upon a second girl to tell
of the work of one of the denomination's leading industrial missionaries. This girl repeated incidents she had read for the occasion. There followed a hymn, and a woman in her late twenties
read for ten minutes an account of the station in Africa where
this particular Middletown church supports a missionary. A boy
of eighteen who had attended a Y.M.C.A. summer conference
related some incidents regarding the life of one of the industrial

[12] There has been no selection among the following services. They are
literally set down here in the order in which they were visited, the visits
being a random sampling with no previous knowledge of the subjects to be
discussed.

missionaries. Then came sentence prayers asking that God would "bless and help the missionaries," and the service ended with another number from the quartet.

The young people's group in a South Side church composed entirely of workers, including, as is usual in working class churches, many older people, met to discuss: "Abolish War? Why? How?"—a program evidently sent out with full references from denominational headquarters. The leader, after giving out printed slips for various people to read, the latter usually seeking to avoid being thus pressed into service, opened the meeting by reading from her own slip and then called for "the first Scripture reading." An elderly woman who had been given that slip arose and fumbled through her Bible for the passage. Other slips and Bible passages were then read. A man of forty, who announced that he "always had something to say," said he had never thought of war as particularly evil until he had come there that evening, but he was ready to go on record as feeling that war could be abolished "just as soon as we Christians care enough to band together and make the will of God prevail." At this point an elderly man, called on for "any remarks," said if he knew his Bible there wouldn't be any peace "until Jesus comes, and there'll be wars and worse wars before that time comes." Here the minister was called upon and seemed embarrassed as to how to answer, saying that the Bible said there could be no peace till Jesus comes, and he certainly believed the Bible, but still he thought it might be worth while to oppose war as much as possible. The discussion ended with one faction feeling that "they," "Congress and the others in charge," ought to stop war or that "Christians" ought to stop war, and the other affirming that wars are inevitable.

A young people's meeting in a church catering particularly to students in the normal college was devoted to "Our Missions in China," led by a boy of about nineteen. Several hymns were sung and a boy and girl were called on by name for prayers, each thanking God for the blessings that we in this country enjoy. The leader went through a long recital of facts about China with considerable difficulty in recalling the names of the places and the details of the incidents he was describing; he followed this by the reading of a long newspaper article on "Impressions of China" by a Middletown citizen, during which his hearers showed a good deal of restlessness. At the close he asked if any one else had anything to offer on the subject. No one did. The minister who stepped in for the last few minutes said that he particularly wanted to emphasize the fact that there are "80,000 Chinese soldiers, all perfect physical specimens of manhood and no one of them diseased or in any way unfit. It gives us pause that although the Chinese are not Christians they have such a fine group of men. If we want to Christianize them we, too, must make ourselves fit."

The testimony of those who have followed Middletown's churches over a period of years is unanimous that there is less spontaneous interest in these young people's services today than thirty-five years ago. Typical of the programs of both periods apparently is the press notice of a young people's service at the most prominent church in 1890, concluding, "Let all come and contribute a word, a verse, or something."

Religious training is also carried on under the elaborately organized system of Bible teaching in the schools under the Y.M.C.A. and the Y.W.C.A. In the spring of 1925 over 2,000 boys and more than 1,000 girls were enrolled in these Bible classes, and in nine out of the last eleven years the Middletown Y.M.C.A. has won first place nationally in the number of Y.M.C.A. Bible diplomas received. Taken by themselves, these Bible classes cannot be fully understood; they are inextricably interwoven with other activities of the Association. Free movies are offered weekly by the Y.M.C.A. to boys attending Bible classes; the entire athletic work of the Association centers about these classes, and thriving teams of the various classes compete throughout the year; most important of all as a drawing card is the large number of free one-week trips to the "Y" summer camp at "the lakes" ninety miles away, given each year to boys who have high grades in the Bible examinations. "The announcement made at the opening of the Bible study work that the winners would have all expenses paid during camping trips proved to be a great incentive to better work," says the press. While the classes are nominally optional, informal pressure is said to be used by many of the school teachers to induce children to attend, and according to some teachers, those who are unwilling to exert such pressure are subject to considerable criticism.

Some clew to the things taught in these classes may be gained from the following: In a class of boys on "Athletes of the Bible," taught by one of the Y.M.C.A. secretaries, five minutes of the twenty were devoted to the teaching of the

Bible and fifteen minutes to athletic talk and announcements about the Bible class basket-ball league and the free "Y" movies for boys attending the Bible classes. A college-trained woman teacher, following her instructions to "teach the straight Bible and avoid discussion," "skipped the second creation story in Genesis because it would make trouble" and expressed herself as baffled as to how she would "teach the story of the angels walking with Abraham, because, you know, the children will find out eventually about angels and think I tricked them." Another prominent Bible class teacher expressed herself as satisfied with the teaching in these classes, on the whole, for "the children respect anybody who can speak with conviction, even if they find later they do not agree." The Y.M.C.A. apparently likes to keep this Bible study work in its own hands and is commonly reported to oppose unofficially the Batavia plan of Bible study as, also, it has unofficially opposed the independent development of the Boy Scout movement in Middletown.

Daily vacation Bible classes are conducted during six weeks of the summer in two churches. The activity of these classes centers on the learning of passages from the Bible and is supplemented by games and handicraft work.

Despite all this industry of teaching the Bible, religious leaders in Middletown express themselves as dissatisfied with its results. "One simply cannot presuppose any familiarity with the contents of the Bible in preaching in Middletown," declared a prominent minister emphatically. Leaders of the Mothers' Council join with certain ministers in urging religious education in the schools under a paid director, "because we all know that children don't spend enough time in Sunday School to get much knowledge of the Bible, and the Sunday School teachers we have aren't equipped to give it to them." A further clew to this dissatisfaction may lie in the uncoördinated character of much of the Bible teaching. The religious training afforded by the Y.M. and Y.W.C.A. is unconnected with that of the churches, and even within the churches much the same situation prevails. In five of the six leading Protestant churches there is no direct effort at coördination between the Sunday School and the Young People's Society; in the sixth the liaison consists in one member of each Sunday School class of Christian Endeavor age being appointed to the Christian Endeavor Look-out Committee, which gets the Sunday School children out for the monthly cocoa suppers before the Endeavor meets. Giving children greater knowledge of the Bible is an avowed aim of religious leaders in Middletown, but it is not always clear precisely in what this knowledge is to consist or how it is to be attained and used.

These five forms of religious observance—Sunday morning and evening preaching services, Sunday School, mid-week prayer meeting, Sunday evening young people's meeting—remain, fundamentally unchanged since the nineties, the outstanding features of the worship of Protestant Middletown. In some churches children's sermons and paid choirs have been introduced into the Sunday morning services; greater variations occur in the evening services in the holding of an occasional forum in one church, an annual service for railway men in another, and in a third a motion picture service, which lived a brief life before being killed by the strong opposition of one group of church members; various innovations described above have at some points altered the Sunday Schools; the use of "outside speakers," stereopticon lectures, and new subjects or methods of discussion from time to time mark an effort to quicken flagging interest in prayer meeting and the young people's societies. But, in the main, such changes have been grafted on to the traditional religious observances without greatly altering their content or form.

Likewise, the Missionary Societies, Ladies' Aid Societies, and Flower Missions in the various churches have altered their essential character as little as have the purely religious services during the past generation, save for more organization in their study programs. Attendance at missionary societies is confined to women, almost entirely middle-aged women, and in the larger business class churches runs from about twenty-five to fifty. The position occupied by these groups appears in the remark of the minister's wife in a church of nearly 1,000 members to a meeting of the missionary society in late October: "We're going to start a mission study class. Of course this next month will be too busy and the holidays will be too busy, but we're going to have a class, and some time in January we'll start it even if there aren't but two people in it. I am ordering five copies of the study text from the Board, and we'll expect that many anyway." At a time when denominational mission obligations are mounting rapidly, local concern for missions appears for most people to assume the prevalent impersonal short-hand form of a money contribution.

But, while Middletown churches have made few modifications in their purely religious activities, they have made fresh departures in incorporating some of the secular interests of their members. Thus, as the informal social life of the churches has declined, the churches are attempting to fight fire with fire by matching the plethora of new outside clubs with new batteries of clubs inside the church. Indeed, church social clubs [13] have apparently multiplied more rapidly than any other one group of clubs in the city since 1890, increasing to 101; of the seventy-six adult church clubs three are men's clubs, fifty-six women's clubs, and seventeen clubs of both men and women. It is in these newer varieties of organizations, organized Sunday School classes, quasi-social clubs, and athletic clubs, that the greatest changes appear. To the devout merchant of 1890 whose diary is quoted frequently in these pages, a Young People's Society social meeting or an all-day sewing meeting of the Ladies' Aid would be a natural part of church activities, but he would probably blink in surprise at the 1924 newspaper headlines: "Methodists Are Jolted 16 to 13"; "Christians and Disciples Are Victorious." The new spirit permeates even the older observances; the greater part of the annual meeting of a leading business class church was given over to the presentation of athletic sweaters to the boys of the Sunday School League by the director of the Y.M.C.A. According to a young business woman belonging to one of the most successful Sunday School social clubs, "seventy-five per cent. of the Sunday School classes in Middletown are organized socially or they couldn't keep going. We have a party every month with a regular program, games, and a feed." Another girl, a popular high school senior belonging to another such class in a fashionable church, remarked, "Our Sunday School class is *some* class. Our teacher is Mrs. ——, and she gives us some slick parties out at her place. Two of the girls in our class got kicked out of their clubs a year ago for smoking—*that's* the kind of a class we have!"

Men's church clubs are not only fewer in number but have generally proved less successful than those among the women

[13] These are clubs, organized Sunday School classes, and so on, meeting at least once a month for purely social, non-religious meetings.

and young people. Furthermore the multiplication of church clubs has occurred chiefly not among the working class, where clubs are relatively less frequent but among the business class, which is already confessedly "clubbed to death" and striving to maintain a "family life that is vanishing because there are so many things to go to"; seventeen of the 123 working class wives and twenty-three of the thirty-nine business class wives giving information on this point, none of the working class husbands and seven of the business class husbands belong to church clubs. This heavy club program of the churches, which adds a third to every two clubs in the city, meets with some protest and much indifference. The perplexity expressed by one business class woman appeared constantly in the statements of others who, while still giving religious allegiance to their church, prefer, except for attending Sunday services, to use their time for home, children, civic, and socially selective extra-church interests:

"We still go to the —— church, because our people always have, and I like the things it stands for. But the question of church social life is becoming more and more of a problem. I want to keep it up, and it's surely a good thing to spend part of one's time with people whose interests are not quite up to one's own, but it takes so much time that one is deprived of forming contacts with people whose tastes and interests would be more congenial in other things one does here in town and who are the contacts my husband needs in his business."

The minister of a leading church complains:

"It's hard to get men to stay for more than an hour's talk at a church men's club supper, though they'll stay till midnight at their civic club dinners."

"South of the tracks," in the working class churches, however, a greater poverty of possible other choices makes church-going, even without the aid of many organized clubs, still a social event. Here newcomers are shaken by the hand by every one present, and people come early to service and stay late to "visit" with their friends. One working class woman remarked enthusiastically, "I like church and lodge work just fine, because you see and get to know so many kinds of people!" Here there persists something of the kind of social life that characterized business class churches in the nineties; today one still reads of "a program of music and orations" at an Epworth League gathering in a South Side church, just as one reads of a leading North Side church in 1890, "Christian Endeavor will give a supper, twenty-five cents for two, with a program of instrumental solos, papers, selections, and recitations, the proceeds to be used to send delegates to the state convention of the Endeavorers." [14] The greater prominence of the church as a place for first meeting and for seeing one's friends among the working class, noted in Chapter XIX, is simply one example of the greater persistence of social life in these churches.

The attempts of religious organizations to retain their hold by incorporating secular activities appears still more in the organized club and athletic groups of the Y.M. and Y.W.C.A. In general the boys' groups with their heavy athletic programs are successful, while among the girls the clubs have a greater struggle. At the "annual reward day" of the Girl Reserves Department of the Y.W.C.A., rewards are given for attendance at the Sunday Social Hours.[15]

[14] It is difficult to recover today the prominence of the churches in the social life of 1890. Christmas entertainments were held on Christmas Day and were enthusiastic whole-church affairs, instead of "parties for the kids," crowded into some other evening today. "In every particular Christmas Day among the churches was a success," concluded the press of 1890 after listing the many afternoon and evening socials and entertainments. Church fairs were frequent; also such parties as "a grand social at Web. R——'s house last evening in the interest of the organ fund—$46 raised," "Presbyterian trolley ride this evening—five carloads at $0.15 each," "Episcopal 'dime party' at Tom's," "Christian Endeavorers gave a taffy-pulling."

[15] At one point the churches and the Christian Associations make little attempt to follow the young, in their predominant new leisure-time interest—dancing. Although the church ban upon dancing is unofficially almost entirely relaxed today so far as the participation of church members is concerned, the throwing open of church buildings for dancing is generally looked upon as mildly scandalous. Only one small business class church of the forty-odd churches in the city has utilized church property for parish dances, and even this met with much criticism. The Y.W.C.A. gives occasional dances and during part of the year conducts a small dancing class.

In connection with the moral protest against dancing and the local agitation for stricter regulation of public dance halls, an interesting sidelight is thrown upon the complex nature of such a "social problem" and upon the piecemeal methods by which people are wont to attack it; although most parents try to forbid their children's attending the two public dance resorts outside town because of alleged "immoral conditions," the owner of one of these resorts is a member of the leading civic club whose slogan is "service" and his resort advertised a Sunday night dance with a picture of a partly clad female dancer and the headline, "Plenty Hot!"

The above account of activities applies primarily to the central core of Evangelical Protestant churches which dominate the religious life of Middletown. In other religious groups certain variations occur: in the Catholic and Christian Science churches the sermon is relatively less important in the Sunday services, with correspondingly greater emphasis on ritual; in these churches there is greater emphasis on the religious instruction of the young than in the main group of Protestant churches. Both the Christian Scientists and the small but ardent group of Spiritualists make a strong appeal through their emphasis upon the here and now benefits of religion in health and prosperity. The small Jewish Reformed Temple without a resident Rabbi differs from other religious groups by its nonadherence to the prevailing Christian tradition and a certain austere simplicity in its service of worship.

As in trying to discover the underlying meaning to Middletown of its getting-a-living activities, noted in Chapter VIII, so in these religious observances one may be puzzled by apparent contradictions. On the one hand, Middletown is building its religion in its own image; there is a tendency to appraise the fruits of religion by the same tangible, material measurements which it applies to its other activities:

"I believe in prayer because of experiences I've had," remarked an active member of Rotary. "A couple of years ago everybody told me not to buy [a material essential in his industry]. Experts said to me, 'Trim your sails. It sure is a time to sit tight.' Well, my judgment was against it, but just the same I went ahead and bought fifty carloads—and in ten days it went up 50 per cent. in price. Now, I don't mean that God said, 'There's Jim in Middletown. I'll help him out,' but somebody guided me, and I think it was Him. That's why I believe in prayer."

When the high school Bearcats were playing for the state basket-ball championship, those unable to attend the game assembled in the high school auditorium to hear the returns. A minister conducted the meeting, opening it with prayer, and as the tension grew during the game, a senior class officer prayed, "Oh, God, we must win. Jesus, wilt thou help us!"

In reporting "the most successful year from every standpoint" in the history of a Middletown church, the press says: "The reports showed the church to be in good financial standing with a good balance in the treasury, to have a building fund of ——, to have had a net increase of membership of ——, to have had an average attendance of —— in the Sunday School and to have paid into missions the sum of ——."

In explaining the withdrawal of a leading family of manufacturers from the small liberal church, a shrewd citizen remarked, "You see, they're used to having everything they touch go over big; they'd scrap an entire plant to improve it. When this church didn't go here, they wanted to let it die and merge it with one of the active, growing churches; when the minister blew up over it they just got out."

A prominent minister urged the claims of the church prior to Go-to-Church Sunday by the following practical appeal directly to the pocket: "The church is the 'reason why' of America. . . . The church has made America prosperous. . . . It is no mere happening that church people become well to do. 'Godliness is profitable' even from a business standpoint. . . . Coöperation with the church program is the greatest contribution any man can make to his community."

On the other hand, there is a tendency to value religion for its very remoteness and difference from the affairs of every day, its concern with "another world, another life." [16] Says one of Middletown's most respected ministers:

"The Christian realizes that the world of things, in which we live and move, is not the real world. . . . You can not but be refreshed by an hour spent in the midst of that awe and reverence with which the believers draw near to their God."

In other aspects of its life Middletown is involved in change. But it values its religious beliefs in part because it is assured that they are unchanging. One page of an 1890 issue of a Middletown paper announced the subject of the meeting of the Ethical Society, "The Life of Man Is One Series of Experiments," and the sermon subject of a leading church, "Our Unchangeable Religion." "My aim is to establish your belief in the assured permanence and power of the Kingdom of God," says the minister of the largest church in the city today. "Every season brings something new—change, change!" exclaimed a speaker to the Federated Women's Clubs. "The one way to keep

[16] One minister lamenting his failure to put over a "Boy Friend Campaign" said, "The trouble is people don't regard that sort of thing as religion and won't respond to it."

our sanity is to have something eternal. God is unchangeable." When the Methodist Episcopal Conference of Middletown's section of the state reaffirmed "its long-standing doctrine of opposition to card-playing, dancing, drinking, and gambling," a Middletown paper commented editorially:

"And somehow, we doubt not, if the day should come when the Methodist Episcopal Church should involve itself in discussions over doctrines and discipline that now are harassing some Christian organizations, it would lose much of the indefinable but sure thing that we call Methodism and that its influence would wane. Others may have doubts about certain things that the Methodists regard as fundamentals of Christianity and Christian living, but not the followers of John Wesley. They and the church stand today as a strong rock of spiritual refuge for those of all faiths that are torn by doubts and distress over material and spiritual things."

This emphasis upon the unchanging character of essential religious beliefs is perhaps not unrelated to a tendency to regard them not as practical concerns, something to be put into practice, like a linotype machine or a new antitoxin, but rather as things hoped for but unattainable. The most marked changes in other sectors of the city's life have been predominantly changes in material culture, adaptations to new tools and techniques involving concrete experimentation, the tracing out of a discoverable sequence of events. Teachers of religion, on the other hand, regarding the solutions of the problems with which they deal as largely discovered and established, sometimes tend to subsume concrete details under categories, to blanket the network of factors involved in a situation under a general denunciation of "sin" or an appeal for "righteousness" or "service." Thus, while doctors and the Social Service Bureau are working at questions of public health, a leading minister tells Middletown that "what Middletown needs is not eugenics, not better public health, not better education, but spiritual regeneration." Another minister instructs his hearers:

"Jesus meant a kingdom of Righteousness, not of earthly power, and all his followers must work to bring it to pass. What sort of work? International commerce will not do it. The progress of science will not do it, for though it can change men's environment it cannot change their spirits. Secular education will not do it—the only thing that will do it is the *Christian* religion; it alone, because it represents religion at its highest point."

The minister of the largest Protestant church exhorted his congregation, "If God has given you wealth, be happy; if he has given you poverty, be happy." [17] Even in churches which do not take the extreme position of some Middletown sects that "these evils must continue until Christ comes" or "no good thing we've ever done can merit any consideration at the hands of God; not by good works of righteousness but by mercy alone are we saved," the gospel of faith is predominantly preached in contrast to the gospel of works followed in everyday living. "The trouble with church people is that they leave everything to the Lord" was the emphatic summary of one church member. During a period of dire unemployment the minister of the most fashionable church in the city prayed, "O Lord, care for the 3,000 unemployed in this city and raise up friends to them. Set the factory machines going that [they] may again have work." In many cases the church seems to serve Middletown not as a method of meeting situations, but, somewhat like Rotary and similar groups, as a repository or safeguard for a set of ideals, whereby the whole puzzle makes sense, and permanence is given to the slippery business of living. It is accordingly noteworthy that in matters of belief the churches apparently retain their most complete dominance over the lives of their members in certain groups of the working class who, on the one hand, have less opportunity for other approaches to problems than the business class, and, on the other, have fewer enjoyments in this life and more urgent need that "it will be

[17] Two of Middletown's favorite poets voice this same mood of religious contentment with one's lot. In his syndicated "Just Folks" column in the leading paper, "Eddie" Guest says:

> "Teach me contentment, Lord, I pray
> With all the joys which now surround me.
>
>   .   .   .   .   .
>
> To bear my share of good or ill,
> The sun of June or April's raining—
> To see through both the weal and woe
> And know that souls are fashioned so."

The program of one of the Women's Federated Clubs quoted from Riley:

> "It ain't no use to grumble and complain,
>   It's just as cheap and easy to rejoice,
> When God sorts out the weather and sends rain,
>   W'y, rain's my choice."

made up to us in Heaven." The relatively more positive affirmations of faith in the answers of working class women cited
in Chapter XX and the tendency to greater questioning noted
among the business class reflect this difference.

As changes proceed at accelerating speed in other sections of
the city's life, the lack of dominance of religious beliefs becomes
more apparent. The whole tide of this industrial culture would
seem to be set more strongly than in the leisurely village of
thirty-five years ago in the direction of the "go-getter" rather
than in that of "Blessed are the meek" of the church; by their
religious teachers Middletown people are told that they are sinners in need of salvation, by speakers at men's and women's
clubs they are assured that their city, their state, and their
country are, if not perfect, at least the best in the world, that
it is they who make them so, and that if they but continue in
their present vigorous course, progress is assured. Meanwhile,
secular marriages are increasing, divorce is increasing, wives
of both workers and business men would appear to stress loyalty to the church less than did their mothers in training their
children, church attendance is apparently less regular than in
1890, Rotary which boasts that it includes all the leaders
of the city will admit no minister, social activities are much
less centered in the churches, leisure time is increasingly less
touched by religious prohibitions in its encroachments upon the
Sabbath, more and more community activities are, as the press
points out in regard to questions of disease and health, being
regarded not as "acts of God" but as subjects for investigation.
In theory, religious beliefs dominate all other activities in Middletown; actually, large regions of Middletown's life appear
uncontrolled by them. Said a member of Rotary, leading prayer
meeting in Middletown's largest church:

"We talk about 'believing in God the Father' and about 'the
church and religion being more worth while than anything in life.'
Now suppose somebody could follow our every thought and act
for just two short days—how much of this would he discover?
I'd hate just to try to say how little. I went to [the state capital]
for two days last week and shouted myself hoarse for our Bearcats, but would I do that for this church? No, you bet I wouldn't!"

When the first rush of the gas boom was over in Middletown and the panic of '93 and the smallpox of that summer
had put the town on its back, many industries failed and others

moved away. A meeting of business men was called. They filed in and took their places in silence. The meeting was opened by the quiet quotation from the Old Testament: "If thou faint in the day of adversity, thy strength is small." Then from another came another verse, then another, and another, and a leading young manufacturer, now one of the city's millionaires, rose and repeated the parable of the house built upon the rock. "And," to quote the story as retold simply by one of the city's veterans, "at that meeting $200,000 was raised to put Middletown back on its feet industrially." To one familiar with Middletown today, such an occasion sounds almost like an incident in the life of another people.

But if religious life as represented by the churches is less pervasive than a generation ago, other centers of "spiritual" activity are growing up in the community. However much the ideal of "service" in Rotary and the other civic clubs may be subordinated to certain other interests, these clubs are nevertheless marked sources of religious loyalty and zeal to some of their members; "civic loyalty," "magic Middletown," as a religion, appears to be the greatest driving power for some Middletown citizens.[18] Some leaders in certain of the women's clubs find in these clubs a similar focus of energy and enthusiasm.

The Ad. Club and the Chamber of Commerce, representing somewhat wider groups, help to translate scattered individual desires into the action of a group. Here, for instance, is a "citizenship" dinner at the Chamber of Commerce on Armistice Day. About fifty new citizens are the personal guests of the business men of the city. A Russian Jewish junk-dealer is speaking in broken English in answer to a speech of welcome by a local lawyer:

"It is not the Chamber of Commerce but Uncle Sam who is our host tonight. I came to this country expecting to find it a land of

[18] There is even a tendency, chiefly noticeable among the business class, to class the church with "civic responsibilities," as appears in the remark of a prominent woman, "I don't go to church, but I contribute to its support, as I pay taxes for the maintenance of the police force." This apparently growing tendency to relax the habit of personal participation in religious activities, accompanied as it is by the feeling that churches ought to be kept up—by somebody—is a phase of the growing specialization and a generally pervasive tendency to leave more and more things to delegated representatives.

gold. But the gold I found in New York was carefully locked up in other people's banks. Slowly I came to realize that America is rich in another kind of gold. I was a peddler with a horse and wagon and one night I asked a farmer if I could stay on his farm. He himself unhitched my horse. I thought he was trying to take it away from me. Then he gave the horse real oats—the first it had ever eaten!—with his own horses; he took me in and fed me with his family—chicken and everything. When I came to leave I asked him what I owed and pulled out half a dollar. He shook his head and I saw my week's savings going. He said, 'Young man, if you want to pay me, give up this wandering life, settle down, and when men and women come to you for help, pass on to them what I've done, if you think it worth anything.'" The little Jew paused in embarrassment, wiped the perspiration from his forehead, and said: "I am sorry I cannot speak better. I honor your schools that are teaching us. But I do want to say that you have gold in America—not paving your streets, but gold of this sort in the hearts of your citizens, the gold which, too, makes each of us able to go all over the world with respect and safety as American citizens. We who have come to your land have left behind us our own ways of living and things dear to us; we gladly take yours and offer you all we possess—our future and our children."

After the dinner a business man, who tells risqué stories and in the next breath quotes Henry van Dyke, a boisterous, successful booster of "magic Middletown," remarked in a husky voice, "This is the greatest meeting ever held in the city of Middletown."

Equally genuine is the farewell luncheon of the Ad. Club to a man affectionately known as "J. T." who is moving away to another city:

The crowd of thirty-five men—merchants, newspaper men—saunter in and devour their luncheons, amid good-natured banter. It is just an ordinary Friday noon luncheon—until the program begins. "J. T." is a Catholic in a bitterly pro-Klan town. And yet one by one the older members, all shades of Protestants, Jews, Christian Scientists, rise, saying simply, as did one man: "Friends go away and often don't know what they really mean to us. 'J. T.,' I want to tell you to your face before it's too late what you mean to me. . . ."

The civic club dinners to teachers and to honor students, the efforts of the women's clubs to foster the Juvenile Court or to bring good pictures to the schools, the memorial meetings, and to some the ceremonies, of the lodges, bear witness to the same thing. To many small boys the boys' work secretary of the Y.M.C.A. is the biggest thing in town. To some people the labor movement or perhaps the Humane Society are centers of work and loyalty. The Community Club fostered by the principal of a school in the factory district brings fresh life to both parents and children. The art and music work and certain classes in the schools stand out as sources of power to individuals. And here, as elsewhere, it is difficult to say how much of this new energy may in fact spring from the older institutional forms of religion.

VI. ENGAGING IN COMMUNITY ACTIVITIES

Chapter XXIV

THE MACHINERY OF GOVERNMENT

Every one in Middletown runs absorbed in keeping *his* job or raising *his* wages, building *his* home, "boosting" *his* club or church, educating *his* children. Now a member of this group, now of that, he shuttles his intent course amid the congeries of jostling groups that make up the larger group which is Middletown. And ever and again he finds his particular business af٧ fected by things which are the corporate business of that larger group. Since the nineties these things regarded as everybody's business have multiplied; more officials and administrative agencies are needed to care for them, more money is spent in operating them, and they involve more laws.[1] But it is perhaps noteworthy that, despite the enormous importance Middletown attaches to money, it chooses the representatives to whom it delegates the work of looking after the $61,000,000 city of today with its 38,000 inhabitants and its $3–$400,000 annual operating expenses in substantially the same way that it chose them for the $5,000,000 city of 1890 with its 11,000 people or for the village of 1850 with its 700 people and annual operating expenses of less than $1,000. One explanation of this constancy of procedure in the midst of radical changes in the number and kinds of activities required of the corporate group is to be found in the fact that the operation of the city's business is dominated by one of its most cherished ideals—"democratic government."

This system of selection is based upon the theory that, if

[1] Among the new activities assumed by the group as a whole or by the state since 1890 are: regulation of working conditions in industry, supervision of a worker's compensation when he is injured, care of health (cf. Ch. XXV), supervision of food, extension of supervision of children, etc., etc.

New group officials in Middletown since 1890 include: The Board of Works, Board of Safety, Sinking Fund Commission, City Planning Commission, City Judge, Judge of the Juvenile Court, City Comptroller, Street Commissioner.

periodically given an opportunity to express itself, the choice of the majority of adult citizens, including today women as well as men voters, will fall upon the person best qualified to fill a particular post. Accordingly, every few years the city's administrators are automatically turned out of office and their reappointment or the election of their successors is turned over to the people.[2] Actually, however, the voters choose between the candidates put forward by one of the two "political parties" which dominate the entire procedure, and the selection of group representatives thus becomes a matter of lining up on one side or the other of an either-or situation. The issues involved in supporting the either's or the or's have become somewhat more blurred since the nineties when the political battle raged about such national policies as the tariff, although both parties even then united in promises of helping the working man, securing economy, and obtaining Civil War pensions. In the national presidential election of 1924 the tariff was scarcely mentioned in Middletown, and differences between the two parties appeared largely in the specific scandals in the record of each which the other singled out for attack. "Conservatism," "stability," the "principles of our forefathers and the spirit of America," "the God-given rights recited in the bill-of-rights amendment to the United States Constitution," served as battle cries for both parties.

Despite this tendency toward obliteration of distinctions, however, party loyalty is none the less passionately urged upon the voter, and the emergence of any third party, differing from the two older ones, passionately denounced by both. It would seem to be the belief of most Middletown citizens that, as stated by a nominee for office, "the inalienable rights of American citizens can only be secured through the existing [two] party

[2] Some citizens in Middletown are now beginning to question this habit of entrusting the complex affairs of the modern city to fresh groups of raw officials periodically turned in and out of office "under what Kipling calls the American idea of versatility—the idea that any man can do anything." A very few go as far as Roscoe Pound in his summary of the study of *Criminal Justice in Cleveland* when he says, " 'Transient administration,' as Mr. Fosdick well puts it, 'is fatal to success in any complex technical enterprise.' The public business is the only sort of business in which it is tolerated in the United States of today. . . . It is one of the legacies of pioneer America. . . . The pioneer notion of short tenure and selection from among the voters of a politico-geographical area is out of place in the city of today." (*Op. cit.*, p. 616.)

method." "Third parties," says a Middletown editorial in 1924, "will have nothing constructive to their credit. Ours is a two-party country and seems likely to continue as such." "The candidates," says another editorial, "are mere instruments through whom the parties exert influence in affairs of government. The party is more important than its representative." In the 1924 election both parties paused now and again in their hurling of recriminations at each other to unite in urging the voters to "preserve the Constitution" and put down the "mad dog" in the person of La Follette and the radical program of the third party. "There usually are two sides to every question, but not where the Constitution of the United States is involved," declared a Middletown editorial.

A person's political party is usually determined, like his religion, by his family, and it is difficult for any one unfamiliar with this culture to picture the intense emotional concern that follows the accident of birth into one or the other camp. A leading club woman, asked whether she was going to vote for a Republican candidate for office, a Klansman, accused of political corruption, replied with a touch of asperity, "I am going to vote a straight Republican ticket. I have always been a Republican and that's the way I always vote." An equally prominent woman justified her membership in the opposite camp, a minority group in Middletown, by saying, "It seems perfectly natural for me to be a Democrat. My family were always Democrats, and so it doesn't seem strange to me." And a Middletown editorial avers, "A man is a Republican or he is not, a Democrat or he is not, and the test of his partisanship is the support he gives his party."

In view of the fact that very few of the city's business leaders are Democrats, it is decidedly "good business" to be a Republican, and this consideration in certain cases overrides the accident of birth. Economic considerations, likewise, prove the most effective barriers to the rise of a third party. In 1924 it was considered such "bad business" to vote for the third party that literally no one of the business group confessed publicly either before or after the election to adherence to this ticket. "If we could discover the three people who disgraced our district by voting for La Follette," declared one business class woman vehemently, "we'd certainly make it hot for them!" And both the old parties united in saying, "The working man who votes for La Follette is voting to put himself out of a job." "Be radical if you wish to be," said a preëlection editorial, "but have sense when it comes to the point of caring for yourself and family." "Timid capital is more afraid of politics than anything else," declared another, and "when capital becomes frightened badly enough the load falls upon the wage-earner." And still another concluded, "A vote for Coolidge is merely a vote for your own safety. You will vote tomorrow for or against your job and your neighbor's job. That's about all there is to it . . . ; maybe you will be one of those who will have none if the election goes wrong." [3]

And yet, great as are Middletown's emotional and financial stakes in politics, elections are no longer the lively centers of public interest they were in the nineties. In 1890 Middletown gave itself over for weeks before each election to the bitter, hilarious joy of conflict. The predilection for "a real good speech" was indulged extravagantly; enthusiastic followers of either party turned out "armed with implements of noise" to march in torchlight processions, drum corps at their head; horns, anvils, "a boiler on wheels and lots of pounding" were a prelude to "addresses by Middletown's best orators." "Twenty-five hundred at the Rink to hear ——, Republican campaign speaker," is a report not unusual in the papers. Not until November 24, twenty days after the close national election of '88, does the day-by-day diary of a leading citizen note, "Everything quiet after the election." Today torchlight processions and horns no longer blast out the voters or usher in the newly elected officials, and, although speeches persist with something of their old vigor, new inventions offering a variety of alternate interests are pressing upon politics as upon lodges, unions, and churches. "I am not going to hear —— speak tonight," said a Middletown business man, turning the knob of his radio in search of an evening's diversion on the air. "It'll be in the paper tomorrow. And anyway, it doesn't make very much difference. I agree with what it says in *Collier's:* This is a good old country. No matter who is in, it gets better every year under any old administration. We don't need to worry." [4]

[3] Cf. Ch. VIII for discussion of the bewilderment of the workers at continued "hard times" after the Republican victory.

[4] Cf. in this connection the following from an article on "The Vanishing Voter" by Professor Arthur M. Schlesinger and Erik McKinley Eriksson:

Not only election preliminaries but the actual choice of the candidates fails to arouse today the interest it did in the nineties. Eighty-seven per cent. of the eligible population of the county voted in the presidential election of 1888 and 86 per cent. in the election of 1892, as over against 53 per cent. in 1916 and 46 per cent. in 1920.[5] In 1924 the Middletown press said after a primary election, "Perhaps one-third of the qualified voters . . . went to the polls in Tuesday's primary election, or at any rate 40 per cent. would cover the number that took a sufficient interest in their local government to vote. This was considered a fair representation for a primary." At least two further factors, in addition to other competing interests, are apparent in this tendency for Middletown citizens to concern themselves less today about the operation of community business. The first is the wide gap existing between what the rank and file of the group actually know about its affairs and the amount they are assumed to know by the nature of their public institutions. In the city of today it is far less possible than in the eighties and earlier for a voter to know personally each of even the two or three dozen local candidates from whom he is expected at a given election to choose the "best men." He is increasingly dependent upon the two political parties for such information as they may give him through newspapers, campaign documents, or speakers.

In the election of 1888 the Republican Party, then as now dominant in Middletown, freely distributed sample ballots listing the names of the candidates of all parties and had them printed in full four times prior to the election in each of the two Republican newspapers. In the election of 1924 no sample ballots were distributed or printed in either of the two leading

"Along with this artificial party situation must be considered a factor of even greater importance: the multifarious interests and multiplied opportunities of contemporary American life. For a generation after the Civil War the United States was predominantly a nation of farms and crossroads villages. Life flowed along pretty nearly on a dead level, and interruptions of the routine were few and far between. To such a people a political rally had all the romance and dramatic interest of circus day. Voting was a diversion as well as a duty." (*New Republic,* October 15, 1924.)

[5] Middletown's experience parallels that of the country at large where the percentage of qualified voters who actually voted dropped from 90 per cent. in 1896 to 73 per cent. in 1900, to 66 per cent. in 1908, to 62 per cent. in 1912, and to 51 per cent. in 1920.

papers, both Republican, and repeated requests for sample ballots at the city clerk's office and at Republican headquarters were met only by statements that there were no ballots and that one could vote for all the candidates by putting a cross over the Republican eagle. "They're afraid to let the ballots get out," explained a leading business man and loyal Republican, "because there's so much talk of not voting straight this year." [6] "The newspapers won't print the ballots," said an experienced Middletown newspaper man, "because they're afraid of the situation here and keep out of it." The leading paper advised that "No outstanding excuse exists for any one to scratch the ticket and there is always risk in that process, particularly in districts using the voting machines."

When the electors went to the polls in 1924 they were told to "vote quickly" from a list of several dozen candidates for the most part unknown to them. Those using the voting machines were in some precincts told to vote "four to a minute! Only fifteen seconds to a person on the voting machines." [7]

"I had planned to vote for —— for governor and one other Democrat," said one woman, "but I couldn't get the list to look it over before, and it was so long and they hurried me so that I just gave up trying to select my candidates and voted straight Republican. I guess it's just as safe in the long run. You don't know much about any of them and in dodging one man you don't want, you are just as likely to pick another as bad if you only knew more about him."

Campaign documents issued by both parties contain such appeals as:

"Vote for our Daddy," with pictures of the candidate's two children, or the picture of the candidate himself, and "He served overseas," or "World War veteran, seven years in service."

[6] A frequent explanation in Middletown of the suppression of ballots was the control of the powerful Republican machine by a well-known political character who, according to the local Democratic paper, "boasts that he named the Republican county ticket and that his strangle hold cannot be broken."

[7] One reason for the speeding up of voting was that the electoral districts, already crowded before the coming of woman suffrage, have not been changed since. Great congestion results. Some people waited an hour or more to cast their votes. The Republican machine is said to oppose any change, preferring to handle the voters as at present.

"Dear Friend and Voter:

"I am seeking the nomination on the Republican ticket for justice of the peace. I have lived in [Middletown] for over thirty years and have been as good a friend and neighbor as I possibly could.

"I am mindful of the benefits of doing justice or extending mercy and know how to be a justice of the peace. . . .

"Thanking you in advance for your support, I am, very truly yours,

"(*Signed*) ——."

Party speeches are both of the informal kind made by a city tax collector to a member of the research staff after assessing her personal property—

"I hope you're going to get out to vote for —— for mayor. There is no reason why this town should not be 100,000 instead of 40,000 if it were not for the ——s [the most wealthy family of business men]. They run the Chamber of Commerce and now they run the town. They want to keep factories out and keep wages down. Henry Ford thought of locating here at one time, but the ——s were so against it that he did not come. Now —— [the Republican machine candidate] is a fine young fellow, pleasant, friendly, nice personality, has a nice friendly smile, too. You'd better vote for him"—

and of the more formal kind represented by campaign speeches. The latter, as Lord Bryce pointed out, "are not directed towards instruction, but towards stimulation." "Ridicule, sarcasm, satire, and bitter condemnation," was the press summary of one 1924 campaign address. "Questions of importance," to quote one candidate, "will all be reduced to questions of right and wrong." It is possibly significant of the way Middletown regards political speeches that less than 5 per cent. of 556 high school juniors and seniors wrote "true" after the statement, "Voters can rely upon statements of fact made by candidates in campaign speeches," while 87 per cent. answered "false"; the others were "uncertain" or did not answer.

The churches, too, give some instruction to voters. "Be sure that you vote to make America a Christian nation," urged the minister of the largest Protestant church on the eve of election, while the minister of the most socially prominent church told his hearers, "Other things being equal, one naturally votes for one's friends. There are several members of our church running for office and I shall vote for them." One church in 1925, as noted above, tried the experiment of a Sunday evening forum on "What Kind of a Mayor Does Middletown Need?"

Another factor in the attenuation of Middletown's interest in choosing its public officials is the fact that, in the minds of many citizens, politics is identified with fraud. "It is probably safe to say that there is not conducted in this state a city election without some sort of fraud, the buying of votes, the introduction of floaters or repeaters, or the falsification of the returns in some manner or other," say the Streightoffs in their study of this state. "This time has so thoroughly disgusted me with politics," remarked a young Middletown business woman after the presidential election in 1924, "that I don't feel like there's much use trying ever to vote again." The business class and many of the working class are prone to regard most of their local officeholders with disinterested cynicism. "There's so little to choose between the two, I'm not going to vote," is a statement not infrequently heard in regard to local elections. A candidate in the 1925 primaries for nomination for city judge, a man who had six months before completed a six months' sentence at the State Penal Farm, jovially addressed a meeting of voters at the court house as follows: "I admit that up until a couple of years ago I was a bootlegger. I sold more white mule than any man in the county except ——. Now here's —— [another candidate], he tries to keep the county dry by drinking it all. Which of us is the better?" "I am going to vote for ——," said one business class woman emphatically in discussing the three leading candidates for mayor. "He's a crook and he's perfectly open about it and makes no pretenses. I think the better of him for it. —— [one opponent] is a crook and doesn't admit it. I'm not sure about —— [the other opponent], but I believe he's just as crooked too."

Not the least significant feature of this political corruption is the fact that it is so taken for granted by most citizens and by both parties. To the Democrats' charge of the Teapot Dome oil scandal in the presidential election of 1924 the local Republicans found it necessary to reply only "Of course, any party is corrupt, and the Democrats are just as bad." Nor is the family skeleton hid shamefacedly in the closet. One of the most widely used and eagerly applauded campaign songs of Middletown Republicans ended with the refrain:

> "When we get to Washington, home, sweet home,
> We won't give a darn for the Teapot Dome!"

In view of the apathy or repugnance with which many Middletown citizens have come to regard politics, it is perhaps not surprising that the "best citizens" are no longer to be found among Middletown's public officials. In theory the kudos adhering to public service through political office is sufficient to attract the ablest members of the group, and a generation ago this was somewhat more the case. For ten years in the nineties and early in the present century, two leading local families warred for political leadership. Today the mayor, the judge, and the councilman are no longer dominant figures; their position of prestige has tended to be taken by business men, and "business and politics don't mix." The activity of getting a living has come to dominate the time, the energy, and the habitual attitudes of the natural leaders among the population to a degree undreamed of in the quiet days before the gas boom. In 1925 the mayor received a $3,000 salary, no sum to tempt any leading business man, while the local press said the city's mayor "ought to have a salary of at least $12,000 a year." The city judge receives $2,100 and the city attorney $3,000, neither sufficient to command the abler lawyers of the city. In so far as the tradition still lingers that group "spirit" ought to prompt a man to step aside from his regular getting-a-living activity to "serve" his community, business men simply reply, "I can't afford to," and the abler professional men say, "Competition is too keen. If a man once steps out of line and gives up his practice, he's done for." Furthermore, the corruption in politics reacts upon itself; brought about in part perhaps by the withdrawal of the "better element," it also deters them from participation in public affairs. "Our politics smell to heaven," said a business leader; "elections are dirty and unscrupulous, and our better citizens mostly don't dare mix in them." "No good man will go into politics here," said another. "Why should he? Politics is dirty. I wouldn't mix in it here. Maybe if I was rich and independent I would, but I'd certainly think twice about it even then." In a civilization in which the health of the community is gauged by its financial pulse, preoccupation with private rather than public business on the part of its ablest citizens is increasingly regarded by Middletown as not only a normal but a desirable state of affairs.

Indeed, it does not consider either marked ability or special training essential for the administrators of public business. The observation of Lord Bryce is at least as true today: "Special knowledge, which commands deference in applied science or in finance, does not command it in politics, because that is not deemed a special subject, but one within the comprehension of every man," a fact which "is the more remarkable because nowhere is executive ability more valued in the management of private concerns." Rarely do campaign speakers mention the special ability of a candidate for office; they extol him as "a man of the people," "four-square," a "real American." "Some persons," said a Middletown editorial of a certain office-holder, "say that ———'s ability has been vastly overrated . . . but in the popular mind those things don't amount to very much. . . . The one thing that they are interested in, whether a President or a ditch digger is asking for their votes, is whether he is 'square.'" "People are rather tired of great ability," said an outstanding business man in commenting upon a candidate for office; "they've seen enough of that sort of thing in the Jews. What they want is a good, plain, common sense man of the people. That's the sort of person we need." [8]

Under this system of choosing representatives the public business goes on: laws are made and more or less enforced; offenders are "brought to justice"; money levies are collected; sewers are laid; streets are paved; institutions for training the young are built and operated; and, unless his particular interests are interfered with, a Middletown citizen does not concern himself greatly about these things.[9] And yet, people of both classes

[8] Cf. Ch. XVI for the attitude of Middletown toward education and intelligence.

[9] It is perhaps characteristic of the prevalent attitude towards the conduct of group affairs that a local doctor, who as mayor is reputed to have "run the town wide open" and was sent to the Federal penitentiary with his prosecuting attorney in 1919 for fraudulent practices, returned to Middletown after serving his sentence and was almost immediately awarded the first prize over some of the ablest local citizens in a city-wide "popularity contest" conducted by one of the newspapers. This doctor had been widely known during his two terms as a "friend of the working man and opposed to the Chamber of Commerce crowd."

Cf. Ch. VII for discussion of the handling of the Middletown employment bureau.

complain of the way they are done. Some clew to their dissatisfaction may be found in the record of the "Republican reform government," put in in 1924, as it appeared from time to time in the Middletown press: [10]

"Mayor indicted on three counts before grand jury. Petition presented to the city council asking for Mayor's impeachment." "When the Board of Public Safety voiced its objections to the mayor on his selection of —— for a place in the police department, the mayor politely told the board that he was under 'personal obligations' to Mr. —— and that he had to be taken care of." "Board of Safety resigns." "——, who has played such a prominent rôle in city administration squabbles during the last two years, and who one month ago began duties as a humane officer by appointment of Mayor ——, is at last about to grasp a place on the city's payroll. The [new] Board of Public Safety . . . assigned Mr. —— to duties on the fire department." (January, 1924—Rep.)

"The appointment of —— [the county boss of the Republican machine] as postmaster naturally comes as a reward for services rendered the Republican party during his two terms as county chairman." (January, 1924—Rep.)

"Malfeasance and incompetency in the conduct of the office of superintendent of highways . . . are charged against ——. During these two years, it is charged Mr. —— kept . . . from five to seven assistant superintendents without any authority and at the expense of thousands of dollars to the county." (March, 1924—Rep.)

"Never before in the history of [this] county have the taxpayers been swindled as they are at present. . . . We dare the county commissioners to investigate the charge that Sheriff —— was interested in the contract for the —— Pike, let by the commissioners to ——. We dare them to ask why the juvenile court bills in 1923, under Judge ——, were more than double the expenditures in 1922, under Judge ——. We dare them to inquire why it cost [this] county $8,000 more to run the poor farm in 1923 than it did in 1922 and why the salary expense at the Orphans' Home has doubled since Judge —— took charge and installed

[10] The Republican papers quoted are morning and evening papers, both widely read and respected locally. The Democratic paper is a vitriolic, anti-Klan, personal organ, published weekly. Business men jeer at it in public, but privately admit that "that fellow —— has the nerve to publish the stuff other papers don't."

Mrs. —— and Mrs. ——, after causing the dismissal of Mrs. ——." (March, 1924—Dem.)

"A threatened break exists between the council and two members of the safety board." (April, 1924—Rep.)

"The various departments of the city have been going along as best they could, with each one practically in the dark as to what the other departments were doing." (June, 1924—Rep.)

"Residents of Seventh, Eighth, and Ninth streets protest assessments for street paving, saying street improperly graded." (June, 1924—Rep.)

"City loses $101,800 suit. Sewer found defective after it was accepted by the board of works and the contractors had been released from their obligations. Bricks and brickbats used instead of cement in interstices." (July, 1924—Rep.)

"—— retires as chief of the police department after serving for three years under the most unfavorable condition that ever confronted a [Middletown] chief of police. During the entire three years there has been nothing but trouble in the police department, and —— has had the pleasant job of trying to serve Mayor —— and three or four different boards of safety during all the upheavals." (October, 1924—Rep.)

"The Council is not on terms with the Mayor." (December, 1924—Rep.)

"There is friction between the sheriff's office and the police department." (January, 1925—Rep.)

"The two Republican members of the school board are evidently hard pressed by the politicians. First ——, Democrat and Catholic, for many years custodian of buildings, was removed to make way for a Republican, and now the announcement is made that Superintendent —— has 'resigned' his position. Mr. —— has the reputation of being one of the best school superintendents in the Middle West, but he is guilty of the unpardonable sin of being a Democrat." (February, 1925—Dem.)

" 'Because of her criticism of the policies of the administration' was the reason given by Mayor —— for disposing of —— as a member of the department of health." (March, 1925—Rep.)

"Office-holders have to spend so much money getting into office that they all try to get it back as best they can," was the succinct summary of the conduct of public affairs voiced by one of the city's ablest business men.

Even in the nineties the growing city was beginning to feel the strain upon certain of its institutions framed to function in a simpler culture. By 1890, the boom community that had grown from some 6,000 in 1886 to over 11,000 was creaking noticeably and the local press complained that there was "no system" in the police force that was replacing the earlier "town marshal"; "no officer has any regular location or beat, but is supposed to do his duty as he finds it or as it runs up against him." By 1900 the population had nearly doubled again and the leading paper says scornfully: "What [Middletown] needs: A mayor, or police judge, who will be fearless in the administration of justice; a mayor, or police judge, who can distinguish between a gambling device and a parlor lamp." In a list in the daily press of 1890 of "A Few of the Things That Would Benefit the Entire Community" appears, "A big stir in city politics and make a change therein."

Today certain parts of the public business are even more cramped by being caught between conflicting political pressures and forced into the molds of an earlier day. A clear example of this appears in the public library, whose policies and annual book purchases of 2–3,000 volumes are controlled, as were the "five new books a month" added to the library in 1890, by a board of citizens appointed by the local office-holders.[11] At one meeting of the Board a reference work on political economy was stricken off the librarian's list of recommendations because there were already a music and an art reference book on the list and that was held to be enough "books of that sort." A group of books on week-day religious education, selected by the librarian because of local interest in the subject, was challenged on the ground that "we already have plenty of books on religion, and the old ones are the best anyway." Meanwhile, a half-score of local citizens bent on keeping down local expenses managed to force a 40 per cent. decrease in income upon the

---

[11] Two members are appointed by the township trustee, two by the City Council, two by the School Board (which is in turn appointed by the Council), three by the Circuit Judge. In 1924 this Library Board contained no college-trained person or outstanding business or professional man in the city, and did not take into account the fact that book-buying for a modern city library from the nearly 9,000 titles published annually has become a highly skilled technique.

library at a time when every other second-class city in the state was advancing its library tax rate.

Various groups in Middletown advocate different solutions for this "problem" of city government. "There simply are not enough good people to outvote the bad element in town," said one business man in discussing the chronically turbulent state of the city's public business. Many would agree with this diagnosis.[12] A state Senator, addressing a great gathering of Middletown people, assures them, "Let the average man, the common citizen, fill full his mind and heart with love of his kind, remember the Golden Rule, and obey the behest to love his neighbor as himself, and social and economic and political ills will fade away like an untimely frost before a summer sun. One so filled with love of his fellows will be an ideal champion of Democracy." A similar solution was advocated by the minister who preached on "The Christian Faith—America's Hope."

To the editor of the small Democratic weekly paper, on the other hand, the "problem" of local politics is a question of securing officials who belong to no man or group and "can tell a whole lot of people to go to hell."

To many, particularly to the newly enfranchised women voters, the "solution" lies in "getting out the vote." "Vote as you please, but vote!" was the slogan of the League of Women Voters in 1924. Coupled with this faith went a demand for more intensive political organization. There was wide local endorsement at the time of the 1924 election of the sentiment expressed by Elihu Root: "All you have got to do is to wake them up, have some one take the head of the crowd and march them. Tell them where to go, whether Democrats or Republicans, I do not care . . . and the organizers . . . will welcome them and set them to work."[13] Lessened interest in the "sacred institution" of the ballot is frequently blamed upon the willfulness of the citizenry. The press in 1924 urged that "the good

[12] This differentiation between "good" and "bad" people as the community regards them does not apparently include all the factors in the situation, as the prosecuting attorney sent to the Federal prison in 1919 was a Sunday School superintendent, and the largest young married people's Bible class in the city is conducted by a city judge who barely escaped impeachment by the state Legislature.

[13] Here again one observes one of Middletown's most characteristic reactions to the failure of one of its non-material institutions to operate satisfactorily—a redoubling of emphasis upon the questioned ritual and a cry for more loyalty to it.

American who will not vote is unfaithful only in less degree than the bad American who tries to vote twice."

It is from the successful business executives of the city that pressure to adopt a new system of city administration is coming; this group recently sponsored a movement to adopt the "city manager plan." The plan was defeated "because for the first time in Middletown's history both Republican and Democratic machines got together and worked to kill the same thing." At the Sunday evening church forum in 1925, on "What Kind of a Mayor Does Middletown Need?" the first of the appointed speakers began by saying, "The first qualification for our mayor is that he must be a Christian." The next speaker, an officer of the church, began brusquely, "In the first place, we don't need a mayor. We need a city manager to run a proposition like this along business lines. And in the second place I don't care whether that man is a Christian or what he is so long as he can deliver the goods."

One thing emerges from all these various attempts at adjustment: changes in the life of the city have put increased strains upon institutional devices originally framed to operate in a simpler culture. As Roscoe Pound said of criminal justice in Cleveland, "Here again the pivotal point is that institutions originally devised for rural or small-town conditions are failing to function effectively under metropolitan conditions." [14]

A large part of this city business consists in the preservation of those group sanctions and taboos which have become solidified into laws and therefore depend for their maintenance not only upon the support of public opinion but upon official enforcement. Protected and to a degree constrained by this body of law and public opinion, all the activities of living go forward; a Middletown citizen does not operate a factory, build a home, drive his car, or even take a drink of water or empty his garbage without both depending upon and being limited by these group regulations.[15] The activity of prosecuting, judging,

[14] *Op. cit.,* p. 627.
[15] Of 1,497 arrests in Middletown in 1924, 324 were for intoxication, 196 for violating the liquor law, 148 for violence to persons, 163 for speeding, eight for driving while intoxicated, eighty-six for being suspects, seventy-eight for sex offenses, seventy-two for being fugitives from justice, fifty-two for gambling, twenty-three for failure to provide, twelve for illegal voting, 242 for violence to property and ninety-three for "other causes."

and trying offenders against these laws constitutes traditionally the most imposing and revered department of Middletown's public business; the most prominent building in the city is the domed court house; the first public institutions in the state for the care of special classes were prisons for law-breakers.

New laws are continually becoming operative in Middletown, made either by the city itself or by the larger geographical units of which it is a part, as new occasions for violation of persons or property emerge. Meanwhile, with the notable exception of the development of the juvenile court, the machinery for their enforcement has remained substantially the same as in the simpler community of thirty-five years ago. The efficacy of this system theoretically depends upon at least six factors: knowledge of the laws on the part of the people; apprehension of offenders by the police, except in less overt cases, such as breach of contract, infringement of patent, and so on, where charges are brought by the offended parties; a fair trial before a jury of twelve "good men and true" selected at random from the citizens; the wisdom and justice of the judge conducting the trial; the emergence of the truth in the clash between lawyers, each riding his case like a jockey to win; the efficacy of punishment in deterring the offender from repeating his misdeeds.

Actually, however, at each one of these points Middletown is experiencing some difficulty in making its traditional, relatively unchanging system of law and justice work under present-day conditions. In the first place, although every one is assumed to know every law and ignorance is no defense in case of infringement, the constantly growing mass of local laws has not been assembled since 1905, and one of the city's ablest lawyers remarked that "no one knows what ordinances are on the books." Furthermore, although Middletown depends for its detection of law-breakers primarily upon the vigilance of its police force,[16] these men, like other political appointees, are not chosen primarily for fitness for their posts, and are not infre-

---

[16] The following statement in a local paper in 1900 still represents fairly accurately Middletown's view: "Many theories are advanced nowadays as to the most available and efficacious means for the prevention of crime; but among all the remedies suggested, none will go so far toward accomplishing the end desired as alertness and efficiency on the part of patrolmen." As over against this attitude may be noted the statement of Raymond Fosdick in his *American Police Systems* that "even were we to secure 100 per cent. efficiency in our patrol and detective work, most crimes would still go unhindered." (New York; Century, 1920), p. 355.

quently, as noted earlier in this chapter, bandied about between mayor and Board of Safety in a struggle for political dominance.

Likewise the device of preëmpting a man's time for jury duty was better suited to the leisurely life of the eighties and nineties than to the more crowded life of a modern banker or merchant, or of a machinist in a shop where overhead costs are closely watched and a machine cannot be allowed to stand idle for a fortnight or so during a worker's absence. One result is an increasing recourse to court-room hangers-on having "no occupation." "Jury-dodging is now one of the favorite sports of some of our so-called 'best citizens,' " said the local Republican press in 1925. And again, "The city's most prominent banker, the most prominent merchant, the most prominent professional man, have as much right to serve upon juries as the 'pick-ups' that so often are chosen." According to the weekly Democratic paper:

"Jury Commissioners —— and —— met Monday and drew the grand and petit jurors for the January term. The petit jury is composed of six men and six women and the grand jury is also evenly divided, three men and three women. The law requires that the names of citizens shall be placed in the jury box and drawn out by the jury commissioners. The law doesn't say anything about drawing out an equal number of men and women. This has happened so often that it has ceased to excite comment.

"What would a jury panel look like unless ——'s name appeared on it, or ——, or ——. Strange how the same old names keep coming out of the box, isn't it?"

"Juries drawn by —— and —— are made up almost wholly from men and women connected with the —— [local Republican boss] outfit—precinct committeemen, deputy road superintendents, deputy assessors, school hack drivers, ditch commissioners, relatives of Judge ——, Sheriff ——, Deputy Sheriff —— and other beneficiaries of the system."

It is perhaps not surprising that a few local people are questioning the adequacy of submitting the complexities of crime to untrained juries in the hands of rival lawyers.

In regard to its judges, Middletown's attitude varies from individual to individual. It has great confidence in the judge of the superior court whose judgment has been reversed by a higher court only once in the course of seven years' service. Its other judges, however, many people tend to regard either with wary tolerance or active distrust. Chosen like other officials by the friction of party politics, and paid salaries insufficient to attract the abler lawyers in a day when corporation law offers the big local prizes, Middletown's judges are often men of little training or local professional standing.[17] The judge of the circuit court was recently tried, as noted above, in impeachment proceedings before the state Senate for irregularities in the conduct of his office; the lower house voted unanimously for impeachment, and the Senate acquitted him by two votes. The Democratic paper, referring to a "notorious colored gambler and bootlegger" who was "promised and received protection for political activities," and who "just laughs at the law," says:

"This time it was ———, late Klan organizer in ———, who, as special judge, gave him another clean bill of health. The last time it was Special Judge ———, another Kluxer, who sent him on his way rejoicing, to sin some more. City Judge ——— always dodges when he is arrested. We don't blame him. [This judge] and [the negro] belong to the same political machine, therefore he has a word or two to say in the matter. Every time he is turned loose by some 'special judge' especially selected for the purpose, he goes back to the 'red light' fortified with the knowledge that he can do as he pleases. That is a bad thing for him and a bad thing for [Middletown]."

Finally, the reliance upon punishment as a deterrent to infringements of group regulations is being somewhat shaken in the city today. The predicament of this traditional machinery of punitive justice in the face of the changing industrial life of the city appears in such a representative case as the following:

[17] Speaking of the difficulty of adapting political institutions of the middle of the last century to the more complex city of today, Roscoe Pound says, "Every one could and probably did know the character and qualifications of the few conspicuous lawyers who were candidates for judicial office in the judicial district or of the rising young lawyers who sought election as police magistrates of the small town of 1850. Under the circumstances of that time the greater the number of citizens that voted, the more intelligent the choice was likely to be. Today, when the average citizen of Cleveland can know the lawyers and judges only from what he chances to read in the newspapers or as he chances to meet them in the course of litigation or in social activities, it is often true that the greater the number of citizens who vote, the more unintelligent the choice." (*Op. cit.,* pp. 628-9.)

A family of six was haled into the juvenile court, the husband, a machinist of twenty-six, accused of gambling and violating the law against passing bad checks; the wife was a neatly dressed, earnest-looking girl of twenty-three; the oldest of the four children was six and the youngest a baby in the mother's arms. They were newcomers to town, lived in a dilapidated house, and had "had only beans and bread for Christmas dinner." It was a time of unemployment and the husband had been long out of work. He had been catching rabbits in the country, selling them for $0.25 apiece, and gambling with the money in an effort to make it more. He had given a grocer bad checks for $1.68, $3.24, and $2.10. "Why don't you get a job?" demanded the judge. "None to get," answered the man. The judge seemed stumped and rather inclined to insist that he get a job anyway. "Why did you pass these three bad checks?" he asked, shifting his attack. "I had to get groceries." Again the judge seemed stumped: "I know, but you can't do that, you know." Again shifting his attack to avoid this impasse, the judge turned to the white-faced wife: "Do you know what this means, to have your husband passing bad checks?" Her only reply was the four quietly spoken words, "We have to eat." Stumped squarely at this, the judge shouted, "Here, Mr. Bailiff, call in the sheriff. If that's the way you feel, we'll just lock your husband up." The girl wilted under this, and putting her head down on the table, wailed softly, "Oh-h-h-h-h!" The sheriff entered and stood beside the husband, whereupon the judge, having frightened the couple, a common practice with him, asked, "Do you say you don't approve of your husband's passing bad checks?" "Ye-e-es," came the meek reply from the wife. "Well, then, you people go home and behave, and, Henry, you get you a job." The family trailed out of the court room, the tow-headed baby looking back with wide, uncomprehending eyes.

The "social problem" of "the increase in crime" is being met most commonly by new state laws augmenting the punishment: changing the punishment for robbery, for instance, from two to fourteen years in prison and $1,000 fine to ten to twenty-one years and $5,000 fine; for rape, from two to twenty-one years to five to twenty-one years, and so on. Some evidence of a growing realization that paying a fine or spending a term of years in prison does not necessarily change a person's conduct for the better appears, however, in the use of the new device of the "suspended sentence" and in attempts by the judge of the circuit court to expose an offender to better influences by "sentencing" him to a period of church and Sunday School attendance. A man, with a wife and four children, charged with "desertion" and "contributing to the delinquency of a minor," was, for instance, sentenced to one and one-half to seven and one-half years in prison, but the sentence was suspended on condition that he "abstain from smoking cigarettes and attend church and Sunday School regularly." A forger was given a suspended sentence of two to fourteen years and sentenced to "attend church and Sunday School services regularly." A similar sentence was given to a man of forty charged with "failure to support" his wife and child, and is regularly given to all children. A system of report cards is used to check up on attendance. On March 1, 1924, forty adults and 110 children were out on probation and all were attending church and Sunday School as part of their probation. The printed probation card given each of these persons warns that "each Sunday missed without an excuse extends your probation one more month." [18]

As observed elsewhere, Middletown tends to be at once more conservative and more experimental in those regions where its children are concerned.[19] Thus these new substitutes for punishment appear chiefly in the children's court, itself an important innovation since the nineties. But this new court has, after the usual manner of social change, been incorporated into the traditional system, a new cog in the old machine. The judge of the circuit court took on this extra duty and the same institution operates with relatively little change in the simpler setting of his office, although the introduction of a woman probation officer, the wide use of the suspended sentence and probations, the informality of the court procedure, and the absence of publicity are significant departures. This tendency to handle young offenders against group prohibitions more flexibly

---

[18] "Well, Olive," the judge of the juvenile court remarked in a fatherly tone to a bobbed-haired girl of sixteen sitting on the edge of her chair watching the proceedings like a cat ready to spring, "we're to dismiss you today. Your record shows six months of regular reporting here and attendance at Sunday School. You've been a good girl. You're not going to do this thing you were sent here for any more, are you? Nor nothing like it? Well, we wish you well. We want you to speak to us on the street and come up and see us whenever you will. Good-by."

[19] Cf. e.g., the persistence of sending children to Sunday School among parents who do not themselves care to attend church regularly but are "afraid the children may miss something they ought to have" (Ch. XXII), and on the other hand the inauguration of group care for teeth and health in the public schools although no similar preventive system is in operation for adults (Ch. XXV).

must fight for its life in the midst of, now helped and now entangled by, a welter of institutional heritages of varying emotional weighting. On the one hand, the court exhibits the evangelical enthusiasm of its judge, who teaches each Sunday a Bible class of over 200 in the largest church in the city; on the other, it reflects the equally imperative loyalty of the judge to his political associates, which is simply another way of saying his livelihood. Although Rotary looks askance at him and the other men's clubs pay little attention to his work,[20] the women's clubs, liking his enthusiasm for children, give his work some support. But immediately following his election, a woman widely regarded as one of the most capable business class women of the city was dismissed from the Board of Children's Guardians and a political henchman substituted, to the dismay of the city's social service agencies; the supervising visitor of the Board was reported dropped for a political appointee, and the head of the local orphans' home was said to have been dismissed for another political appointee, whereupon the remaining two business class women on the Board resigned in disgust. There are disputes between the police and the probation officer and between the Social Service Bureau and the court over their respective jurisdictions; the police charge the probation officer with "keeping things back"; the court attempts to save the people's money by taking local children out of public institutions, while the Social Service Bureau tries to have a larger number of needy children cared for by the city or state.

In addition to use of the suspended sentence, further modification of the assumption that law-breaking is a "voluntary" act on the part of a willful wrong-doer appears in the tendency to include somewhat more of the population in the group whose infringements of taboos are regarded as not "their own fault." Thus, to cite but one instance of this extension of the period of helplessness before circumstances, the "age of consent"—when a girl is supposed to be able to decide for herself as to the

[20] Two days before the Advertising Club acted, as noted in Ch. VII, at the request of Middletown manufacturers to request the local press to suppress advertisements for machinists from out-of-town factories, the local press reported fully the warning of the juvenile court against advertisements for girls as "models" appearing in the local press, but no suggestion was made at the Advertising Club that these advertisements be suppressed. The judge of the juvenile court complains that he can get no help from the civic clubs in finding jobs for boys who come through the court.

sex approaches of a member of the opposite sex not her husband—was raised in 1893 from twelve to fourteen years, and in 1907 from fourteen to sixteen years.

So justice in Middletown gets itself done amid many of the same conflicting currents that condition the rest of the city's public business. As private business looms larger, draining off the best legal talent to serve its ends, the machinery of justice tends to become less independent and imposing and somewhat more an adjunct to the city's dominant interests; one not uncommonly hears the statement from both business men and workers that, by and large, a rich man or a corporation stands a better chance in Middletown courts than a poor man. "Science" in the form of the medical man and intelligence tester is beginning sporadically to displace some of the omniscience of the court, although, except in extreme cases, neither juvenile nor adult offenders receive even a medical examination as a part of the trial procedure. The religious atmosphere still clings strongly to the activity of dispensing justice at certain points, though the ceremonial of administering the oath has lost virtually all of its earlier meaning. Save in the case of the judge of the superior court, neither juries nor judges have the confidence of the city. Meanwhile, in their government, as in training their children and in their religious practices, Middletown people are tending increasingly to delegate their interests, while they busy themselves with more pressing and immediate concerns.

Chapter XXV

KEEPING HEALTHY

Around the central core of the city's business are various other general concerns not conducted as public or total-group affairs but widely supported as civic movements. Many of the latter are in process of evolution into the former. Among these activities, gradually developing from individual into group responsibilities, is the care of health.

In the main, the health of Middletown is cared for by a group of more or less well-trained doctors operating under the competitive business system. To understand the status of this activity, however, one must recall how close is the pioneer background of Middletown with its cures of "nanny tea" and bleeding. Measuring children for "short growth" is still practiced by a few people. A number of citizens have gone in recent years to "the old nigger out in ——," an outlying village, who is alleged to drive disease down through a patient's feet into the ground by waving his hands before him. Some people still treasure incantations for curing erysipelas and other ills, carefully passed down from generation to generation, always to one of the opposite sex. A downtown barber regularly takes patients into a back room for magical treatment for everything from headache to cancer. Some people still believe that an old leather hatband wrapped about each breast of the mother at child-birth will prevent all forms of breast trouble. An old leather shoe-string wrapped about a child's neck will prevent croup; "Our little boy had croup," said a working class woman, "and I'd forgotten about this cure, but I got a leather shoe-string, and the boy got well." Asafoetida worn about the neck prevents the catching of contagious diseases; "Our little girl wore one bag and played day after day with children who had whooping cough and never caught it," said another worker's wife. If one rubs a wart with a bean picked at random from a sack of beans and then drops the bean back into the sack the wart will disappear. Some people believe in curing a stye by rubbing a wedding ring on the eye, in carrying a copper wire about the wrist or a buckeye in the pocket to prevent rheumatism—copper rheumatism rings may still be purchased—or in the magical potency of flannel.

Such methods are largely confined to the working class, but some business class families still treasure such books as:

DR. CHASE'S

Third, Last and Complete

RECEIPT BOOK

and Household Physician

or

Practical Knowledge for the People.

From

The lifelong observation of the Author, Embracing the Choicest, Most Valuable and Entirely New Receipts in Every Department of Medicine, Mechanics, and Household Economy; including a Treatise on

THE DISEASES OF WOMEN AND CHILDREN,

in fact

THE BOOK FOR THE MILLION,

by A. W. CHASE, M.D.

Author of "Dr. Chase's Receipts; or Information for Everybody"; also "Dr. Chase's Family Physician, Farrier, Bee Keeper and Second Receipt Book."

Why conceal that which Relieves Distress?

This my third and last Receipt Book is most respectfully dedicated to the twelve hundred thousand families throughout the United States and Dominion of Canada who have purchased one or both of my former books.

1890

Among the counsels of this household guide are the following:

"It is claimed by many scientific men that it is best to always lie with the head to the north, on account of the fact—a supposed fact, at least—that there is an electric current passing through the system when one is lying down, whether awake or asleep, and that its influence is best with the head to the north. Invalids, at least, had better do it, if the situation of their room will allow it."

"DIPHTHERIA—CLOSING REMARKS UPON.—The author leaves the subject with his readers, believing that he has presented a larger number [twelve] and more reliable remedies or recipes for the cure and prevention of diphtheria than are to be found in any other publication whatever; . . . if these recipes are well studied, and one or more of them adopted by the heads of households containing young children, . . . nothing like the fatality will hereafter take place from diphtheria, as has heretofore been the case. . . . Then the responsibility rests with each one who shall have this knowledge. . . . The author has done his duty, which is a great consolation to him."

Among a similar list of optional treatments for "Rheumatism, Spinal Affections, Cancers, etc." is listed a "California Cure" consisting in allowing "wild parsnip roots . . . to simmer on the stove until they assume the consistency of paste; then spread on chamois skin and apply to the cancer . . . and the cancer will contract and loosen, until it may easily be extracted with its roots."

Akin to these time-honored methods of treatment but far more potent in shaping Middletown's habits today are the numerous and constant newspaper advertisements of "patent medicines." A locally made "Winter Pep Cough Syrup" is publicly recommended by the president of a highly respected Middletown life insurance company, by the judge of the circuit and juvenile courts, the judge of the city court, and the chief of police. In the leading Middletown paper one morning in January, 1925, there were sixty-eight display advertisements of one inch or larger, exclusive of amusement, agricultural, financial and classified advertisements; thirty-seven of the sixty-eight concerned various salves, soaps, and remedies.[1] In January,

[1] The complete list was: Tonsiline ("The National Sore Throat Remedy") ; Corona Wool Fat ("Old Sores Quickly Healed") ; Marmola Prescription Tablets ("Reduce Your Fat Without Dieting") ; Pyramid Pile Suppositories ("Quick action for piles") ; Dr. Platt's Rinex ("Kills Colds—*quick!*") ; 666 ("Colds, Fever and Grippe. It is the most speedy remedy we know, preventing pneumonia") ; Dr. James' Headache Powders ("Druggists sell millions . . . because they are safe") ; Mother Gray's Sweet Powders ("Benefit many children complaining of Headaches, Colds, Feverishness, Worms, Stomach Troubles and other irregularities") ; Zemo ("For itching torture") ; Dr. Bell's Pine Tar Honey ("Coughs") ; Cuticura Soap and Ointment ("Pimples—healed") ; Resinol Ointment and Soap ("At the first sign of skin trouble apply") ; Tanlac ("If your system is run down," etc.) ; Dr. Edwards' Olive Tablets ("Bad breath—Gets at the cause and removes it") ; Nature's Remedy ("For constipation") ; Musterole (usually gives prompt relief from [nineteen ailments]—it may prevent pneumonia") ; Sys-Tone ("For not only building up your strength but for steadying your nerves") ; Bromo Quinine ("Cold and Grip Tablets") ; Poslam ("Often ends pimples in twenty-four hours") ; Pinex ("Ends stubborn coughs in a hurry") ; Haley's M-O ("Mineral oil users try this!") ; Peruna ("The recognized treatment for catarrh") ; Epsonade Salts ("World's finest physic") ; Pope's Cold Compound ("the quickest, surest relief known for colds") ; Kondon's ("Get rid of that catarrh") ; Jad Salts ("A glass clears pimply skin") ; The Williams Treatment ("To prove the Williams Treatment conquers kidney or bladder disorders, rheumatism, and all ailments caused by

1890, the leading paper carried on one day in one column six advertisements of "cures" for "weak men." A local diary records in 1889, "Elixir of life is all the talk. Claims to make old people feel young, etc. Great many have experienced bad effects of it, sore arms, abscesses, erysipelas, and blood poisoning." The printed membership directory of the largest Middletown church in 1893 advertised "Blitz's Instantaneous Toothache Cure and Filler. A sure cure for Toothache, and a filling that will last for six months." Then as now the advertisements were reassuring to the doubtful: "It is not always perfectly safe to soothe the baby with opium preparations, but you can rely on Dr. B——'s Baby Syrup; it contains nothing injurious." Though group action has curbed somewhat the content and unlimited claims of these "cures" since 1890, their number appears to have increased.[2] In October, 1890, 25,004 lines of medical advertising, more than any other kind of advertising, appeared in the leading Middletown paper as against 45,451 lines in the same month of 1923.[3] Supporting and supported by this heavy advertising is the widespread local propensity for "taking things." Marion Harland remarked a generation ago the American habit: "In case of indisposition, the first question in this country is: 'What shall I take?' 'What shall I do?' is left unasked." And according to the head of the Social

'too much Uric Acid,' no matter how chronic or stubborn, we will give you one 85¢ bottle (thirty-two doses) free"); McCoy's Cod Liver Oil Compound Tablets ("Tens of thousands of thin, run-down men," etc.); Mentholatum ("It heals, soothes, softens"); Creomulsion ("Money refunded if any cough or cold . . . is not relieved"); SSS ("To regain strength"); Pape's Diapepsin ("Correct your digestion"); California Fig Syrup ("Harmless Laxative"); Konjola ("Stomach, liver, kidney troubles, catarrh of mucous membranes, nervousness, constipation, awful neuritis and rheumatism"); Lydia E. Pinkham's Vegetable Compound ("An operation avoided"); Baume Bengué ("When in pain"); and a three-column advertisement of many remedies, including Tanlac, Pepgen, Trutona, Vigorlac, Ozomulsion, Father John's Remedy, Pierce's Golden Medical Discovery, etc.

[2] There were 825 proprietary items listed in the catalogues of leading wholesale drug houses in 1871, 2,699 in 1880, 17,780 in 1906, and 38,143 in 1915. "A recent authority states the number of manufacturers of proprietary medicines in 1879 to be 563, and in 1914 this number had increased to 2,903, and we know a much larger proportional increase has taken place since that date." W. A. Hofer, "The Economic Value of the Jobber and Distributor" (*Standard Remedies;* Vol. VI, No. 6).

[3] See Table XXII.

Though this represents more than an 80 per cent. absolute increase, the relative percentage of this type of advertising fell off from 25 per cent. of the total advertising space in October, 1890, to 8 per cent. in October, 1923.

Service Bureau in Middletown, "People of the poorer class here like to keep taking things and going to see the doctor." [4]

A further source of diffusion of health habits is the newspaper advertisements of so-called "quack" doctors. There are in Middletown two main groups of medical men: the regular group whose "professional ethics" forbids them to advertise for patients, and a scattering of other men outside the local Academy of Medicine and usually less well trained who flood Middletown's newspapers with prestigious large space advertisements. These advertisements utilize the price appeal Middletown is habituated to respond to in buying dresses, shoes, and other articles, and drive straight at the popular belief that "it costs an awful lot to go to a doctor":

"Where a Multitude Go to Get the Worth of Their Money Is a Fairly Safe Place for You to Go—Be It Lawyer, Merchant, or Doctor. . . . I have treated thousands . . . ; therefore, frankly and candidly, I do not hesitate to say to you that I can give you more actual service for your money than you will get anywhere else in this whole country."

"I care not what you are worth in dollars and cents. During the past twenty-five years I have successfully treated and cured thousands. . . . Some paid me with money, but many who were financially unable to do that paid me in gratitude. . . . Therefore, I now, again, extend to you, one and all, rich and poor alike, the opportunity to make one more honest effort to regain your health."

"Ten Dollar Examination Absolutely Free," beckons over the authoritative-sounding name of a "medical institute." And another two-column advertisement crowded with pictures of cured persons proclaims:

"They say Prof. ——, Noted . . . Magnetic Masseur and Foot Correctionist, Does the Work. WILL BE IN [MIDDLETOWN] FRIDAY. Good News for All Chronic Sufferers. *Absolutely No Charge for Glorious Services.*"

[4] One reason why they "like" to do it may conceivably be found in the fact that for those whose lives are somewhat cluttered by many baffling affairs—job, children, ill temper, unemployment, bickering with neighbors, a surly husband—psychological comfort and a social justification may be afforded by "being sick." It is an incidental detail that the thing "taken" frequently has no more direct efficacy to better the specific condition than have the burns on the shoulder with a hot stick which the Todas inflict "to cure the pain arising from the fatigue of milking."

Two other appeals are used in their frontal attack upon sickness by these doctors: They reassure sick people—"A Road Traveled by Thousands Without Accident is Fairly Safe for You"—and they offer avoidance of the much-feared operations—"Why Endure the Knife when Dr. ——'s Natural Method will Cure You." One of them even goes so far, in assuring women how to "avoid the operating table," as to say:

"My name is as firmly connected with the treatment of diseases of the pelvic organs as is Edison's connected with electricity. . . . My office records show that 90 per cent. of the patients cured by my treatment were unsuccessfully treated elsewhere before they came to me."

Occasionally, as in the following, appeals to desire for health and the powerful and closely associated religious tradition of Middletown are combined:

WHAT POWER DOES
DR. M —— POSSESS?
Some say Magnetism. Some say Faith Healing.
Others say Divine Healing. Here are the facts.
During his boyhood Dr. M —— had the gift of healing. . . .

M —— GOES TO COLLEGE
Desiring to be fortified with exact knowledge, M —— in 1912 entered the —— Institute of Suggestive Therapeutics. Here he obtained a very comprehensive knowledge of Suggesto-Therapeutics and supplemented this with a special course of training in applied psychology. . . . Dr. M —— studied under Prof. ——, conceded by thousands to be the world's greatest Magnetic healer. . . . The United States Government has recognized his success and has officially authorized his School and Graduates.

HE CURED A BAD CANCER OF THE FACE
John W ——, ——, had just about given up all hopes of recovery from a bad cancer of the face. Having heard of Dr. M——'s remarkable cures, he came to [Middletown]. After only twenty-one treatments the once raging cancer vanished. He returned home well, shouting Dr. M ——'s praises from the house tops. . . .

LARGE THRONGS SWARM THE OFFICES
The throng of sick folks that swarm the offices and Sanitarium at —— keeps this healer busy from early morning until almost midnight. Then he slips away . . . only to take up his duties early next morning of helping suffering humanity's footsteps back into the pathways of health and happiness. . . .

One such doctor strikes squarely at the traditional medical ethics in a column advertisement:

"Why do I advertise? When I know that by so doing I debar myself from all the privileges of the so-called Regular Medical Societies? Answer: For humanity's sake. . . . I determined to be a benefactor to the races by making it my life study to keep people away from the unnecessary use of the knife. . . . But how was I to inform the public? I might have sat in my office from now until Doom's Day and very few people would have known about my special qualifications. . . . So my only recourse as a Humanitarian and God-Fearing Man was to spread the news through the newspapers."

The attitude of many business class citizens toward all such health advertisements is probably represented in the comment of one prominent woman, "People ought not to be so foolish as to be taken in by such advertisements, but then, of course, the papers need the money and it would be a worse tragedy not to have papers than to have the patent medicine advertisements."

The attitude of the members of the local Academy of Medicine is one of silent condemnation not unmingled with envy.[5] The State Medical Association, following the disclosure of a prohibited method of issuing medical diplomas in another state, put forth in 1924 a dignified statement, reprinted in the press, on the danger of being treated by persons not properly trained: "There are thousands of unqualified persons [in this state] who treat the sick who are not licensed or under the jurisdiction of any responsible state body"—but this bulletin did not talk the language of the workers and the inconspicuous news note was a weak rival of the smashing type and full-page space of the irregular practitioners. The local Academy of Medicine passes resolutions, but beyond that their plight is described by one of the ablest of the younger men: "We're helpless. We

[5] The situation is the more galling to them in that, according to a leading doctor, "Dr. —— [one of the quacks who advertises] has openly boasted to some of us that the practice of medicine in [Middletown] can be made to yield a million if properly handled, and he has already cleaned up more than half of his million." The practice of this man was described by a local physician as "among ignorant South Side people whom he keeps coming back week after week at $2 or so a visit. Just last week he told a woman who had been coming to him for a year that her trouble was cancer of the womb and he couldn't help her. She came to me as a surgeon, but the cancer had been allowed to run too long. Had she come to me even six months earlier, an operation probably would have saved her."

ought somehow to assume more community leadership because we are the people who know the facts. But the trouble is that as soon as we did so, people would begin saying we were acting from selfish motives." The Ad. Club has passed a set of mild resolutions condemning the practice of local papers in running medical advertising, but the club's more active interest in "Truth in Advertising" is in less locally controversial affairs.

Meanwhile,

"The doctor, as popularly conceived by the uneducated, is a mystery at best, something of a magician from whom the patient dares expect only a small part of the truth and no explanation of it. . . . He [the uneducated person] finds as wide a choice of doctors as of religions, and as he hesitates, bewildered, the more watchful and aggressive forces find him." [6]

The claims of the medical profession itself to "scientific" methods are of fairly recent growth. As late as 1885 a state law regulating the practice of medicine was headed, "Requirements to Practice the Healing Art." Up until 1897 the state law governing the admission of a person to practice medicine was so lax that the clerk in one county is said actually to have issued a license to an applicant who submitted a Chinese napkin with the claim that it was a diploma issued by a Chinese medical college. "It was entirely possible for any one to secure a license under the said law by submitting any sort of instrument and stating that the same was issued by a reputable medical college." [7] Since 1897 the educational requirement for doctors has risen first to a high school diploma and then to two years in a recognized college, including specified scientific work, and in addition in both cases graduation from a recognized medical school, while closer supervision of medical schools by the state has weeded out all but one medical school, and it is a "Class A" institution.[8] It is perhaps indicative of the refinement of

[6] Mary Strong Burns, "Quacks and Patent Medicines," from *Education and Practice in Medicine, Dentistry and Pharmacy,* Part VIII, *Cleveland Hospital and Health Survey* (Cleveland; Cleveland Hospital Council, 1920), p. 673.

[7] Quoted from a letter of May 9, 1924, from the Secretary of the State Board of Medical Registration and Examination.

[8] "Since 1900, through a campaign carried on by the American Medical Association, greatly improved conditions have been secured in the medical schools of the United States. . . . At that time this country had 160 medical schools—more than in all the rest of the world—but only a few of them were equipped to provide a satisfactory training in medicine, and only two

the medical technique of Middletown's doctors even since 1890 that the death records for the city in 1890 reveal but fifty-one kinds of "causes" of death as against 160 in 1920. For the last half-dozen years a Middletown Academy of Medicine, reputed to be one of the strongest in the Middle West, has met weekly to hear specialists from many large cities discuss new developments in their fields.

With the increase in medical skill have inevitably come vested interests. Says one of the city's leading medical experts, "Our Medical Association refuses to recognize the right of an outside medical man to come here and earn money raised in this county. We don't believe in 'socialized medicine'—no, sir! ——— was sent here by the State Venereal people, but he wasn't one of us. Now we've arranged to have the work done by local men." An outstanding doctor, asked about the observance of the state law requiring doctors to report to the state authorities all cases of venereal disease, shrugged his shoulders and said, "Doctors must live—like everybody else. A fellow comes to a doctor and says, 'Am I going to be reported?' It's only human nature to answer 'No,' rather than lose the case." A responsible social service worker with every facility for knowing the facts said of the tuberculosis clinic, "We have no good chest man in town, and a man was sent in from the state capital one day a week to hold a clinic. But the local medical men raised such a protest we had to stop, and now we use local men." [9]

It begins to be apparent that the profession of medicine, like the occupation of running a drill press or selling real estate in Middletown, swings around the making of money as one of its chief concerns. As a group, Middletown physicians are devoting their energies to building up and maintaining a practice in a highly competitive field. Competition is so keen that even the best doctors in many cases supplement their incomes by putting up their own prescriptions. Meanwhile, one observes the situation of some fifty local doctors spending much time sitting

required more than a high school course for admission." At the present time seventy of the eighty schools in the country are rated in "Class A." (From ~ press bulletin issued by the State Medical Association.)

[9] This opposition by local doctors to large-scale group care of these two groups, the venereal and the tubercular, assumes even more meaning in the light of a statement by one of the best-trained of the younger medical men that "the best doctor is the man who understands syphilis and tuberculosis, since they underlie so much of all the rest of ill health."

in their offices waiting for patients to come in and proffer the requisite money for treatment, "needing," as one of the ablest of them expressed it in a professional memorandum to a group of associates, "a chance to grow in practice as well as professional attainments," while at the same time 38,000 people, most of whom have some physical defect great or small needing correction,[10] are in only relatively few cases having these defects treated by the best medical skill the city possesses. A boy haled before the juvenile court was an obvious adenoid case, so obvious that the court recognized it although no medical examination was made. The judge said to the parents, "Your boy needs his adenoids out. Who's your doctor?" In answer to their mention of a local medical man, the judge rejoined, "He doesn't do adenoids. You go out and find some one to take your boy's adenoids out." Apparently noting their shabby appearance and blank looks he added, "You can get somebody who'll do it if you pay a few dollars a week. Come back and report to me next week." And the father, mother, and son were shooed out of the court room, bewilderment still on their faces. Some people in such a predicament drift in to the Social Service Bureau, but there each case must be handled separately and the head worker protests that she is turned into "a high-class errand girl," going about trying to arrange now with this doctor and now with that for the treatment of such charity cases as wander in to her. The principal of a school said:

"I have a child who has lost his hearing in one ear trom adenoids, and the other ear is going. Another child needs glasses badly. The father is out of work, and the mother has asked if I could find a doctor who could fit the girl for glasses and let the family pay later as they can. I don't know what to do with these cases. If I send word to the Social Service Bureau they may get around after a while with all they have to do to investigating the

[10] According to Eugene Lyman Fisk of the Life Extension Institute, an analysis of 10,000 American industrial and commercial workers engaged at their tasks, and supposedly in good physical condition, showed that 83 per cent. showed evidences of nose and throat defects (17 per cent. marked or serious); 53 per cent. showed defective vision uncorrected; 21 per cent. flat feet; 56 per cent. defective teeth; 62 per cent. of mouths X-rayed showed root infection; 12 per cent. of those examined showed well-marked changes in the heart, blood vessels, and kidneys; 9 per cent. showed marked lung signs requiring observation for possible tuberculosis.

family, and then after more delay they may be able to do something, but both these cases need help right away."

One of the working class wives interviewed, a woman of thirty-nine, the mother of four children, replied to the question, "What use would you make of an extra hour in the day if you had it?"

"I'd go to bed. They tell me I have chronic appendicitis and gall-stones. Our doctor and the surgeon he took me to wanted to operate last year, but they said it would cost $150 for the operation and two weeks in the hospital, and we can't afford that much. I'd go to —— [the state capital] where it could be done cheaper, but I just can't bear to leave the children to run wild that long." [11]

Neither the physicians nor the people are satisfied with this situation in which medical skill is engaged part of its time in a game of teeter-totter with the city, the institutional devices of "price," "competition," and "professional ethics" being the fence over which the two groups see-saw up and down.

So much for the care of health by individuals. Since 1890 the city has interested itself in the matter. In July, 1890, the first health officer was appointed, serving as a part-time officer at a salary of $250 a year. His services were evidently somewhat casual, for in 1900 a newspaper asks, "What has become of the health officer?" Today Middletown still has a part-time health officer, paid slightly more than half the salary of the lowest grade policeman. The city health office is in this doctor's home and he is assisted by his wife, who, according to the newspaper, "when not doing some sanitary or humane work

[11] Royal Meeker, United States Commissioner of Labor Statistics, speaking before the Employment Managers' Conference in Rochester in 1918, said, "I am convinced that very few, if any, laborers' families can expend enough for doctors, dentists, and medicines to meet the requirements of a minimum standard of health under present conditions." (Bureau of Labor Statistics Bulletin No. 247), p. 47.

In his study of *Medical Education in the United States and Canada,* Abraham Flexner says, "Society reaps at this moment but a small fraction of the advantage which current knowledge has the power to confer. That sick man is relatively rare for whom actually all is done that is at this day humanly feasible—as feasible in the small hamlet as in the large city, in the public hospital as in the private sanatorium. We have indeed in America medical practitioners not inferior to the best elsewhere; but there is probably no other country in the world in which there is so great a distance and so fatal a difference between the best, the average, and the worst." (New York; Carnegie Foundation for the Advancement of Teaching, Bulletin No. 4, 1910), p. 20.

. . . is busy baking cookies." The Health Officer is assisted, also, by a part-time policeman, who, according to the press, "has been serving as humane and sanitary officer for several months, when he was not assigned to other police duties." Two political appointees are paid $100 a year each as members of the Board of Health; in 1924-5 they were the wife of a policeman and the wife of a foreman. The method by which they were selected is suggested by the comment of the Democratic weekly on the occasion of the displacing of one of them by the mayor:

"Because of her criticism of the policies of the administration, ——, female Republican politician, . . . was this week removed by Mayor —— as member of the city health board for pernicious political activity, and Mrs. ——, wife of the . . . fifteenth precinct committeeman, was appointed in her place. —— was at once given a job as deputy township assessor, under Assessor ——, along with ——'s [the local political boss] sister, Mrs. —— [the wife of the county superintendent of schools], and Mrs. —— [the wife of the Republican candidate for mayor]."

The report of the field staff doctors of the American Child Health Association, which included Middletown among the eighty-six cities studied in 1924, concluded that Middletown "is still in its infancy as regards organized health work." [12] The significant point to recognize here, however, is not that Middletown has not adopted a theoretically complete program of public health work, but that during the last four decades it has come to regard health as a matter in which the public should interest itself. In view of Middletown's strongly individualistic traditions and the comparative recency of this new attitude toward health, it is far from surprising that public control of health has met with some opposition and is unevenly diffused throughout the city.

Different levels of health habits appear, for example, in the handling of such a contagious disease as typhoid fever. In the nineties it was so common that, according to the health officer, "it was the doctor's principal work during the summer." A diary notes in July, 1890, "Considerable typhoid fever in town," and records two or three deaths within the following week. Since the State Board of Health began, about 1904, to analyze city water supplies throughout the state and condemn

[12] *A Health Survey of Eighty-six Cities* (New York; Research Division of the American Child Health Association, 1925).

about 50 per cent. of those examined, better water has been secured; town pumps have vanished and the early city water that "looks like cider" has given way to acceptable water analyzed weekly in the state laboratory. But, despite a law requiring city water and sewerage connection, in January, 1925, as pointed out above, approximately one home in four in Middletown depended upon private backyard sources of water.

"Last year," according to the office of the Health Officer, 'there were twelve cases of typhoid in the Isolation Hospital which we traced directly to the use of water from two backyard wells. But we get little coöperation in enforcing the sanitary laws. For instance, there are many influential citizens who own property and just won't clean up some of it but are content to rent it for something like $10 a month. One of the ——s [a wealthy family] owns some property in town that rents for $8 a month and we have tried and tried to get him to clean it up, but he begs for time and says he will do it in the spring, and we let it go, and then summer comes and complaints immediately start coming in, and he doesn't do a thing about it. This is all made harder for us because the health officer is paid so little he can't afford to antagonize anybody. Some of his best patients are people who own property that ought to be cleaned up, but if he goes and orders these changes made and they object and he ejects their tenants, he will probably lose them."

A somewhat similar situation exists in regard to other contagious diseases; recurrent epidemics are taken for granted. In 1890 Middletown factories had at times 25 per cent. of their workmen laid up with grippe, and the press would report "schools and churches are both crippled. The *Times* has been obliged to cut its reading space because so many printers are ill, and the flour mill has been forced to close because it has not enough experienced men able to work." In 1893 schools, churches, and general business houses were closed for two months because of smallpox. It is perhaps indicative of the acceptance of recurrent sickness as a perennial characteristic of this culture that when the quarantine was lifted in the fall of 1893 the popular mood was not one of inquiry into the factors lying back of such an untoward interruption of the ordi·· nary business of living, but: "General jollification, assembling of people on streets, blowing of horns, burning of red fire. On the following day church bells call the people to divine service, and on November 6 the public schools are opened. Mercantile business revives at once." Thirty-one years later the report of the American Child Health Association states, "Although [Middletown] has had an epidemic of mild smallpox for about two years, there has been no wholesale vaccination effort." [13] In 1888 a prominent citizen notes in his diary that "John has the German measles but is out all the time," and thirty-five years later the papers are still speaking of many children who have measles but are "allowed to mingle with other children by their parents." A visiting health expert notes that free immunization for smallpox, diphtheria, and typhoid are not available, and the Health Officer is quoted as reporting that "the doctors are only fairly good about reporting cases," while physicians counter with the reply that their reports to the Health Officer are sometimes neglected.[14]

In recent years the city has assumed further responsibility for care of health through the public school system, which conducts brief periodic dental and medical examinations of all school children. This work is independent of the Health Officer save that a child absent from school more than three days must secure a permit from the Health Officer to return; both school nurses and Health Officer are said to "stand on their rights," each insisting that the other keep within his own province.[15] The schools, also, give to all children a course in

[13] *Op. cit.* The report notes, moreover, that there are no records of vital statistics before the current year or of deaths from communicable diseases for the current year and no spot map studies of the incidence of disease.

[14] It is perhaps significant that only sixteen diseases are recognized by the local health department as "contagious and quarantinable," as over against thirty-six "infectious diseases" recommended as "notifiable" by the Eleventh Annual Conference of State and Territorial Health Authorities acting with representatives of the United States Public Health Service in Minneapolis in 1913. Cf. I. S. Falk, *The Principles of Vital Statistics* (Philadelphia; W. B. Saunders Company, 1923), pp. 102-5. A few of these thirty-six, such as Asiatic cholera and Rocky Mountain fever, are not common enough in Middletown perhaps to be listable, though the Middletown list does include such locally rare maladies as leprosy, cholera, and yellow fever. Such far more common maladies as malaria and mumps are not reportable. Although the above list of notifiable diseases recommended in 1913 also includes in addition to the thirty-six "infectious diseases" cancer, pellagra, syphilis, gonococcus infection, eleven "occupational diseases," and "any other disease or disability contracted as a result of the nature of the person's employment," none of these is reportable in Middletown. "There is 'not much' reporting of tuberculosis," according to a responsible public health expert who studied the local situation, "and reporting of pneumonia and opthalmia neonatorum is 'not done.'"

[15] On one occasion, e.g., when a child in one of the schools was believed to be "breaking out" with an infectious disease, the teacher sent emergency

hygiene; in this department of school health work alone, among the eleven types of health work classified, was Middletown ranked among the upper third of the eighty-six cities studied by the American Child Health Association in 1924. It is noteworthy as a commentary on the process of social change that this new habit of diffusing free preventive diagnostic service to an entire age group of the population has emerged first as a measure to protect the children. Chief among the factors influencing this differential diffusion are, probably, the appeal and recent popularity of child welfare work, the popular habit of thought that the health of an adult is "his own funeral" unless he has a contagious disease, while a child's health is somebody's else responsibility, as well as the fact that preventive child health work is in Middletown largely an unworked field in which there are few vested interests. Care is taken that concern for the health of the children shall not interfere with "private practice," for children are not treated by school doctor and nurse but simply given a card to their parents pointing out defects and suggesting that they consult their private physician; dental work is done free, as in the case of the adult venereal clinic, only for those too poor to pay for private treatment.

As in this public school health work, diffusion from without is responsible for the rise of "food inspection" in Middletown. The happy-go-lucky conditions prevailing in 1890 appear in the *Seen and Heard* column of the leading paper:

"I was strolling near the river and saw one of our industrious dairymen coolly drive into the dampness, dip up a bucket full, and thicken up the contents of his can."

Not until after 1890 was the first tentative pure food law passed by the state.[16] Today there are explicit state laws covering milk

calls first to the school nurse and then to the school doctor; failing to locate either, she called the Health Officer. The latter refused to see the child, saying it was the business of the school doctor.

[16] The first Federal meat inspection law was passed in 1890. It is an interesting commentary upon the incidental nature of "progress" that this first move was made "not because Congress had in view the protection of the people of this country" but simply to enable American meats to gain entry into foreign ports with more strict food supervision than the United States had yet learned. This early meat inspection concerned only meat for export. Not until 1906 did the idea of protecting its own citizens spread sufficiently throughout this country to prompt action to that end. (Cf. Carl L. Alsberg, "Progress in Federal Food Control" in *A Half-Century of*

and milk products, meat, canned goods, and certain other forms of food, and state inspectors visit Middletown from time to time.

In Middletown, however, although the new habit of food inspection has spread, it presents a ragged array of institutional practices. For years the State Board of Health has tried to get the city to pass a milk ordinance. The City Council finally passed an ordinance, but, according to a public health expert who studied the city, "without getting expert advice, and the result is a very complicated arrangement for recording milk sellers, with much red tape, and with no one to enforce it, so that it is entirely ineffective." The office of the local Health Officer reports, "We try to enforce the act as well as we can, but we ought to have a full-time policeman to help us as the present part-time man doesn't have time to do all he ought to." [17] And another region of conflict appears in the statement of the Health Officer that "the papers often will not coöperate in publishing the names of firms who are fined by the state health authorities for selling illegal foods such as bad meat or lard substitutes, because they want advertising from these firms." When the Woman's Club agitated against improper branding and weighing of food-stuffs, local dealers became uneasy, and according to a woman active in the movement, a leading wholesale grocer, a prominent churchman, telephoned the president of the club and protested, "Don't you think this business has gone far enough?" [18] A leading butcher in Middletown

Health, the Jubilee Historical Volume of the American Public Health Association, and also A. D. Melvin, *The Federal Meat Inspection Service,* United States Department of Agriculture, Bureau of Animal Industry, Circular 125.)

[17] According to the report of the American Child Health Association, "Milk dealers are registered with the Health Department, but no effort is made to make sure that all are registered. There is no dairy farm inspection and no laboratory analysis of milk being done by the city. A few inspections have been made by the state milk inspectors. About 70 per cent. of the milk is pasteurized, but there is milk being sold raw in Middletown from cattle not tuberculin tested, and the pasteurizing plants are not equipped with recording thermometers so that some one in authority may have definite information that pasteurization is properly done." (*Op. cit.*)

[18] Cf. the observation of Prof. A. B. Wolfe: "The business men who 'run' the average American town are very likely to oppose any exposé of conditions calling for more effective public health administration. . . . Druggists are commonly found to be hostile toward any movement for limiting the manufacture and sale of fake patent medicines." *Conservatism, Radicalism and Scientific Method* (New York; Macmillan, 1923), p. 74.

stated that while a number of local meat dealers have been arrested and fined for selling adulterated meats, many others continue to sell such "doctored" meat. "I guess politics is to blame," he concluded with Middletown's usual good-nature in such matters.

The decline of the venereal and tuberculosis clinics noted above is indicative of the cross currents which surround phases of health work in their transition from individual to social handling. Although the city reads in its papers that "syphilis and its effects on posterity are the one big problem before us today," and that "it is a conservative estimate that today the care in state institutions of those directly wrecked by syphilis costs [the state] over one-third of $1,000,000 a year; while those in these institutions suffering indirectly from this disease cost the state a much larger amount," [19] the doctor in charge of the clinic states:

"The doctors here hang on to patients as long as they can pay, since venereal treatment pays well, and only after they can't pay any longer send them to the clinic. Attendance at the clinics has fallen off by 50 per cent. Although the important thing about venereal work is regularity of treatment until a cure is effected, the police give us no help. The result is that a great many of the cases we treat stay only long enough to get superficial cures."

Since the above interview, the county has dispensed with the services of the state doctor, and the clinic has been suspended save for the treatment of those too poor to pay a local doctor.[20]

The County Tuberculosis Association, started in 1917 as a charity fostered by a few interested individuals, has now entered its second stage as a semi-public agency, and will probably eventually be taken over by the group as part of its corporate business. In 1924, 14 per cent. of the population sup-

[19] *Mental Defectives in [this State]* (Third Report of the State Commission on Mental Defectives, 1922). Only 6.7 per cent. of the state's feeble-minded, 25 per cent. of the epileptic, and 79 per cent. of the insane, were in these state institutions at that date.

[20] In the year ending June 30, 1923, 3,140 treatments were given in the clinic to 148 patients, and 106 old patients were discharged as non-infectious, as against 1,410 treatments in the year ending June 30, 1925, to eighty-two patients, with only four continuing treatments long enough to be discharged as non-infectious. There is no reason to suspect any diminution in venereal disease; the falling off reflects rather the failing local support of the clinic. The year closing June 30, 1925, includes the last months under the old régime and the commencement of the new plan of treating only those too poor to pay a doctor.

ported the Association, and in 1925, with the inauguration of the Community Chest, its support was still more widely distributed.

In response to the educational work of this and other agencies, some change has been made since forty-five years ago when a writer in the annual report of the State Department of Statistics and Geology for 1879 notes in "court-room, theater and places of public resort large areas of filthy saliva upon the floors, festooned with tobacco quids varying in size from that of an apple-dumpling to that of a tree-frog," but some women in Middletown still say that they walk on the inside of public sidewalks because of the constant spitting. In the seventies and eighties a diagnosis of "consumption" was a sentence to slow death. Middletown has today shifted to a basis where it is, according to the local Tuberculosis Association, "arresting and improving many cases, but with its limited facilities curing few." Under this semi-public agency a free tuberculosis clinic is conducted, open-air shacks furnished free to the sick, educational campaigns and drives carried on, and summer nutrition camps conducted for selected groups of both boys and girls. Two full-time trained nurses and a permanent paid secretary are now employed in place of the volunteer workers of five years ago. It is characteristic of the shifting status of much of this health work that it is through this semi-public agency and not through the Health Officer that Middletown coöperates with the State Infant and Child Hygiene Division and state and sectional conferences on health and hygiene. Indeed, the Health Officer did "not know anything about the conduct of the venereal clinic" and did "not know much about the work of the Tuberculosis Association."

The Visiting Nurses' Association is yet another health agency which, beginning in 1916 as a private charity of a few women, is rapidly assuming the status of a semi-public institution through the payment in 1923 and 1924 of $1,200 from the city funds for its support. Its private charity status is being attenuated further through the large share of support received from the Metropolitan Life Insurance Company for the care of its patients and the increasing tendency of local industries to assume responsibility for their employees by paying for their care at the hands of the visiting nurses. This agency is entering the home and breaking down one of the most cherished regions of "ancestor wisdom" in its diffusion of modern health habits in pregnancy and childbirth. In 1924 its seven full-time nurses made 12,217 visits, three times the number made in 1920.

The Middletown hospital, likewise, has ceased to be a purely private institution, as the county contributes to its support for the care of the poor. But it remains largely a hospital for private patients with no clinical facilities and so small as to be "too expensive for the poor and not good enough for the rich." Middletown has fifty-three hospital beds for its 38,000 of population.[21]

According to Sir Arthur Newsholme, "Infant mortality is the most sensitive index we possess of social welfare and of sanitary administration, especially under urban conditions." [22] In the four years 1910-14, ninety-six out of each 1,000 babies in the state died before reaching their first year; in 1915-19, eighty-five; in 1920, eighty-one; in 1921, seventy-one; in 1922, sixty-eight. Presumably, some roughly similar trend obtains for Middletown, although local vital statistics happen to be too loosely kept to invite reliance upon their use. A veteran Middletown physician reports of the obstetrical work of forty years ago: "As I think back in the light of our present knowledge, the things we used to do then were awful! I used to go out here in the country and deliver a woman and go off and leave her with a servant girl. And if God was good she got well. By golly, they *did* get well, too, most of them! They were healthy folk." Says a "doctor book" in Middletown homes in the nineties, "Among the working classes it is still too much the custom for women to be confined in their everyday dress." Middletown diaries of 1890 report the birth of a child as a semi-gala event: "Next day mother and son had lots of visitors." Such conditions are by no means uncommon today. A recent survey by the Division of Infant and Child Hygiene of the State Board of Health of 6,809 mothers in the half of the state in which Middletown lies, revealed that 10 per cent. of the mothers reported

[21] According to S. S. Goldwater, President of the American Conference on Hospital Service, five beds per 1,000 of population is usually considered an average for general hospital purposes.

A new 150-bed hospital was being built in 1927-28 through the philanthropy of wealthy citizens.

[22] Quoted by Falk, *op. cit.*, p. 72.

hospital births, 13 per cent. care by a registered nurse, 37 per cent. care by an experienced nurse, 40 per cent. no care by nurses, 34 per cent. no pre-natal or post-natal supervision, and 67 per cent. no rest from heavy duties during pregnancy.[23] Medical men state that even this amount of care was unknown locally in the nineties.

In general, then, Middletown people individually and as a group are much more concerned about health than in 1890, so concerned that the city is more and more assuming care of health in cases where individuals may be apathetic. But the strong pioneer individualism which clings to health as a private matter persists, and various aspects of this single activity of caring for health appear in every stage from assertion of complete individualism to complete public control. On the same city block live a mother who is eagerly seeking to follow modern suggestions for her health and that of her baby, and another who rebuffed the visiting nurse with the emphatic declaration, "You can't tell me nothin'. I've had seven kids. And I don't want no doctors around, either; one came and put some stuff in one of my babies' eyes that made them sore. All I put in their eyes is tea one day and castor oil the next." Even the preventive work among school children meets with opposition from a religious group in Middletown, composed principally of members of the business class who deny reality to disease. A mass meeting was held at the Middletown high school in 1924 to boom a State Society for Medical Freedom seeking to oppose "compulsory methods which disregard the inalienable rights of the citizen."[24] For the most part sickness to Middletown means, as in 1890, "too sick to work." Minor creakings in the human machine many people tell themselves are "just notionate"; "I never go to a dentist until my tooth aches so badly I can't stand it any longer," said one business class

[23] Annual Report for 1923.
[24] "This society," asserted the speaker of the evening, "seeks to protect people. You can't stop an epidemic by vaccinating the school children who are only 5 per cent. of the population. School medical examinations are financed by public money, often without authority. Public nurses should not be hired. 'Preventive medicine' is just another excuse of the doctors for getting work. It's ridiculous to frighten people by talking about these millions of germs in our food, air, and water. I went to school in a small room with lots of other children, and we were all rosy-cheeked and healthy and didn't know a thing about germs. There wasn't any of this foolishness of weighing children and frightening them to death because they may be underweight."

woman. "If you go to a doctor or dentist he is sure to find something wrong with you." [25]

The congeries of conflicting powers and jurisdictions in care of health, consequent upon its shifting status, appear in the following summary of the local situation in 1924: The Health Officer (or Board of Health) has specific power to—

(a) Condemn and destroy foodstuffs, but there were no prosecutions in 1924; he tried to prosecute one case but the court did not support him. He has to depend upon the State Board of Health for anything that gets done.
(b) Condemn building for use to which put; has placard to put on house but doesn't prosecute.
(c) Destroy condemned buildings.
(d) Condemn private water supplies believed to be menace to health. (No prosecutions in 1924.) He does *not* have power to compel connection with the public water supply.
(e) Compel sewer connections. (Prosecuted one negro and tried to persuade several others.)
(f) Compel abolition of cesspools if a nuisance.
(g) Require vaccination of those in a group exposed to smallpox who cannot show evidence of vaccination. (Has persuaded some, but not really required.)
(h) Bring action for failure to report births, deaths, diseases.
(i) Enter premises to examine health hazards.
(j) Compel hospitalization of communicable disease where successful quarantine impossible. (For smallpox only, but really not compelling.)
(k) Examine, isolate and treat carriers of contagious diseases. (None remembered by Health Officer.)

The following activities are handled as follows:

(a) Plumbing inspection: by the City Engineer.
(b) Garbage collection: by the Board of Public Works.
(c) Street cleaning: by the Board of Public Works and a Street Commissioner.
(d) Public scavenger: ditto.
(e) Nuisance inspection: by the Board of Health through a sanitary policeman.
(f) Contagious disease hospital: by the Board of Health (smallpox only).
(g) Housing: by the Building Inspector.

[25] Cf. Ch. VII for new health measures introduced into Middletown's factories.

(h) Food and milk control: by the State Board of Health.
(i) Control water supply: ditto.
(j) Vital statistics: by the Board of Health.
(k) Health prosecution: by the Board of Health and city attorney.
(l) Pre-natal service: by the Visiting Nurses' Association.
(m) Infant welfare: ditto.
(n) Distribution of antitoxin and immunizing agents: by the Board of Health (for indigents only).
(o) Medical school inspection: by the Board of Education. All school health work is under the general supervision of the State Board of Health.
(p) Tuberculosis clinic: by the County Anti-Tuberculosis Association.
(q) Mental clinic: by the Board of Education for school children. (This work was abandoned in 1924 as unnecessary.)
(r) Dental clinic: by the Board of Education for school children.
(s) Venereal Clinic: by the County Commissioners
(t) Nursing:
  Maternity: by the Visiting Nurses' Association.
  Child Hygiene: by the Visiting Nurses' Association and Board of Education.
  Tuberculosis: by the County Anti-Tuberculosis Association.
  Bedside care: by the Visiting Nurses' Association.
  School: by the Board of Education.

According to a trained local health worker:

"There is little coöperation and considerable friction between the Visiting Nurses' Association and the hospital. The head of the hospital thinks the work of the Anti-Tuberculosis Association is good and the head of the Visiting Nurses' Association thinks it poor. This last organization has little use for most of the doctors in town. Social Service Bureau, Red Cross, hospital, Anti-Tuberculosis Association, and Visiting Nurses' Association all overlap and duplicate in jobs at some points."

To cap it all, the Health Officer was even blocked by a fellow doctor in trying to compel the installation of sanitary improvements in a local dwelling during 1924, on the ground that the owner of the property was a friend of the doctor's. The whole scene still resembles the situation in 1890, when, shortly after the appointment of the local Health Officer, the Board of

Health, in a tiff with the City Council, climbed on its high horse and resolved solemnly:

"Whereas, The City Council refuses to give the Board of Health authority to have the filth cleaned from the alleys, therefore

"Resolved: That in the future all complaints concerning alleys will be referred to the Council and we will not assume any responsibility in such matters."

In this picture of rival jurisdictions and lag and friction in the performance of a common activity we see in operation the sprawled process of cultural diffusion by which Middletown is inching along into the business of keeping healthy. Back of local health work lies usually a more active policy of a state board. Diffusion frequently proceeds from national public or private agencies, to state private agencies, to state public, to local city private, and finally public, agencies. A period of years frequently elapses between the adoption of a given activity by the state group and its adoption by the city group.

A final trend worth noting is that towards the secularization of health matters. As suggested at various points above, in the eyes of many, treatment of health is in large part a religious matter. A paper on health given before one of the women's federated clubs in 1924 combined a discussion of "the body as the temple of the soul" with a matter-of-fact description of some of the national work of the American Child Health Association; prayers for the health of ailing members are offered regularly in connection with Middletown religious ceremonies; finally, one religious group, while denying reality to sickness, gains much of its strength through healing. The growing acceptance of a more secular attitude, however, is clearly reflected in such a statement as the following from an editorial in the local press:

"Within twenty-five years, infant mortality has been cut at least 60 per cent. through scientific care. Not so many years ago when a baby died the parents mustered whatever resignation they could assume and attributed the death to 'the will of God.' It has been learned, however, that more often than not the death was due to ignorance of parents, the lack of sanitary and medical equipment and knowledge on the part of physicians, and the absence of organized societies whose business it is to see that babies have a chance to live."

Chapter XXVI

CARING FOR THE UNABLE

Throughout three-quarters of the geographical city there is a fairly even sprinkling of people who are at any given time unable to secure the necessary food, shelter, and care of health under the economic system by which people live in Middletown.[1] Their numbers tend to increase when "times are bad" and diminish when "times are good." Throughout a fourteen-month period divided equally between prosperity and depression, an average of nearly 200 appeals for help each month, involving an average of nearly sixty-three fresh breakdowns a month, came before the "Social Service Bureau."[2] Additional breakdowns are handled by the Salvation Army, Red Cross, and other agencies. With scarcely a single exception these come from the working class. Something over half of the families and individuals seeking aid from the Social Service Bureau applied only once during the year; the others made repeated calls for help.

Sickness and unemployment are the most frequent reasons for appeal to the Bureau, accounting for 35 and 32 per cent. respectively of all appeals. "Inadequate income" brings 12 per cent., and old age 9 per cent. The remaining 12 per cent. include those unable to make adjustments because the wife is left widowed, or "stranded" or "deserted," or with the "breadwinner in jail"; because the individual is "handicapped" or "incorrigible," and so on. When "times are bad," according

[1] Based upon a spot map of families appealing to the Social Service Bureau.

[2] The figures here given are for the fourteen months from November 1, 1923, to December 31, 1924. Owing to a change in the fiscal year, these data of the Bureau are for a fourteen- instead of a twelve-month period.

During these fourteen months there were 2,696 appeals for aid coming from 877 different units, 828 of them white and forty-nine colored. Eighty-three per cent. of the cases involved a family, and 17 per cent. single persons.

See N. 8 below, regarding the work of the Salvation Army, Red Cross, War Mothers, and other agencies.

to the head worker, "unemployment outruns all other causes." None of these appeals can, of course, be taken to indicate a single antecedent of maladjustment; they are surface indices of a long sequence of events which lead at some particular point to collapse.

This chronic inability of a certain number of the group to operate under the existing system requiring the proffering of money at every turn is taken for granted; indeed some Middletown persons point to the statement, "For the poor ye have always with you" in support of this attitude. Middletown has developed four kinds of institutional devices to deal with this recurrent overwhelming of certain of its members: (1) person-to-person individual giving; (2) giving through voluntary groups such as churches, clubs, lodges, and labor unions, which do not exist primarily or exclusively for this purpose, but which give a semi-personal assistance to the "unfortunate"; (3) giving through voluntarily supported, semi-public social service organizations existing solely to alleviate the condition of the "needy"; (4) giving, through elected or appointed representatives of the entire group, funds raised from the entire group through taxes.

The first of these, the face-to-face giving of money and time and services to meet an immediate emergency of some one "less fortunate," is traditionally the commonest form of "charity." Its roots reach down indefinitely into the religious and rural neighborhood habits of the group. When the home of a family burned in 1890 the press would note that "the family deserves the aid and assistance of the citizens in getting a lodging place and a little start in the world again." Or when the New England Home for Little Wanderers sent out twenty-five orphans who "sang at different churches Sunday and were placed for inspection at the Methodist Church Monday," the citizens opened their hearts, and "ten of them were provided with homes before noon, others being distributed in the afternoon." [3] Today, likewise, there is felt to be a peculiar merit in personal giving: "Charity should have the personal touch," was

[3] "Between 1865 and 1891," according to a veteran social worker who became head of the State Board of Charities about 1891, "the Children's Aid Society of New York, alone, had placed more than 2,000 children in [the state]. . . . I made an estimate in 1891, from the best data obtainable, that the total number brought from states [to the] east . . . from the time the New York society set the example was probably more than 6,000."

the stout assertion of a member of Rotary, and another commented, "We should keep too much organization out of welfare work. It's the healthy and normal thing for people spontaneously to help their brethren in trouble." Little children in the wealthiest Sunday School in town are given envelopes and told to "deprive yourself of something you want this week and bring the money in the envelope for the poor next Sunday." So, according to the press at Christmas time:

"While the young trip the merry fantastic, night after night, in gayly lighted ball rooms of hotels to the soft strains of appealing orchestras, there is a feeling in their young hearts that they are not the only ones that are happy, and more than one little sweet young thing can close her eyes while gliding from one end of the ball room to another, dreaming of the pathetic family she has visited during the afternoon to which she had delivered a basket of good substantial foodstuffs, or the memories of the Christmas party at which dozens of little tots danced in glee about a huge Christmas tree laden with toys and all sorts of goodies."

The continuing prominence of the neighborhood in the lives of the working class and also their somewhat more literal religious life are reflected in a greater tendency to continue neighborhood face-to-face charity. The social calling of certain working class housewives is still not uncommonly the "visiting the sick and needy" engaged in by all classes in 1890, and women of this class appear to receive house-to-house beggars more sympathetically; it is even common for neighborhoods to "pass the hat" for some unlucky member. Of the 100 working class families for whom income distribution was secured, fifty-six reported contributions to charity; the amounts ranged from $0.50 to $255.00, averaging $18.72.

The more impersonal group methods of giving are, on the contrary, diffusing more rapidly among the business class. Civic clubs have exalted the sporadic semi-personal group form of giving and are enabling it to flourish in the face of the advancing longer-term planning and control represented by the inauguration of the Community Chest in 1925. Men's, women's, girls', and boys' clubs periodically, chiefly at Christmas and Thanksgiving, become civic over the "needy": each Christmas, as noted above, Rotary goes to the local poor house with gifts, and the Dynamo Club of the Chamber of Commerce banquets 150 "poor kids" and gives each of them a watch; the

Eagles visit and give gifts to the inmates of the orphans' home; the boys of the Hi-Y club give their annual "cheer party" to the orphans' home; the ladies of one of the federated clubs "observe their annual all-day charity meeting" when "as a result of the efforts of fifteen members, the humming of sewing machines and a jolly spirit, eighteen garments and a comfort were completed and twelve dollars' worth of provisions and some articles of furniture provided for the comfort of a worthy but unfortunate family." "At Christmas time," said the head of the Social Service Bureau, "we are deluged with baskets. We have almost more than we can use. If only some of these clubs would remember that people need help at other times of the year!" An editorial on "Watch Your Giving" in the press at Christmas time, however, urged the use of organized agencies "in touch with the situation" rather than the earlier more or less random giving "at this season of the year when hearts are easily touched."

Thirty-five years ago lodges were a prominent source of charity. Conspicuous in the statement of purpose of each of these secret orders are such statements as this: "Our object is to visit the sick and the afflicted and relieve their necessities; to educate the orphan, care for the widow, comfort the sorrowing." Lodge charity plays a relatively smaller rôle today as other charitable agencies have developed, as the fraternal spirit declines, and as lodges are more preoccupied in using their resources in building competitive club houses. The local Elks, e.g., gave a little more than a dollar a member for all kinds of charity, donations to needy members, the widows of members, and so on, in 1923—only 4 per cent. of their total annual expenditures. In two or three lodges additional money for charity is obtained by exacting a "rake-off" on the gambling in the club houses. In lodges frequented chiefly by workers the popular 1890 habit of "raffling off" gold rings or watches for charity is still not uncommon, but there is a growing disposition to relax the sporadic giving of incidental aid to "needy brothers" in favor of a more impersonal and systematized insurance system; many workers today keep up their lodge connection "only for the sick and death benefits." The habit of fraternal calling upon sick members is becoming so attenuated that in some lodges it is maintained only by means of rigid fines upon those who fail to call when such duty is delegated to them.

The churches of Middletown, also, are less conspicuous in local charity today. As in the care of health, the trend toward secularization is an outstanding characteristic of Middletown's charity in the last thirty-five years. Speaking of the "benevolent institutions of the state," the state yearbook for 1889-90 boasted that they "challenge the admiration of all who comprehend and appreciate the animating spirit of our Christian civilization." In 1890 the churches shared the care of the poor only with lodges, unions, the Free Kindergarten for poor children, and the Blue Ribbon Temperance Society, an offshoot of the churches. During the "bad times" of 1893 the first local "Associated Charities" was organized "under the control and auspices of the churches of the city," with the name "Associated Church Charities." So closely was charity connected with the churches that when a "charity ball" was announced by a club of young ladies, the Associated Church Charities caused a press notice to be printed, that "the proposed charity ball is in no way connected with the Associated Church Charities, as we believe that a public ball is not in keeping with the spirit of the work of the churches, and we cannot permit the churches to even have the appearance of endorsing the dance." The Central Executive Committee of the Associated Church Charities was composed largely of ministers. Not until 1900 was the word "Church" dropped from the title of the Associated Charities. No aid was granted to a family even then until after relatives, neighborhood, and churches had been canvassed; all cases in any way affiliated with a church were turned over directly to the church. So successfully was this policy pursued that the report of the Associated Charities for 1906 states that out of a budget of $1,082.52 only $24.59 was spent for "relief," nearly the same amount for carfare, and the remaining $967.71 for rent, salaries, and office supplies. The cover of the annual report in 1906 bore the words, in "Bible type," "Organized charity is organized love." So far has local charity been secularized in Middletown today that of the nearly $10,000 spent by the Social Service Bureau in 1924 (before the coming of the Community Chest), less than 3 per cent. was contributed by churches.[4] The first Community Chest budget, raised early in

[4] No church had the Social Service Bureau on its budget, though the Missionary and Benevolent Society of one church regularly contributes $20 or $30 a year, another church gives its Thanksgiving offering of about $100, and the Sunday School of a third, whose superintendent is treasurer of the Social Service Bureau, contributed in 1924.

1925, included $40,560 for local welfare work of all kinds, exclusive of an additional $37,050 in the budget for "character building" work through the Y.M.C.A. and similar local agencies. During 1924 five leading churches disbursed for all local benevolence work respectively $410.00, $361.23, $229.68, $212.26, and $0.00. This means, at a conservative estimate, that today the churches of Middletown *qua* churches are spending on local charity something well under one dollar for every four being raised in the annual collection from the city at large.[5] One of the largest churches in the city, which gives $5.00 a year to the Visiting Nurses' Association, had 554 visits to various of its members by nurses of that organization during 1924, 113 of these being visits to people too poor to pay the small fee and therefore a direct expense to the Association. A similar situa-tion exists among the members of other churches. Of the min-isters of five leading churches, the head of the Social Service Bureau said that three had not been in her office since she took over the local work two and a half years earlier, one had called once, and the other "three or four times." [6] She did not recall that the head of either Y.M.C.A. or Y.W.C.A. had been in her office, though the predecessor of the latter had. "The churches rarely report their charitable work among their members to us," she said, "and the only time they take an interest in our work is at Thanksgiving and Christmas, when some of them send to us for names of needy families." The "giving of alms" appears to be today largely a secular activity.[7]

[5] The money of these churches is being heavily diverted to activities out-side Middletown. One leading church of 922 members, e.g., contributed $5,931.23 to all missions and benevolences in 1924, of which $361.23 went to Middletown; in another church of 1,600 members, $4,486.57 was spent for all missions and benevolences, of which $410.00 was spent in Middle-town. Another church of over 1,200 members contributes over $7,000.00 a year to missions and benevolences, of which slightly over $200.00 goes to local charity. These figures are statements by the ministers. Occasional small contributions by church societies may possibly be omitted from these sum-maries.

[6] The director of the Bureau is *ex officio* a member of the Ministerial Association.

[7] During 1924 a movement was launched by the Married People's Class of the largest Protestant church in the city to raise a permanent fund to help finance poor boys and girls of the church in college. A small fund was raised and an organization incorporated to manage it. Five boys were helped during the first year. It is perhaps enlightening as to the conflicting institu-tional devices of the group as well as to the shift in interest of organized church finances that there was much grumbling throughout this church on

While a peculiar merit still clings to personal and semi-personal giving, the rapid growth of more organized, impersonal giving in recent years is significant. Even as late as 1905 "less than eighty persons" out of the entire city of 22,000, according to the Annual Report of the Associated Charities, contributed to its support, and in 1924 only about 350 people in a city of 38,000 contributed to the corresponding agency, the Social Service Bureau, while all local public welfare work of this organized kind was supported by only 500 or 600 persons. In 1925, with the inauguration of the Community Chest, the load was spread over no less than 6,402 contributors.

This new agency for financing the needy extends to charity work some degree of centralized planning and control; it involves "one annual war-dance to whoop up enthusiasm and funds" instead of thirteen. A "drive" of this sort, like a revival or a high-pressure sales campaign, furnishes an interesting study in forced-draft diffusion and suggests possibilities for bringing about other changes in habits. Given the right group of men behind it, such a drive appears almost irresistible in Middletown. The World War taught Middletown the technique; the method used is simply to invoke again this technique, and the spread of the desired patterns follows. The first step was to enlist the big men in town, the Rotary crowd, as the responsible heads upon whom success or failure depended. There was the minimum of Christian *caritas* about it, no zeal for a particular emergency or needy family; in fact the movement took shape despite the opposition of some of the thirteen local charities who feared they might "lose their individualities and power of separate appeal." The attitude of the "drive" workers in 1925 was simply, "Here's a mean civic job that has to be done. Money must be raised for these institutions, and we, the big business men, are tired of being pestered every few weeks to help in a new drive. Let's pitch in and get the thing off our chests." Men's and women's teams were organized with "captains" for each; the men were further

the ground that the minister would not coöperate actively in the movement since it diverted funds from the published denominational statement, on which alone the church's standing in the denomination is based. The idea of starting this fund grew out of the enthusiasm aroused by a visit to the city of one of the city's former poor boys who had been educated through the generosity of a local citizen and is now a powerful bishop of the church in the "foreign field."

organized as "non-coms" and "buck privates," while a "colonel"
was placed in charge of these "shock troops." First, meetings
were held to "whoop up" these workers: "It is absolutely imperative that every worker be present. The various team captains will be held responsible for full attendance." "When we
say 'everybody out,'" said the "Colonel" at the first of these
meetings, "we mean just that: there's no room in this city for
civic slackers. Let's go!" And when the Rotary crowd says that,
Middletown goes. "Men," the leader of the campaign declared,
"this city needs a revival. We've been lying back too much on
the things we did during the war, and we've got to snap out of
it! You remember the Fourth Liberty Loan? We'd always
taken a week or ten days on the loans, but that time we decided
a day was enough, and we raised a million and a half in a single
day. We're going to raise the $77,000 we need for this Chest
in three days." The city was districted name by name, publicity
workers turned loose—and the city "went over the top."

Under this centralized, city-wide massing of resources Middletown supports nine welfare and four other agencies, as over
against the struggling church mission sewing school and the
charity kindergarten supported by "penny collections" and
similar devices which constituted the local organized charity of
1890.[8]

---

[8] These thirteen new activities supported by voluntary charity from two
out of three of all families in the city include: *Welfare agencies:* Day
Nursery for children of working mothers (3,407 child-days of attendance
in 1924)—$2,600; Anti-Tuberculosis Society (1,959 instructive calls in 1924
and 153 new patients attended)—$9,750; Social Service Bureau (in January,
1925, alone, 435 calls for help, involving 861 individuals, were cared for)—
$9,750; Visiting Nurses' Association (746 nursing visits and 140 instructive
visits during January alone)—$8,450; Jewish Welfare—$390; Humane Society (five thousand school children, members of the local Junior Humane
Society, the largest membership in the United States)—$650; Salvation
Army ("The Army gave aid in 669 cases, made 2,040 calls, distributed
Christmas presents to over 500 children in 1924")—$5,850; War Mothers
("—who mother invalid 'Buddies' of the great war, watch over the families,
widows and orphans of former soldiers and 'keep alive in the minds of the
people the service rendered by America's boys during the war'")—$520;
Red Cross ("—to aid afflicted ex-service men . . . to render first aid to the
community in a great emergency. The local chapter successfully handled 503
appeals in January alone")—$2,600. *Character-building agencies:* Boy
Scouts—$6,500; Playground work—$1,950; Y.M.C.A. ("The attendance of
[Middletown] boys at all the activities of the 'Y' Boys' Department in
1924 reached the astonishing total of 118,177")—$19,500; Y.W.C.A. ("The
Y.W. reaches 15,000 girls annually through its various activities")—$9,100;
total, $77,610.

As in care of health, religious education, and various other sectors of Middletown's life, there appear here a variety of institutions all engaged in carrying on the same activity, but the pressure of the city's business men is making for the reduction of overlapping spheres of work despite the threat to vested interests involved.[9] Not only does conflict arise between the newer and the older forms of giving and among the various agencies concerned, but, also, between this annual all-inclusive charity and the traditional church tithing. In 1890 one's church contribution covered much local charity. Today Middletown families support local charity through one big subscription made once a year as a civic duty often under considerable social pressure. It is not surprising, in the face of such local competition, that ministers report tithing greatly on the wane.

Another characteristic emerging in connection with the centralization of charity and the increased use of paid executives is the crowding out of the volunteer business class women who formerly directed the charitable enterprises; many of these women, having turned from "church work" to "charity work," are now turning to "civic work" as a substitute.[10] They continue to do the less organized sorts of charity and civic work through their clubs: placing pictures of Shakespeare in schools, library, and the rest room at the Court House; working for better local pure food supervision; sponsoring free twilight concerts in a

[9] The confusion existing at present, owing to multiple agencies engaged in a common activity, is apparent in the providing for children unable to attend school because of lack of clothes. The child's teacher notifies the Vocational Department of the schools, which notifies the Social Service Bureau, which eventually calls to look over the family in question, and if it is sufficiently poor, recommends the purchase of the needed clothes; the Township Trustee is then notified, and he furnishes the money, an assistant in the school Vocational Department buys the clothes—and the child gets to school. The same procedure must be gone through for school books for poor children; according to one teacher, there were children in her room from September to February without books, waiting for the system to unwind for them.

[10] No little opposition to the consolidation of welfare agencies in Middletown comes indirectly through the opposition of their boards to relinquishing their work. The extent to which these boards have been social affairs is indicated by the fact that the professional worker of the local Visiting Nurses' Association is not allowed to attend the meetings of the board and was not even asked to be present when a worker from a large neighboring city was called in to describe her methods of work. At a meeting of the local Federated Clubs at which the work of the Visiting Nurses' Association was discussed, it was not the professional worker in charge who made the talk, but a society woman on the board.

local park, and so on. But the loss of their earlier activities is perhaps reflected in the eager way they organize a new club, such as a local chapter of the American Association for University Women, then cast about for something to do, and end by offering fifteen dollars in prizes to school children writing essays on "Heroes" or holding meetings with papers on various scattered civic subjects.[11]

In this trend away from predominantly informal, face-to-face giving to the more organized forms of charity, the fourth and most completely socialized type is the routinized, city-wide giving through tax funds. Since 1795 public charity has existed throughout the state side by side with personal giving, in the form of public doles handed out to the poor by township trustees.[12] Next to this public dole method, the oldest form of tax-supported charity in Middletown is the county poor asylum "for those persons who by reason of age, infirmity, or other misfortunes may have a claim upon the aid and beneficence of society." This institution was in 1890 the county catch-basin in which all sorts of human sediment collected: "the insane, the feeble-minded, the epileptic, the deaf, the blind, the crippled; the shiftless, the vicious, respectable homeless, poor and bright

[11] Since the completion of this study the local chapter of the A.A.U.W. has adopted the educational program of the national organization by launching a child study group among its members. This is an example of the way new habits are being diffused locally through yet another channel: the educational program of a national organization financed by a New York philanthropic foundation.

The type of voluntary participation in charity work that is being superseded by centralized Community Chest financing is vividly shown in the fact that in 1923 Middletown women raised sums for the local Visiting Nurses' Association under the following catch-as-catch-can methods of pre-Community Chest financing: "White Elephant" sales $779.05; card party $171.50; charity concert $152.53; rummage sales $352.10; winter's aid sale $1.75. The Day Nursery has been financed since its founding in 1917 by a sorority of young business class women through similar devices, supplemented by dances.

[12] The care by county and state agencies, in part supported by tax levies from Middletown, for those overwhelmed in life, shows trends somewhat similar to those apparent in the city itself. Between 1890 and 1895 more than half a million dollars was spent annually by the state through sums doled out as emergencies arose by township trustees. Without supervision of any kind these officials simply gave as they saw fit. In 1889 state charities passed from scattered to organized administration and after 1895 the caring for the poor by this public dole method system began to be curtailed, though it is still somewhat in force. The drift toward concentration of control is apparent today in the anticipated change from nineteen boards controlling state charitable and correctional institutions to two.

young children." It was contrary to the traditions of this pioneer community that anybody should be habitually dependent upon the group; if he was, it was certainly "his own fault." Accommodations, accordingly, were not such as to encourage one to live on the county. The press in 1890, summarizing the report of the newly appointed State Board of Charities on the local "asylum," described it as having "a capacity of thirty and a population of thirty-six; one room twelve feet square holds three beds in which five men sleep every night." It went on to report a weekly cost per inmate of $0.64. In 1903 the county Board of Charities and Correction found conditions "a disgrace to the county," and two years later the State Board declared it to be "slovenly kept" and its inmates "ill-clothed and fed." As recently as 1915 conditions were described as "shocking and deplorable"; "diseased persons were admitted and allowed to mingle freely with the healthy"; on petition of the Humane Society the official in charge was "allowed to resign." [13] The multiplication of new group agencies since the organization of the State Board of Charities in 1889 has drained off from the poor asylums to special institutions many of those unable to provide for themselves because of feeble-mindedness, epilepsy, tuberculosis, and other infirmities. The steady movement towards a wider and wider assumption of group care for the unfortunate is evidenced by the passage through one of the houses of the state legislature in 1925 of an act to abolish the "poor asylum" and establish an "old age pension" for certain eligible classes.[14]

Other manifestations of the drift towards administering welfare agencies, like fire protection or street cleaning, as part of the city's business, appear in the establishment since 1890 of an orphans' home and the recent transfer of the kindergartens from private charity to the public schools.

As Middletown thus reduces "Christian charity" increasingly to a secularized business proposition, the roots of emotional

[13] From minutes of the Humane Society for 1915.

[14] It is noteworthy that this bill was introduced by the senator from a recently heavily industrialized section of the state and was denounced by a senator from a more agricultural region as "socialistic," since "the government owes nothing to any man." This illustrates the tendency noted elsewhere for changes to occur first in regions where the old culture is subjected to greatest pressure through the advent of new material tools.

conflict remain. It has been pointed out that "there is no touch-stone, except the treatment of childhood, which reveals the true character of a social philosophy more clearly than the spirit in which it regards the misfortune of those of its members who fall by the way." [15] Middletown's philosophy is essentially personal—a philosophy of pioneer, and more recently of democratic, independence in the present, and of personal salvation in a hereafter. A deep strain in its religious traditions has "lent a half-mystical glamour both to poverty and to the compassion by which poverty was relieved, for poor men were God's friends. At best, the poor were thought to represent our Lord in a peculiarly intimate way." [16] Over against this is the equally insistent religious tradition of "individual responsibility" which has so far departed from the primitive Christian attitude as to convert thrift into a Christian virtue. This last tendency is re-enforced in Middletown today by the overshadowing importance of industry, of getting ahead with the compact, intergeared world of modern business in which one's life seems inextricably fixed. Split thus at the very core of its philosophy between an emphatic feeling that "personal charity," "giving to the poor," "helping those less fortunate," is somehow inescapably "right" and necessary, and an even more emphatic, because more urgently present, feeling that one must "get on" financially and that in this competitive world "people get pretty much what's coming to them," Middletown wavers in its handling of practical situations. It extends ready sympathy and help to individual unfortunates whose plight is immediately before the community, and yet it is not particularly concerned with preventing the recurrence of similar situations, because there are bound to be some sick, unemployed, or otherwise miserable people in the world, and no change in the present social or industrial system could presumably prevent this unfortunate condition, nor should the community seem to encourage dependence. Furthermore, granted that one should love one's neighbor as oneself, who *is* one's neighbor in this city of 38,000 people in which neighborhoods are disappearing?

[15] Tawney, *op. cit.*, p. 268. An excellent picture of the conflicting points of view touching such a matter as charity that historically have got themselves woven into the religious philosophy of a modern Protestant community like Middletown is presented in this book.

[16] Tawney, *op. cit.*, pp. 260-261.

And so, not surprisingly, Middletown's working philosophy in the matter tends to be: People in actual need must be helped, because "you wouldn't let a *dog* starve," but we must not make it too easy for them, and by all means let's get the unpleasant business over with and out of sight as soon as possible!

Chapter XXVII

GETTING INFORMATION

Mention has been made of the printing of election news and other accounts of group business in the press. Middletown's dependence upon some such artificial diffusion of information grows as the city grows. The city's political devices, for instance, assuming as they do first-hand familiarity with collective affairs, candidates, and elected officials, grew up in a culture more akin to the Middletown of the eighties, when fewer things were done by the group as a whole, citizens were somewhat less pressed by the competitive urgency of getting a living, and the community's comings and goings were more exposed. Today, when even a professional politician in Middletown has been known to confuse the "city plan commission" and the "city manager plan," and the average voter does not know many of the men on a Middletown ballot even by sight, this early assumption of full knowledge by each citizen is an anachronism. Hence the press becomes more and more an essential community necessity in the conduct of group affairs.

Every business man in Middletown wakes in the morning to find a newspaper at his door. Workers, rising and getting to work an hour or two earlier, seldom have time to read papers in the morning, but there was no family among those interviewed which did not take either a morning or evening paper or both. It is, however, left to the whim and purse of the citizen whether he ever reads a newspaper. The local morning paper distributes 8,851 copies to the 9,200 homes of the city, and the afternoon paper 6,715, plus at least half of an additional 785 sold on street and news-stand.[1] In addition, the circulation of out-of-town papers, which, according to the older citizens, was negligible in 1890, now totals 1,200 to 1,500 a day.[2]

[1] Figures for the morning paper are for April 15, 1924. An additional 316 copies are sold on news-stands and the street, probably most of them to transients. Afternoon figures are an average for February, 1924.

[2] These include an average of 800 copies of one paper from the state capital and 125 of another, eight copies of the *New York Times* (thirty-nine

Twenty-three (three out of five) of the thirty-nine business class families giving information on this point and only three of the 122 working class families took an out-of-town paper in addition to one or more local papers, six of the business group and thirty-one of the workers took both morning and evening local paper but no out-of-town paper, ten (one-fourth) of the business group and eighty-nine (three-fourths) of the workers took only one local paper. This may seem merely an odd bit of data until it is recalled that such differential diffusion is the stuff of which the habitual reactions of the group are to no inconsiderable extent formed. The fuller news content of the large city papers should be borne in mind.[3]

The ostensible purpose of Middletown's newspapers is to present an accurate array of the "news." Nearly two-thirds of the morning paper bought by the great bulk of Middletown families is composed of advertising:[4] first, in point of quantity, things to wear, next things to eat, and next things to restore health.[5] The rest of the paper, 800–1,000 inches of space daily, is devoted to reading matter of various kinds. About 20 per cent. of this reading matter concerns local happenings, another 10 to 20 per cent. county and state affairs, 25 to 50 national news, 10 international, and the remaining 20-odd per cent. non-geographical features.[6] A count for the first week in March, 1923, showed 18 per cent. of this reading matter in the leading paper to be devoted to public affairs, 16 per cent. sports, 10 per

on Sunday), 170 copies of the *Chicago Tribune* (800 on Sunday), sixty-three *Chicago American,* forty-two *Christian Science Monitor,* sixty *Cincinnati Enquirer* (240 on Sunday), one New York *World,* fourteen *Detroit Free Press* (61 on Sunday), four *Cleveland News-Leader* (21 on Sunday), etc.

[3] This differential diffusion from outside sources becomes even more significant when combined with periodical subscriptions: a third of the 122 working class families interviewed take one or more newspapers but neither subscribe to nor buy regularly any periodical, as against none of the business class group interviewed; an additional one in ten of the former and again none of the latter take only a women's periodical or periodicals in addition to one or more newspapers. Cf. Ch. XVII on periodical reading.

[4] Sixty-three per cent. of Middletown's local morning paper and 47 per cent. of the afternoon paper during the first week in March, 1923, were composed of advertising, as against 63 per cent. and 43 per cent. respectively of the two leading papers for the corresponding week of 1890.

[5] See Table XXII for a complete distribution of the advertising contents of the leading Middletown papers for October 1890 and 1923.

[6] The ratio of news content of Middletown's papers from a geographical point of view has not altered greatly since 1890. See Table XXIII.

cent. business, 10 per cent. personal items other than social functions, 8 per cent. police and court news, 6 per cent. social items, 3 per cent. religious, with smaller percentages of space devoted to health, women's interests, radio, agricultural news, "Eddie" Guest's daily poem, and other material. As compared with a similar count for 1890, the current paper shows a decrease in news space devoted to agriculture, education, and politics, and heavy increases in sports (300 per cent.), women's news (200 per cent.), business items, and cartoons.[7]

Middletown is served with some three and a half times as much news in its leading paper each week as it was a generation ago, and much of this news is based upon more adequate reporting. Only one Middletown paper in 1890 had a daily telegraph service, and the column of Foreign News might contain stray items such as "Pipe ignites clothing—man in London burns to death," or "Two Berlin policemen killed by bomb." Today, wireless, radio, the cable service of the Associated Press, and syndicated features, reënforced by pictures, including radio photographs, bring the world to Middletown breakfast tables.[8]

[7] See Table XXIV. These figures are based on counts for the same weeks as Table XXIII. Counts for single weeks cannot be taken as indicating more than general tendencies. Since, however, the week in 1924 was characterized by no unusual occurrences likely to distort the picture, and since a second week was counted in 1890 as a check, the contrasts indicated probably represent roughly actual shifts.

[8] Just how much of the world reaches Middletown, however, may be gauged very roughly from a comparison of the front pages of Middletown's leading paper and of the *Christian Science Monitor* containing the news of the French recognition of the Soviet Government in October, 1924, the *Monitor* appearing the afternoon before the Middletown morning issue. Whereas the *Monitor* ran a two-column headline over its front-page, three-fourths column despatch, the Middletown paper carried this piece of news in a three-inch note at the bottom of its first page. The other major headlines in the *Monitor* that day were: "Wang Invited to Be Premier" (Pekin); "Tax Law Policy Is Vital Topic Before Cabinet" (Washington); "Anglo-Turkish Dispute Before League Council" (London); "Vocational Survey Clarifies Economic Status of Women" (signed story of a report just issued by the Bureau of Vocational Information in New York); "War Mothers of World May Outlaw War" (interview with Indianapolis head of War Mothers); "Tory Plot Seen in Alleged Red Note to Britain" (London); "New Movement Toward Russia Pleases Borah" (Washington). The Middleton paper carried the following on its front page under major headlines: "Campaign Funds Quiz Brings on Heated Debate" (Washington); " 'Little Rody' Blasts Hopes of Democrats" (Providence); "Davis's Speech Is Assailed by Organ of Klan" (Atlanta); "4,000 at [a neighboring small city] Hear [a Republican]" (state); "Davis Cheered by New York Negroes" (New York); "Coolidge Rests Sure of Result" (Washington); "Hughes Scores Third Party's Aim" (Chicago); in addition a four-inch cut and

Furthermore, the news Middletown reads today is of a more organized, impersonal kind. News in 1890 tended to be chatty and informal. The illness of the British Prime Minister, the visit of "our friend, Jack ――, to see his girl in S――. That's the third time this week. Look out, Jack!", the arrest of a notorious Middletown prostitute, "Nellie ―― on the rampage again," were all reported in the same jovial fashion. The announcement of an arrest and conviction frequently took the form of "―― is pounding stone for the county again." The first column of the front page consisted sometimes of foreign news, sometimes of advertisements, and again of a "Seen and Heard" column of personals. Sensational headlines, pictures, and cartoons were absent. In the same column editorial notes, jokes, the International Sunday School lesson, and advertisements for "Hood's Sarsaparilla" and "Dr. Kilmer's Complete Female Remedy to Prevent Tumor and Cancer" jostled each other. Today the press has become a more organized and routinized agency for presenting the news.

And yet, although the purpose of newspapers, as far as the individual citizen and the community as a whole are concerned, is ostensibly to give information, Middletown's newspapers have three other uses, for a few individuals or limited groups, that sometimes appear to take precedence over keeping the community adequately informed. For politically ambitious individuals or for political parties, Middletown's newspapers serve the purpose of shaping public opinion to their aims; this so-called politically partisan aspect of the press, which colors extensively the news allowed to reach the group, is emphasized by the fact that both dailies belong to the same political party.

To the owners, thanks to the heavy growth of advertising in the past few decades, the newspaper serves an increasingly alluring money-making purpose; in May, 1890, the leading paper carried 108,715 lines of advertising and in October of the same year 101,448 lines, as against 628,856 lines in the lead-

caption showed the new president of the national society of surgeons, and a one-half-column-across-three-column cut depicted certain national figures under the caption, "Baring of Income Tax Secrets Embarrasses Notables."

Of the 6,900 inches of news space in Middletown's leading papers in the first week of March, 1923, 21 per cent. was syndicated or signed news material of a more or less nationally standardized type (Associated Press, David Lawrence, etc.), 5 per cent. was nationally standardized news pictorial matter (not including cartoons), 16 per cent. national features of the Dorothy Dix and "Eddie" Guest sort, 1 per cent. filler material clipped from national magazines, and the remaining 57 per cent. was locally written copy.

ing paper in May, 1923, and 604,292 in October.[9] The advertising columns of the press appear to be run on frankly *caveat emptor* principles. Advertisements of doctors and nostrums condemned by the health experts of the community, as noted in Chapter XXV, as well as dubious offers of building lots in remote states and advertisements for "girls" for models, the latter condemned by the juvenile court and social welfare agencies, appear regularly. Or the credulity of a large section of the population is exploited in such an advertising and circulation-getting scheme as the one in which an individual calling himself "Rajah Raboid" is introduced to the community as "America's Greatest Master Mind," "The only man in the world who can tell people what they are thinking of without having them write it down"; the citizens were asked to send in "questions pertaining to your life—love, courtship, marriage, business, etc. . . . and watch the paper daily for your answers." Various groups in the community oppose different types of advertising, but there are few exceptions to the rule that if a man's money is good, his advertisement appears.

The growing profit in controlling the agencies of news diffusion has developed yet another use of the press—that of buttressing the interests of the business class who buy advertising; more than ever before it is the business class advertisers who are the supporters of the newspapers, rather than the rank and file of readers of the paper.[10] It is largely taken for granted in Middletown that the newspapers, while giving information to the reading public as best they may, must not do it in any way that will offend their chief supporters. Independence of editorial comment happens to be in rough inverse ratio to the amount of advertising carried. The leading paper rarely says anything editorially calculated to offend local business men; the weaker paper "takes a stand" editorially from time to time on such matters as opposition to child labor; while the third paper, the four-page weekly Democratic sheet, carries no advertising except such political advertising as must legally be given to a rival paper, and habitually comments freely and vociferously on local affairs.

[9] See Table XXII. In 1890 the papers appeared only six days a week, while in 1923 a Sunday edition was published.

[10] The receipts from advertising, as reported by the United States census of the publishing business covering newspapers and periodicals, were $793,898,584 in the year 1923. This was more than twice the figure for returns from subscriptions and sales ($361,178,329).

This third paper, in reflecting the editor's personality, is much more in line with the leading papers of 1890 than are the two current dailies. Then, an editor was a person of local authority like the judge and minister, and he did not hesitate to express thumping opinions on any and every subject. In the midst of a front-page column of news, he would appear to comment, "It is a known fact that the *Herald* [a rival sheet] can print more plain lies that have not even a semblance of truth than any other sheet ever known." Today, neither daily is personally edited by its owner, both owners live permanently in distant cities, although one of them maintains a technical citizenship in Middletown for political reasons, and the leading paper is a member of a national syndicate of papers, with most of its editorials actually written by a central editorial board in another city. It is symptomatic of the change that has occurred that the habit of citizens' "writing to the paper" has decreased heavily.

Not only advertising and editorial comment but the actual news presented is not unaffected by Middletown's dominant interests. It is generally recognized in Middletown that adverse news about prominent business class families is frequently treated differently, even to the point of being suppressed entirely, than news about less prominent people. The following from an editorial in the small outlawed weekly, while reflecting obvious bias and animus, nevertheless describes a condition generally recognized locally as containing a large measure of truth:

"DISCRIMINATION

"Three youths belonging to important people in [Middletown] were recently arrested with booze on their persons, but no mention of the matter was made in the two daily newspapers. They never do. Watch the daily newspapers here and you will generally see the word 'laborer' following the names of those who are exposed as violators of the liquor laws. We hope the day will come when it will be a criminal offense for newspapers to protect higher-ups and ruin the reputations of those without influence. . . . When youths belonging to the families of the elect get in a jam with liquor they even escape without paying fines and their names are kept from the public."

It is usually safe to predict that in any given controversy the two leading papers may be expected to support the United

States in any cause, the business class rather than the working class, the Republican party against any other, but especially against any "radical" party.

In some cases, as has been pointed out in regard to pre-election information, partisan politics, official mishandling of public works, violation of the pure food law, and other aspects of the city's business, the influencing of the handling of news by extraneous factors amounts actually to misrepresenting or withholding it. Possibly reflecting this situation are the responses of the 241 boys and 315 girls in the two upper years of the high school in November, 1924, to the statement in regard to Middletown's leading paper, "The ——— presented a fair and complete picture of the issues in the recent election": 32 per cent. of the boys and 36 per cent. of the girls marked the statement "true," 27 and 20 per cent. respectively "false," 37 and 39 per cent. respectively "uncertain," while the rest did not answer.[11]

Here, then, is a community of nearly 40,000 individuals, founded upon the two principles that one adult's judgment is as good as another's and that ignorance is no excuse for incompetency, and increasingly dependent upon information furnished by its daily press. But despite the assumption of the adequacy of the information of each citizen, it is left to the whim and economic status of the individual whether he shall see a paper at all, obstructions, political, economic, and personal, are thrown at many points in the way of the newspapers' gathering and publishing from one day to the next the facts needed by the citizens to carry on a democratic form of government, and the information upon which both individuals and the community depend is left to the outcome of the resulting battle royal. In view of the above, many vagaries which appear in the politics, public opinion, health, and other habits of this community, in which the habits of one section of the population at many points already lag behind those of other sections, become more understandable.

[11] Attention should be called in passing to the answers by the same boys and girls to the statement, "It is safe to assume that a statement appearing in an article in a reputable magazine like the *Saturday Evening Post* or the *American Magazine* is correct": 36 per cent. of the boys and 43 per cent. of the girls marked the statement "true," and 44 and 36 per cent. respectively "false," while 18 and 20 per cent. were "uncertain," and 2 and 1 per cent. did not answer the question.

Chapter XXVIII

THINGS MAKING AND UNMAKING GROUP
SOLIDARITY

"Middletown is an easy place to get acquainted in; I really know every one in town now," remarked the wife of a socially prominent minister.

"Lady Eagles, all eyes in town is upon you!" exhorted the coach of the Lady Eagles' Drill Team.

So the two worlds live; to each one the other is largely out of the picture. The "every one" of the minister's wife included at most the three in each ten of the population who belong to the business class, while the coach of the Lady Eagles almost as definitely excluded the business group. Overlapping and shading into each other as the two groups do, the division between them nevertheless emerges more sharply than any other single line of demarcation in Middletown.

But there are not just two worlds in Middletown; there are a multitude. Small worlds of all sorts are forever forming, shifting, and dissolving. People maintain membership, intimate or remote, formal or tacit, in groups of people who get a living together—factory, department of factory, group within department under the "group system" of production, factory welfare association, trade union, board of directors, Chamber of Commerce, Merchants' Association, Ad. Club, and so on; through one's home activities one belongs to a group of relatives, a neighborhood, a body of customers of certain shops, patrons of a bank, of a building and loan association; a student may belong to the high school, the sophomore class, the class pin committee, the managing board of the school paper, the Daubers' Club, History B class—and so on indefinitely through all the activities of the city. Some of these groupings are temporary—a table at bridge, a grand jury, a dinner committee; others are permanent—the white race, the Presbyterian Church, relatives of John Murray. Some are local—depositors in the Merchants' Bank, the Bide-a-Wee Club, friends of Ed Jones, residents of the South Side; others are as wide as the county, the state, the nation, or the world.

In the main, as has been suggested, the cleavages which break up Middletown into its myriad sub-groups appear to have become somewhat more rigid in the last generation. When one of the grand old men of the city returned from Europe in 1890, "the whole town" was invited through a newspaper notice to meet him at the train and attend an "informal reception" at his home afterward; such an invitation would not be given even nominally to the larger, more self-conscious city today. Racial lines, according to old residents, were less felt in the days before the Jews had come so largely to dominate the retail life of the city and before the latest incarnation of the Klan. Jewish merchants mingle freely with other business men in the smaller civic clubs, but there are no Jews in Rotary; Jews are accepted socially with just enough qualification to make them aware that they do not entirely "belong." The small group of foreign-born mingle little with the rest of the community. Negroes are allowed under protest in the schools but not in the larger motion picture houses or in Y.M.C.A. or Y.W.C.A.; they are not to be found in "white" churches; Negro children must play in their own restricted corner of the Park.[1]

Religious differences, of a doctrinal sort, as noted in Chapter XXI, are less conspicuous today, with the exception of the line between Catholics and Protestants temporarily deepened by the Klan, but they have apparently been succeeded by somewhat more marked social and economic distinctions between churches.

At many points economic cleavages make themselves felt more than in the county seat of the nineties, whose newspapers

---

[1] News of the Negroes is given separately in the papers under the title "In Colored Circles." The sense of racial separation appears in widely diverse groups. At a meeting of school principals held at the Y.M.C.A. to arrange for inter-school basket-ball games, one of the Y.M.C.A. secretaries said that any school having a Negro on its team could not play in the Y.M.C.A. building, but would have to play in the high school. In answer to mild protest he said simply, "Well, you know, it's the sentiment here." And so it stood. There has been some talk of "colored" Y.M.C.A. and Y.W.C.A. buildings. The secretary of the Trades Council has tried to persuade the Molders' Union to take in Negro molders, but they have consistently refused. One is struck by the absence of Negroes at a place like the large tabernacle built for a community revival. They appear to keep very much to themselves, and Klan agitation has emphasized this tendency.

published friendly front-page accounts of labor meetings and wished "all success" to new unions, and whose mayor, super-intendent of schools, and other leading citizens welcomed con-ventions of laboring men to the city.[2] If one mentions "or-ganized labor" either to a man across the luncheon table at Rotary or Kiwanis or to a member of the working class Trades Council, one is likely to get a response charged with emotion. The former may refer to the time organizers pulled the men out of a Middletown glass plant, leaving the glass to harden in the pots at a heavy loss to the owners; while the latter may grow vehement over the time the head of a local union was dis-covered to be a spy reporting all proceedings to his super-intendent, or over alleged trickery during the machinists' strike.

Class distinctions between "rich" and "poor," "North" and "South Side," "East" and "West End," "above" and "below the tracks" there were in 1890; the newspapers frequently singled out "society people," and the baker in his diary refers to the "dudes." Indeed the kudos of belonging to one of the "old families" is probably even less of a factor today marking off one part of the group from the rest, as the money standard of the industrial city displaces old landmarks.[3] But most people agree that this family claim to prestige has been replaced by less impassable but no less real economic barriers. Some newcomers are impressed by the city's democracy and lack of class feeling. "The best thing about Middletown is that it is so democratic," said one woman. "Wealthy families here live simply rather than setting a pace others can't keep up. They don't have uniformed chauffeurs and nurses and all that kind of thing. They tend to live more as I do." Such statements, however, usually come from people who are themselves "our society people." Others dissent sharply: "In order to be accepted in Middletown, you must dress well, entertain, and spend money," said one woman who had come from another cultural background. "When I first came here I knew another woman new to town, a woman who

---

[2] Cf. in Ch. VIII the discussion of the changed status of organized labor in Middletown.

[3] That this cleavage according to length of residence still exists among certain of the business class is apparent, however, at a number of points; e.g., the new minister of an exclusive church was given by one of his active workers two sets of cards, the pink ones inscribed with the names of members of long standing and the yellow ones with the names of "new people in town who might work in well later." Some of these "new people," he discovered, had been in the church for ten years.

was wealthy and entertained, but she used such poor English I always felt sorry for her. Now she has been accepted socially, belongs to the Country Club and all the rest of it." One is reminded again of the genial comment of the old resident, "You see, they know money and they don't know you." [4] Bitterness on the part of the working class toward the "dudes" appears today as it did thirty years ago. The leader of a young working men's chorus, although it met in the Chamber of Commerce Building, addressed it as follows:

"We're training good soloists and a good chorus here, and I want you to work hard. If you hear criticism of the chorus about town I want you to stand up to it and answer it. You know if you had a million and got up and sang, these blatherskites would make a fuss over you because of your social position, but if you haven't money they pick flaws. I'd rather have your praise and listen to your criticism than that of all the rich blatherskites here in Middletown, because nine times out of ten you can't believe them."

The differences that break up Middletown into its many subgroups are in part simply devices by which the growing city facilitates the association of like with like in work and play. The whole process may be seen in little, as noted elsewhere, in the multiplication of clubs and other group-finding devices in the rapidly expanding and increasingly indiscriminate high school population, as over against the much smaller, more select high school of 1890, in which "everybody went around together."

Coming upon Middletown like a tornado, catching up many of these latent differences into a frenzy of activity, the Ku Klux Klan has emphasized, during its brief career in Middletown, potential factors of disintegration. Brought to town originally, it is said, by a few of the city's leading business men as a vigilance committee to hold an invisible whip over the corrupt Democratic political administration and generally "to clean up the town," its ranks were quickly thrown open under a professional organizer, and by 1923 some 3,500 of the local citizens are said to have joined. As the organization developed,

[4] Money is by no means a sole criterion of popularity, however. The most prized citizens in Middletown are the millionaire philanthropists frequently mentioned in these pages. Another family of large wealth but small interest in Middletown has little place in the activities or affections of the city.

the business men withdrew, and the Klan became largely a working class movement. Thus relieved of the issue that prompted its original entry into Middletown, the Klan, lacking a local issue, took over from the larger national organization a militant Protestantism with which it set about dividing the city; the racial issue, though secondary, was hardly less ardently proclaimed. So high did the local tide of Klan feeling run that in 1924 a rebel group in Middletown set up a rival and "purer" national body to supersede the old Klan. Tales against the Catholics ran like wildfire through the city. In a sermon on "The Godlessness of America" the minister of a thriving working class church earnestly passed on to his flock this story

"They say the Pope isn't wanted in Italy. France has been approached and she doesn't want him. The Balkans say no, Russia—'Not on your life!' England, Germany, Switzerland, Japan—all refuse; and they say the Catholics are building a great cathedral in our national capital at Washington which is to become his home." Then, as though half-ashamed at relaying this gossip, the minister added, "I don't know this; it's just talk but that's what they say."

Local Klansmen vowed they would unmask "when and not until the Catholics take the prison walls down from about their convents and nunneries," and the "confessions" of Helen Jackson, "an escaped nun," were widely sold at local Klan rallies. Fed on such threatening rumors, Klan enrollment boomed.

"Lady," exclaimed an earnest woman to one of the staff interviewers, "you have asked me a lot of questions, and now I want to ask you some. Do you belong to the Klan?" To a negative reply she continued, "Well, it's about time you joined the other good people and did something about this Catholic situation. The Pope is trying to get control of this country, and in order to do it, he started the old Klan to stir up trouble among the Protestants, but instead of doing that he only opened their eyes to the situation, and now all the Protestants are getting together in the new Klan to overcome the Catholic menace. I just want to show you here in this copy of the *Menace*—look at this picture of this poor girl—look at her hands! See, all those fingers gone—just stumps left! She was in a convent where it was considered sinful to wear jewelry, and the Sisters, when they found her wearing some rings, just burned them off her fingers!" [5]

[5] This woman was a neat, healthy-looking woman of thirty-seven, the mother of four children and wife of one of the highest grades of worker

To this Catholic hatred was added Negro and Jewish hatred fed by stories that the Negroes have a powder which they put on their arms which turns their bodies white, and that the Jews have all the money, but when the Klan gets into power, it will make a new kind of money, so that the Jews' money will be no good.

"We are charged with being against the Jew," thundered a lawyer from the state capital at a Klan rally. "We are against no man. Jesus Christ is the leader of the Ku Klux Klan, and we are for Him. The Jew is not for Him, and therefore the Jew has shut himself out of the Klan. We are not against the Negro. Rome fell because she mixed her blood. God Almighty has commanded us, 'Thou shalt not mix thy blood.' The *Outlook*—or some other periodical—reported the other day 113 marriages last year in Boston between whites and blacks, and I'm sorry to say it was white women marrying black men. We must protect American womanhood."

At this point a shuddering "Yes, yes" went up from the crowd, like the fervently breathed ejaculations one hears punctuating many working class church services.

"Lincoln said," continued the speaker, "that no nation can exist half slave and half free. My friends, this nation cannot exist with half its children in the great American free school and the other half being taught a different thing in the parochial schools.

"I am a member of the Klan because I believe before Almighty God that it is His appointed instrument. This country needs a Moses to lead this great people of ours. That leadership we have in the Klan. We, the rank and file of its membership, offer ourselves, a living sacrifice, for Christian America and the world."

Still another note of division was struck by another speaker who stormed:

"This great country of ours—bounded on the north by the aurora borealis, on the south by the equator, on the east by the rising sun, and on the west by the hereafter—is American, thank God! We make our boys and girls live here twenty-one years before we allow them to vote, and we ought to do the same with all foreigners."

Klan feeling was fanned to white heat by constant insistence in season and out that "every method known to man has been used

in an automobile plant. The husband earns $40.00 a week, and had had no unemployment in five years; the family owns its home, though it is mortgaged, owns a new Ford sedan, and plans to send the oldest child to the local college.

and is being used by the alien-minded and foreign influence to halt our growth." Social clubs were broken up and church groups rocked to their foundations by the tense feeling all this engendered. The secret of this eruption of strife within the group probably lies in the fact that it blew off the cylinder head of the humdrum. It afforded an outlet for many of the constant frustrations of life, economic tensions and social insecurity, by providing a wealth of scape-goats against whom wrath might be vented; and two of the most powerful latent emotional storm-centers of Middletown, religion and patriotism, were adroitly maneuvered out of their habitual uneventful status into a wild enthusiasm of utter devotion to a persecuted but noble cause. The high tide of bitterness was reached in 1923, and by 1925 the energy was mainly spent and the Klan disappeared as a local power, leaving in its wake wide areas of local bitterness.

In the midst of these divisive elements certain other tendencies act to draw various groups in Middletown together. "Civic loyalty" to "Magic Middletown" is better organized and almost as conspicuous today as in the flamboyant gas boom days when every issue of the press breathed forth boastings and slurs on rival cities. Following a mistaken rumor in 1890 that some of Middletown's ball players were to be given a trip to Europe "to exhibit America's national sport," the press seized the opportunity to exult: "This will be quite a nice trip for the boys and a bigger advertisement than —— can scare up for their dried-up town. [Middletown] is always ahead." "Boom breezes," "natural gas," "the new electric street railway," "50,000 population in ten years," have simply given way to "Shop in Middletown," "Middletown offers more," "Beat 'em, Bearcats," "I'm for Middletown College," as focal points of enthusiasm. The European War, like the earlier gas boom, served as the emergency evoking much of this civic loyalty.

"Some time ago," commented a member of Rotary, "one of the papers asked a lot of us what was the most important change we noted in Middletown. I said it was the change in community spirit. In 1910 nobody would help anybody else. Such a thing as the Presidents' Club over here at the Chamber of Commerce that brings the heads of all the civic clubs together would have been out of the question then—the men just wouldn't have come together. The war did it—forcing men to learn to coöperate."

Two activities promoting civic loyalty deserve special mention. The Chamber of Commerce, the successor of the old Citizens' Enterprise Association, while still opposed by many workers as a capitalistic organization, is becoming increasingly a community center. In 1890 the diminutive predecessor of the Chamber of Commerce existed solely to boost business, but in one month in 1924, 157 different meetings sponsored by forty-one different organizations were held in the Chamber of Commerce building. The spirit of the Chamber of Commerce is expressed by a recent president of the organization, "Merchants and farmers, Catholics and Protestants, bankers and working men—we're all living here together, and as far as I can see we're going to have to keep on, and I'm against anything that splits us up against each other." [6]

An even more widespread agency of group cohesion is the high school basket-ball team. In 1890, with no school athletics, such a thing as an annual state high school basket-ball tournament was undreamed of. Middletown claimed, to be sure, a "world's championship" in polo, but the gate receipts reported at the games were small, and the young baker who took in everything about town reports going to only one polo game in three and one-half winter months and then only because a "date" failed him; baseball received much newspaper space, but support for the teams had to be urged. Today more civic loyalty centers around basket-ball than around any other one thing. No distinctions divide the crowds which pack the school gymnasium for home games and which in every kind of machine crowd the roads for out-of-town games. North Side and South Side, Catholic and Kluxer, banker and machinist—their one shout is "Eat 'em, beat 'em, Bearcats!"

"[Middletown] probably will resemble the Deserted Village Friday night," said the press. "So strong is the spirit upon the citizens who are backing the basket-ball team of the high school that it is probable the greater part will follow the team to N——. But

[6] School houses, also, are increasingly being used as community centers. A flourishing Community Club in one working class section of town centers in the school. School yards serve somewhat as community playgrounds, and the high school auditorium is used for a variety of lectures, concerts, and other meetings.

on Saturday night the city should resemble some large metropolis, inasmuch as the team is entertaining L—— here."

For a critical out-of-town game the morning newspaper announced that it had "installed a complete new telephone system, and on Friday night three special operators will be on duty to give out the returns. . . . A fourth special wire, connected directly with the gymnasium at S——, will be used to get the game play by play." Several hundred citizens of all ages stood in the street outside the newspaper office that night chanting themselves hoarse in a chorus of "Fight! Fight! Bearcats!" The Ad. Club had been forced to set a dinner meeting for that evening in order to secure an important national speaker, but it warned the speaker to stop by eight-thirty, when the game started, and arranged for messenger service every three or four minutes the rest of the evening to bring the score.

"The meanest man in [Middletown] has been found," exclaimed one of the newspapers. "He doesn't care a thing about the basket-ball craze which has enthused the other inhabitants of the Magic City."

After the season, "as an expression of appreciation for what [Middletown's] Bearcats accomplished during the basket-ball season just closed, the High School team will be given a reception by the community in the near future. . . . The reception will be entirely a community affair, to be sponsored by no one organization, but by residents of [Middletown]."

At least two values seem patently involved in these various massed boostings of "Magic Middletown" that crop up recurrently in the Chamber of Commerce, in basket-ball, at noonday luncheon clubs, in the suppression of news judged unfavorable, in the desire which one hears expressed over and over that "Middletown make a good showing." [7] One is assurance in the face of the baffling too-bigness of European wars, death, North Poles, ill health, business worries, and political graft; the big-

[7] "What's wrong with Middletown?" members of the Kiwanis Club were asked at roll call. "The majority of the members declared that there was nothing vitally wrong. . . ." Ten "tests of a city" were submitted to Rotary and the other civic clubs for a grading of Middletown and brought forth mutual congratulations because the grades ran so high, rather than any self-searching and inquiry into the possible shortcomings of the community. "People always hush up any talk of immorality in the high school," said a Rotary member. "We don't want anything bad about the town to get out. When there's anything wrong with Middletown, people don't like to talk about it."

ness of it all shrinks at a championship basket-ball game or a Chamber of Commerce rally, and the whole business of living in Middletown suddenly "fits" again, and one "belongs"; one is a citizen of no mean city, and presumably, no mean citizen. This glorification of the homely necessities of living is apparent in the exhortation of an outstanding Middletown club woman to a group of children:

"You must have community spirit. You must think that there is no finer town in the whole United States than this. There is no finer school than yours, no finer parents than yours, no finer opportunities anywhere than you have right here. People talk of California where there is sunshine all the year round, but I've lived in California, and give me Middle Western rains! I tell you there's no lovelier place on God's footstool than this old state of ours."

The other purpose served by civic boosting is the more tangible one of bolstering the crucial getting-a-living activity by which the city lives and prospers. As the local slogan tersely expresses it:

"United we stick, divided we're stuck.
United we boost, divided we bust."

"What will become of the town if we don't believe in the thing we're going to manufacture here?" Sherwood Anderson has the citizens of Bidwell ask. "Of course we like Middletown—we have our living here," said a worker's wife. When in the early nineties the paper denounced "the traitors among us who say that gas is failing," the venom in the denunciation came from concern over what such a traitorous occurrence would mean for business. When in 1925 citizens wore cards announcing, "I'm for Middletown College!" it did not mean essentially pride in the college as an educational institution, but, as press and merchants proclaimed, "A live college here will mean thousands of dollars annually for local business." Twenty-five years ago a local editor complained: "We all know how thousands of dollars that should be spent at home are spent in other places by Sunday excursionists. Let us find the remedy!" And today the press still urges, "Shop in Middletown. It should be a matter of civic pride with every citizen to spend his money in the home community." And yet, once thus frankly made instrumental to other things, it is not surprising to observe civic pride openly made secondary to the business interests of this or that group—even of the Chamber of Commerce. When the Chamber remodeled its buildings after a fire it let the contract to the cheapest bidder, a firm which buys none of its mill-work in Middletown, i.e., does not "shop in Middletown"; a local dealer protested at this unpatriotic act, and the secretary of the Chamber hotly rejoined by calling him "uncivic" and a "knocker." Speaking of the smoke nuisance, one of the city's civic liabilities, a member of Rotary said, "The trouble is a lot of our prominent men own stock in the Electric Light Company down here that is one of the worst smoke offenders, and if we tried to do something about it over at the Chamber of Commerce they'd blow cold air on it and kill the scheme. Why, look at these grade crossings, too; our traffic is held up sometimes as much as forty-five minutes right down here on the main street, and that sort of thing hurts business. We got a movement started to elevate the tracks, and then some of the manufacturers learned it would inconvenience them in loading their cars and they killed it." Here, as among other peoples, as Malinowski has pointed out, "The unity of the clan is a legal fiction in that it demands—in all native doctrine, that is, in all their professions and statements, sayings, overt rules, and patterns of conduct—an absolute subordination of all other interests and ties to the claims of clan solidarity, while, in fact, this solidarity is almost constantly sinned against and practically nonexistent in the daily run of ordinary life. . . . This, like everything else in human cultural reality, is not a consistent logical scheme but rather a seething mixture of conflicting principles." [8]

National patriotism is civic pride writ large. Its uses are much the same. An account of a meeting of the Woman's Club in 1889 records: "The club closed with singing patriotic songs, and the president remarked that if all citizens had as deep love of liberty as the club ladies the country was safe." "The United States government is not going to the dogs," says the press today, "so long as the movie audiences cheer every appearance of the American flag." "If the coming generation of this country will be as patriotic as the past, the country's future is assured," said a prominent minister addressing the high school

[8] *Crime and Custom in Savage Society* (New York; Harcourt, Brace and Company, 1926), pp. 119, 121.

students. "The United States is the best country on earth and should give her ideals to the rest of the world," said "the most intellectual minister in town" in a Sunday morning sermon in 1924; "England is a good country, too; she has done many wrong things, but many goods things, as well. The two countries should stand together to civilize the rest of the world." "I want to see America first," declared a Rotarian; "then when I go abroad I can tell them all about America and what a fine place it is." "Patriotism for our country is the noblest sentiment any man can possess," was the pronouncement of one of Middletown's doctors to a civic club. The press glories in America's record in such editorial passages as:

"In the military annals of the United States can be found no page blotted by the inglorious record of a war to conquer or oppress. . . .

"Four great wars are written into American history, and for none need even a pacifist blush. . . .

"But we would have you note that this story is marked by a steady advance toward a great goal. Each chapter in it denotes a forward movement in the nation's life. We cannot read it without the conviction growing and deepening that the hand of God had part in its writing and that a purpose more glorious than any yet attained lies in its development."

One of Middletown's outstanding business leaders paid $80.00 for the American Legion history of the World War, so that he could "know the real facts about the war and just why the United States is where it is today."

Not only does this sentiment of patriotism give stability to life in the present, it attempts to bring the vicissitudes of the past into line with the habits of thought and action regarded as desirable today. Thus a senatorial candidate, educated at Harvard and prominent in the American Legion, assured the fashionable annual banquet of members and guests of the local chapter of the D.A.R.:

"The historic event we call the American Revolution was not really a revolution. 'Revolution' suggests to the mind the thought of violent change, a sharp breaking with the past, a complete overturn of traditions and the creation of a new order of life. The American Revolution was no such upheaval. . . . Russia is in the midst of a real revolution, but Americans simply declared their independence by confirming a situation which was already existing. . . .

"Nor was the American Constitution a revolutionary document. On the contrary it was the sum of the accumulated wisdom which had been evolved out of the experience of the thirteen colonies.

"It is because of this condition that the United States today has the most stable government in the world. . . . This is the only country that is settled—in every way. . . . It is because of the firm foundation of conservative liberty under the Constitution that America stands first among the nations of the world. Because America has this government, the best government on God's earth, America can bring salvation to the world." [9]

Since patriotism performs such a useful service to the city,[10] as in the case of civic loyalty, various institutions have arisen to foster it. Aside from the work of such organized groups as the Daughters of the American Revolution, the Woman's Relief Corps, the Military Affairs Committee of the Chamber of Commerce, the War Mothers, and the American Legion, other devices new to Middletown since the nineties are active. Annually an oratorical contest is held in the high school on the Constitution of the United States. "Would a person who criticized the Constitution stand any chance of winning the contest?" the lawyer in charge of the contest was asked. "Oh, no!" was the surprised reply. "That's not the point of it." Girls between fifteen and twenty competed for a trip to Washington by writing essays on "Why a Young Man I Know Should Attend a Citizens' Military Training Camp." Practically every organization in the community, from the public schools to the men's civic clubs with their salute to the flag, actively fosters patriotism. At a meeting of the Matinee Musicale devoted to American music, the speaker apologized because she could not pronounce the names of most of the artists on the New York programs of the fall "because they are foreigners" and said, "We all want to work for having *American* programs with *American* artists."

Civic loyalty and patriotism are but two of the pressures tend-- ing to mold Middletown into common habits of thought and

[9] Cf. the responses of high school juniors and seniors to the statements, "America is undoubtedly the best country on earth," etc., quoted in Ch. XIV.

[10] Cf. A. R. Brown's *The Andaman Islanders* (Cambridge; Cambridge University Press, 1922), Ch. V, especially pp. 233-4, for an enlightening treatment of the utility of the system of sentiments of a group and of the rôle of the group's ceremonials in maintaining these sentiments.

action. Every aspect of Middletown's life has felt something of this same tendency: standardized processes in industry; nationally advertised products used, eaten, worn in Middletown homes;[11] standardized curriculum, text-books, teachers in the schools; the very play-time of the people running into certain molds, with national movie films, nationally edited magazines, and standardized music contests. Certain sharp focal points of this pressure appear in the celebration of a host of annual "days" and "weeks," nearly all of them new since 1890. Among those observed in Middletown between April 1, 1924, and April 1, 1925, were the following:

"Suburban Day"; "Home Sewing Week"; "One Hour Dress Week"; "Ice Cream Week," with a special essay contest on "Why I Should Eat Ice Cream Every Day"; "Truth Week," fostered by the Advertising Club and aiming to "build confidence through truth in advertising"; "Father's Day"; "Mother's Day"; "Boys' Day"; "Boys' Week," "a nation-wide movement to emphasize the fact that our boys are loyal"; "Thrift Week," with its "Own Your Own Home Day," "Savings Day," "Pay Your Bills Day," "Make Your Will Day," "Insure Yourself Day," "Live on a Budget Day," "Share with Others Day"; "Home Beautiful Week"; "Education Week"; "National Picture Week"; "Art Week"; "Music Week"; "Odd Fellows Reconsecration Week"; "Joy Week," in which the ministers were asked to preach on "Does it Pay to Do Good Turns?"; "Week of Prayer"; "Father and Son Week"; "Mother and Daughter Week"; "Go to Church Sunday"; "Labor Sunday"; "Golden Rule Sunday," endorsed by the governor, who "would like to see the Golden Rule enshrined in every heart and believes it the solution for numerous situations"; "Saratoga Day"; "Clean Up and Paint Up Week"; "Child Health Week"; "Tuberculosis Day"; "Non-spit Week"; "Hospital Donation Day"; "Tax Reduction Week"; "Fire Prevention Week"; "Courtesy Week," with an essay contest on the value of courtesy; "Pep Week"; "Constitution Week"; "Defense Day"; "Buddy Poppy Day."

While many of the multiplying social groupings of Middletown tend to facilitate the performance of a specific life-activity

[11] The pressure of advertising is reflected in the assertion by one of the national advertising media that through it "4,500,000 families who eat advertised foods, wear advertised clothes, drive advertised automobiles, use advertised tooth pastes, wash with advertised soaps, buy advertised furniture, and believe in advertised products, may be told and sold effectively." One surmises that Middletown scarcely realizes the extent to which it is being "told and sold."

—e.g., the displacing of frequently unskilled home nursing by the Visiting Nurses' Association—others appear to be in the interest of a sub-group of the community and possibly opposed to the interests of the group as a whole, e.g., the "Shop at Home" campaigns of committees of such groups as the Merchants' Association and Advertising Club, or the programs of certain labor groups.

Such a multiplication of groups performing often criss-crossing activities regarded as vital to business class or to working class, to merchants, to the Middletown Republican machine, to residents of Twelfth Street, to "all Protestants," to owners of Middletown industries, to "every one living north of the tracks," to the local medical profession, present to the individual in Middletown an increasing host of occasions for conflict.[12] Confronted by the difficulty of choosing among subtle group loyalties, the Middletown citizen, particularly of the business class in this world of credit, tends to do with his ideas what he does with his breakfast food or his collars or his politics—he increasingly accepts a blanket pattern solution. He does not try to scratch the "good fellow" ticket, but votes it straight. To be "civic" and to "serve" is to "put over" "Magic Middletown," the church, the party, a get-together dinner, a financial campaign, one's friends; it is to be "a booster, not a knocker"; to accept without question the symbols. Hence one tends to find certain groups of loyalties linked together. A successful lawyer is likely to belong to the Bar Association, Chamber of Commerce, Rotary, the Republican party, one of two or three leading churches, a high order in the Masons, the Country Club, and be a director of the Y.M.C.A., while his wife joins the

[12] An occasion for such conflicts is involved in the entry by various new groups into a region formerly dominated by churches and more recently shared with the lodges, namely, the conscious formulation of an ethical code to which the complete adherence of members is sought. Certain groups like the Bar Association have their separate code of ethics. During 1924 each member of the Middletown Chamber of Commerce was asked to sign a card saying that the undersigned organization, firm, or individual "has accepted the Principles of Business Conduct as passed by the Twelfth Annual Meeting of the Chamber of Commerce of the United States." During the same year the District Governor of Rotary, outlining the code of business ethics adopted by Rotary International, said, "Rotary isn't a club. It is a movement. I see the Rotary ideal, this Rotary way of living with one's fellows, spreading eventually over all the world." Here is a new religion, which as noted elsewhere conflicts with the traditional primacy of church loyalty in the case of some Middletown men.

more fashionable women's clubs, the group directing the local charities, and so on. A successful banker belongs to the same groups, save that the Bankers' Association replaces the Bar Association. This linking of loyalties is easier since the groups which any one individual supports tend to unite in standardizing and fusing the pattern of loyalties: "the lowly Nazarene" becomes "the first Rotarian"; George Washington, the ideal of the "America first" politician, "would undoubtedly have been a realtor. Because that is precisely the kind of a man he was. . . . He joined everything there was to join. If there had been a Rotary (or Kiwanis or Lions or Civitas) or a real estate board or a chamber of commerce, George Washington, the 'regular fellow,' would surely have been a member"; [18] as Rotary and the Country Club go together and both support the Chamber of Commerce, so the church, the law, and the Republican party tend to hold up each other's hands; distinctions are blurred under symbols.

As pointed out above in connection with the disappearance of the Ethical Society, the discredit attached to voting for La Follette, and so on, strong heretics—religious, economic, political, social—in Middletown are increasingly frowned upon. An educator recently moving to Middletown was warned, "No matter what you think, you'll have to become a regular churchman to get along in Middletown." The Teachers' Federation, wanting to bring Ida Tarbell to lecture, was told by a leading Middletown lawyer, "I don't believe you'd better do it. Why, when I was in the university I believed all the teachers told me and went out thinking I could help change things. Now I realize it was downright wrong for them to talk to us the way they did." "It is rather comforting to know," says a Middletown editorial in 1924, "that the great masses of business men, of which there are excellent examples in [Middletown], 'sit steady in the boat.'" A man of wide experience who had grown up in Middletown and is highly respected by the city said that the two things he felt most upon coming back to Middletown from a distant country were its prejudice and superficiality. "These people are all afraid of something," he said. "What is it?" A Middletown writer, almost the only member of Rotary allowed to be "a bit queer," describes this same standardized,

[18] Quoted in the leading Middletown paper from *Collier's Weekly*, February 21, 1925.

patterned avoidance of unusual living in a story about the town in the *Saturday Evening Post:*

"The Rotary Club ate sociably, without attention. Had they ever been really hungry, or thirsty past the point of mild discomfort? Or feared, or hated, or spent the last ounce of their courage or their strength on anything?

"Look at their faces. Oddly, they seemed all of one type. Fat or thin, old or young, one mark was on them all. Not dullness; not exactly; these were successful men. The keenest of them showed the mark most plainly. Moderation—the keying down of all spiritual force to the general level. No deep calm lines of single purpose, no steady driving set of jaw, no eyes of meditation. Rather a harassed and a scattered look, the mark of a thousand small habitual restraints, the price of living comfortably with neighbors. The petty lines of worry—moderate fear.

"What did they care about the feel of earth? They shut it out, fenced themselves in with houses, and played safe."[14]

That is one side of the picture; the other is the occasional lonely individual who does not want to vote the good-fellow ticket straight but dares not break away.[15] One after another of these wistful individuals appears—each hungry for companionship, but knowing no one else in the city who speaks his language. "I just run away from it all to my books," says one of these men who scrupulously keeps up the appearance of conformity. Another essentially lonely man disguises his distaste for the lock-step under a never-failing quizzical smile and external good humor.

"I'm just tired—tired in the legs—tired in the spirit," said another man occupying a substantial position in the city. "I know there are others as lonely as I am. There's ——, he's a fine fellow, and I want to know him better, but I can't get under his skin; we're always just friendly and joking together. Neither of us ever gives himself away."

A veteran who had known Middletown many years exclaimed in speaking of the old Ethical Society, "You know, that's interest-

[14] "The conservatism of active business men, like that of all other conservatives, is motivated by fear and desire for security. . . . More especially in the smaller towns and cities is he careful to conform to the accepted standards of respectability." Wolfe, *op. cit.,* p. 69.

[15] Cf. Charles Horton Cooley's discussion of factors which deter the individual from writing "'whim' on the lintel of his doorpost," thus infringing upon the established conventions to which his neighbors are habituated. *Social Organization* (New York; Scribner, 1909), p. 46.

ing, I never thought of it before, but we've lost all that ability to ponder over life those men had. We're too busy being busy, going to places and to committees, to live that way. We don't think of it unless some fellow stops us as you've stopped me."

"I went a hundred miles to —— the other day," said a worker, "just to get a chance to talk to a fellow who understands the things I think about. Working men here are all alike—a lot of jokes and a little to drink and a few smutty stories, and they're happy."

The foregoing pages have suggested various factors tending to decrease and to increase the cohesion of the community. A final point remains to be suggested. It was noted in an earlier chapter that the working class, particularly the males, appear to be more isolated as members of the group than formerly. In the present chapter indications have been strong that the cohesion and conformity of the business class is proceeding very rapidly. In other words, the sense of "belonging," of fitting their world—none the less real psychologically though possibly based on nothing more substantial than symbols—appears to be growing more rapidly among the business class than among the workers. The one group gives more easily with the stresses to which the group is subject, because its members have built their lives about these dominant stresses at more points; it does the "civic" thing easily, because civic values are its values at so many points. The workers, on the other hand, instead of yielding to and reinforcing the pressures of this organized community life, are more often inert, uncoöperative, and even resistant, as in the case of their opposition to the Chamber of Commerce; the major drives of "Magic Middletown" are not so completely their drives, and only at second hand do they tingle to the exhilaration of some of the things that are living itself to the business group.

Chapter XXIX

CONCLUSION

A number of people, after hearing of the field investigation recorded in the preceding pages, have demanded, "Well, what are your conclusions? Is American life growing better or worse?" Consideration of Middletown in terms of "good" or "bad," "better" or "worse," is, obviously, irrelevant to the purpose of the study; in so far as any data have been presented which may suggest comparison with other communities or with any hypothetical standards, they have been given simply to bring out significant aspects of Middletown's life. The course of the study has not been, in fact, such as to encourage smooth generalizations at the close. On the contrary, the attempt to reveal interrelations in the maze of interlocked, often contradictory, institutional habits that constitute living in Middletown has led to few general conclusions save as to the inchoate condition of this one small modern community and the extent and complexity of the task confronting social science.

The description of the phenomena observed in Middletown has, however, suggested some recurrent manifestations of the processes of social change involved in them and largely responsible for what Middletown considers its "social problems."

Living moves along in Middletown, as we have seen, at a bewildering variety of gaits. Differential rates of adjustment in the performance of the same function have been observed between elders and their juniors and between people living next door to each other, while the females have exhibited greater conservatism than the males at many points, and the males, with seemingly no more coherence or pattern in their adjustments, are more resistant to adaptation at many other points. In many activities, as has been repeatedly pointed out, the working class today employs the habits of the business class of roughly a generation ago; if it were possible to differentiate clearly the gradations by which each of these two major groups shades into the other, it might appear that many changes are slowly filtered down through various intermediate groups. Shifts sometimes diffuse, however, in the reverse direction, from working class to business class, as has been noted, for example, in the use of commercially baked bread and canned foods.

Not only do these variations, in many cases pronounced enough to affect markedly one's capacity to deal with one's world, appear between individuals and between different age, sex, and other groups within Middletown in the performance of the *same* life-activity, but the city as a whole and groups within the city live in different eras in the performance of *different* life-activities. It is apparent that Middletown is carrying on certain of these habitual pursuits in almost precisely the same manner as a generation ago, while in the performance of others its present methods bear little resemblance to the earlier ones. Among the six major groups of activities a rough hierarchy of rates of change is apparent. Getting a living seemingly exhibits the most pervasive change, particularly in its technological and mechanical aspects; leisure time, again most markedly in material developments such as the automobile and motion picture, is almost as mobile; training the young in schools, community activities, and making a home would come third, fourth, and fifth in varying order, depending upon which traits are scrutinized; while, finally, on the whole exhibiting least change, come the formal religious activities.

Thus Middletown, due allowance always being made for wide variations in practice within the city, may be observed to employ in the main the psychology of the last century in training its children in the home and the psychology of the current century in persuading its citizens to buy articles from its stores; it may be observed in its courts of law to be commencing to regard individuals as not entirely responsible for their acts, while in its institutional machinery for selling homes, failure to pay, whether due to unemployment, sickness, or other factors, is regarded as a deliberate violation of an agreement voiding all right to consideration; a man may get his living by operating a twentieth-century machine and at the same time hunt for a job under a *laissez-faire* individualism which dates back more than a century; a mother may accept community responsibility for the education of her children but not for the care of their health; she may be living in one era in the way she cleans her house or does her washing and in another in the care of her children or in her marital relations.

Furthermore, not only does the rate of change vary widely, but the direction of change is highly erratic. Social change and the thing Middletown calls "progress" are not synonymous. There is real reason for suspecting, for instance, that the slackening of the drift towards centralized laundering of clothing outside the home that has appeared in connection with the installation of costly electrical machine units used only one day a week in hundreds upon hundreds of Middletown homes represents not "progress" but a back-eddy in home-making technique.

One could go on listing these variations in Middletown's everyday practice indefinitely. It is not intended here either to endorse change at a given point or of a given kind, or to glorify rapid change as over against more leisurely adaptation. Whether one is temperamentally well disposed towards social change or resistant to it, however, the fact remains that Middletown's life exhibits at almost every point either some change or some stress arising from failure to change. A citizen has one foot on the relatively solid ground of established institutional habits and the other fast to an escalator erratically moving in several directions at a bewildering variety of speeds. Living under such circumstances consists first of all in maintaining some sort of equilibrium. If, as has appeared over and over again, Middletown tends to bear down harder on the relatively solid ground, it is simply exhibiting the reluctance to changing habitual ways common to men everywhere.[1]

[1] As has been pointed out so often, it is characteristic of mankind to make as little adjustment as possible in customary ways in the face of new conditions; the process of social change is epitomized in the fact that the first Packard car body delivered to the manufacturers had a whipstock on the dashboard. A. L. Kroeber, referring to the development of the Greek alphabet, says, "They followed the method which is characteristic of invention in general. They took over the existing system, twisted and stretched it as far as they could, and created outright only when they were forced to." *Anthropology* (New York; Harcourt, Brace and Company, 1923), p. 272.

In Middletown a wealthy citizen was forbidden by the City Council in 1890 to lay a "new-fashioned concrete walk" in front of his home and ordered to make it of brick. Likewise, as W. I. Thomas points out, the iron plow was resisted in the early days as an insult to God and therefore as poisoning the ground and causing weeds to grow, while the first man to use an umbrella on the streets of Philadelphia was arrested. *The Unadjusted Girl* (Boston; Little, Brown, 1923), pp. 228-31.

Since the emergence of "social problems" would seem to no small extent traceable to the ragged, unsynchronized movement of social institutions, better understanding of how such institutional habits develop and change. what ones are learned most readily, and what accompanying circumstances facilitate or retard such learning, would appear to be a prime concern confronting those desirous of applying more effective planning and control to Middletown's living. In case after case the preceding pages have revealed Middletown as learning new ways of behaving towards material things more rapidly than new habits addressed to persons and non-material institutions. New tools and inventions have been the most prolific breeders of change. They have entered Middletown's industrial life more rapidly than new business and management devices. Bathrooms and electricity have pervaded the homes of the city more rapidly than innovations in the personal adjustments between husband and wife or between parents and children. The automobile has changed the leisure-time life more drastically than have the literature courses taught the young, and tool-using vocational courses have appeared more rapidly in the school curriculum than changes in the arts courses. The development of the linotype and radio are changing the technique of winning political elections more than developments in the art of speechmaking or in Middletown's methods of voting. The Y.M.C.A., built about a gymnasium, exhibits more change in Middletown's religious institutions than do the weekly sermons of its ministers or the deliberations of the Ministerial Association. By and large, a new tool or material device, the specific efficacy of which can be tested decisively and impersonally, is fairly certain to be fitted somehow into Middletown's accepted scheme of things, while opposed non-material factors, such as tradition and sentiment, slowly open up to make room for it.

A further characteristic of social change as observed in Middletown is the sidling procedure whereby the innovation first appears in many cases as an optional alternate mode of doing a given thing and then in time displaces its older rival. Such developments as the newer methods of administering justice in the case of juvenile offenders, side by side with the earlier court procedure still used for adults, the public and semi-public care of health as compared with the traditional individualistic methods. and the development of the impersonal Community

Chest paralleling the continuance of separate, personalized charities, are all cases in point. In passing, the question may be raised as to why, in view of man's predilection for changing his ways as little as possible, this relatively painless method of the alternate procedure is not more generally adopted by would-be reformers instead of the widely current method of a head-on assault upon established institutions, the immediate result of which is to stiffen the defensive resistance of the opposed institution. The observed tendency of Middletown to be at once more prone to cling to cherished traditions and at the same time more adventurously experimental in embracing new ways in matters concerning its children suggests the possible strategy of utilizing situations involving the child in developing new alternate modes of behavior.

New habits not opposed by strong emotional resistance have apparently entered Middletown more readily than those confronted by the reverse situation. Thus the automobile was more quickly adopted in Middletown for delivering groceries than for use at funerals, and for driving to business than for driving to church. This suggests the significance of the process of secularization of Middletown's institutions, noted at many points, for social change in the regions affected.

At the same time that secularization is lessening emotional resistance at some points, the very swiftness of the penetration of impersonal institutions like automatic machinery, electrical devices, and automobiles, has, however, multiplied occasions for emotional explosion in Middletown through the discrepancies between these shouldering new ways and the other habits upon which they force themselves. One need only mention the new friction spots between parents and children incidental to the incorporation of automobile and movies into their world, or the emotional ramifications suggested above in connection with the advent and operation of automatic machinery among Middle-town working men and their families. An explanation of the tendency already noted for Middletown to link its emotional loyalties together, to vote the good-fellow ticket straight, may probably lie in its increasing sense of strain and perplexity in its rapidly changing world that can be made to hang together and make sense in no other way.

Among the questions arising from the phenomena observed and inviting further study, the following may be mentioned:

What are the implications for social change of the present tendency for Middletown to subsume more and more of its living under pecuniary considerations? Where is the vicarious living of such a pecuniary community increasing, and where diminishing, the readiness to accede to the demands of a changing cultural environment?

How do the shifting emphases upon present and future, fostered by such urgent new inventions as the automobile and by the marked spread of education and of the institution of credit, load the situation for or against specific kinds of change?

What is the significance for social adjustment of the increased rôle of youth with its greater adaptability?

Under what circumstances may change in process in one sector of an individual's habitual responses to his world tend to make him resistant to or responsive to specific changes in other sectors?

At what points is the shifting relative status of men and women likely to condition the adjustment of Middletown to the new institutional habits being pressed upon it on every side?

What specific types of emotional adjustment and maladjustment are likely to arise as a result of the restraints imposed by the accepted standards of different groups within Middletown? For instance, how does the fact of belonging to business or working class subject individuals to differential strains in the performance of identical life-activities?

Can there be any clearer determination of the kinds of culture traits which are characteristically diffused from business to working class and of those which travel in the reverse direction, and of what specific factors may accelerate or divert the course of a trait in a particular community?

As Middletown has become reluctantly conscious from time to time of discrepancies in its institutional system, it has frequently tended to avoid "doing something about" these "social problems" of "bad times," "the younger generation," "corrupt politics," "housing," "street traffic," and so on, by blaming the difficulty on the "nature of things" or upon the willfulness of individuals. When the "problem" has become so urgent that the community has felt compelled to seek and apply a "remedy," this remedy has tended to be a logical extension of old categories to the new situation, or an emotional defense of the earlier situation with a renewed insistence upon traditional verbal and other symbols, or a stricter enforcement or further elaboration of existing institutional devices: thus difficulties in the business world are met by a greater elaboration of financial devices and by an attempt to apply the familiar individual ethics to corporate activities, increase in crime by an elaboration of the police and court systems or the doubling of penalties, political corruption by a harking back to the Constitution in all the schools, by nation-wide oratorical contests, and by getting more people out to vote, indifference to the church by the forming of more church organizations. The foregoing pages suggest the possible utility of a deeper-cutting procedure that would involve a reëxamination of the institutions themselves.

APPENDIX

Appendix

NOTE ON METHOD

The first intimation that Middletown had that it was being studied was the unheralded opening of an office in a local office building by a group of people who had come to "study the growth of the city." The writers of the report and Miss Flournoy, the staff secretary, lived in Middletown from January, 1924, to June, 1925, Miss Davis for a year, and Dr. Williams for five months.

At the outset no fixed schedules were drawn up; the six main life-activities in terms of which Middletown was to be studied had been agreed upon, and the study began with the attempt to observe in as detached a manner as possible what people were doing in carrying on these activities and the trends of change these processes exhibited over the preceding thirty-five years. Such a procedure involved some loss in time but a gain in flexibility and, it is to be hoped, in objectivity; as schedules and questionnaires were drawn up, they were framed in terms of the specific behavior observed in the city.

As the study progressed, the scope of the inquiry was narrowed by concentrating on 1890 and 1924 rather than attempting a detailed study of the intervening years. The aim here was not to reveal contrasts but a clearer understanding of past behavior as conditioning present behavior. Furthermore, it appeared that, homogeneous as Middletown is racially and in many other respects, it was impossible to describe the carrying on of even its major life-activities in terms of the city as a unit. Various attempts were made to define different groups within the city, and it was finally decided, as pointed out in Ch. IV, especially Notes 2 and 3, that the differentiation which most adequately represented the facts was that into business class and working class. The study was accordingly narrowed further in dealing with certain data by concentrating upon business and working class groups. The result is not a complete structural diagram of the city, but it does, it is believed, indicate the outstanding characteristics of the culture, the chief points of divergence within the city, and the principal regions of movement. Whenever throughout the text, however, Middletown or any group within the city is referred to as if it were a unit in thought or behavior, it should be borne in mind that this masking of the individual under the group is merely a shorthand symbol.

Even as so delimited, however, the investigation obviously could not claim to be comprehensive; some subjective selection of data was inevitably involved. In general, the criterion of selection was that no institutions were studied for their own sake but always with reference to the life-activities which they served. Thus many details which would have been of paramount interest had the operation of a given institution been the chief concern became less relevant.

The various techniques employed included the following:

1. *Participation in the local life.* Members of the staff lived in apartments or in rooms in private households. In every way possible they shared the life of the city, making friends and assuming local ties and obligations as would any other residents of Middletown. In this way a large measure of spontaneity was obtained and the "bug on a pin" aspect reduced. Staff members might dine one night with the head of a large manufacturing plant and on the next with a labor leader or a day laborer. Week in and week out they attended churches, school assemblies and classes, court sessions, political rallies, labor meetings, civic club luncheons, missionary meetings, lectures, annual dinners, card parties, etc. At the end of the period they had access to a kind of information which would have been entirely inaccessible at the outset. This constant interplay of spontaneous participation and detached observation presented difficulties which have not at all points been successfully overcome.

In reporting meetings attended, as in individual interviews or casual conversations, the method followed was to take such inconspicuous notes as were possible in the course of the meeting, service, or interview and then immediately afterward to write them up in detail according to the standardized form adopted. In cases where it was impossible to take notes at the time, the record was made immediately afterward from memory.

2. *Examination of documentary material.* Census data, city and county records, court files, school records, State Biennial Reports and Year Books, etc., were used wherever available.

The two leading daily papers were read in detail for the full years 1890 and 1891 and were supplemented at many points by the Democratic daily and a labor paper which flourished for a time. The two current Republican dailies and the Democratic weekly were read and clipped during the year and a half of the study, and in somewhat less detail for a year thereafter. Frequent use was made of the papers in 1900, 1910, and other intervening years.

The minutes of various organizations were read for both early and current periods and, in most cases, for the intervening years as well. These included the Board of Education, the Missionary Societies of two leading churches, the Ministerial Association, the Federated Club of Clubs, the Woman's Club, the Library Board, the Humane Society, and other groups. School examination questions, of which there were complete files for both grade school and high school for the nineties and today, were compared in detail.

Two unusually detailed diaries, one of a leading merchant and prominent Protestant churchman and one of a young Catholic baker, were read for the years 1886 to 1900. These and various other diaries and "scrap books" of clippings, programs, letters, club papers, etc., when coupled with the memories of older citizens, helped to supply for the earlier period a partial equivalent for the informal person-to-person contacts and folk-talk of today.

Histories of the state, county, and city, city directories, maps, "boom books," Chamber of Commerce publications, high school annuals, and health and other surveys of the city, etc., were used wherever available.

3. *Compilation of statistics.* In some cases where no statistical material was available it was possible to compile data from sources in the city, at the state capital, or from actual counts by staff members. Such data included figures on wages, steadiness of employment, industrial accidents, nearness of residence to plant, promotion, club membership, church membership, contributions, and attendance, library and periodical circulation, attendance at motion picture theaters, ownership and use of automobiles, etc.

4. *Interviews.* These varied all the way from the most casual conversation with street-car conductors, janitors, barbers, or chance associates at luncheon or club meeting to carefully planned interviews with individuals especially qualified to give information on particular phases of the city's life. Among the latter, for instance, were detailed interviews with the six leading Protestant ministers and the secretaries of the Y.M.C.A. and Y.W.C.A., lasting about four hours each and covering with a long schedule intimate phases of the life of their organizations and of their personal problems as religious workers in Middletown. Such questioning, coming at the close of many months of close contact with these men, elicited what is believed to have been almost complete frankness of response.

As the study progressed it seemed desirable to test in individual families certain hypotheses as to trends observed in the behavior of the community. Accordingly, schedules were drawn up on the basis of these observed characteristics, and a group of 124 families clearly of the working class and a group of forty families clearly belonging to the business class were interviewed. The requirements in the selection of these families were that they be native-born, white Americans, that they live within the city limits, that both parents be alive and living together, and that they have one or more children between the ages of six and eighteen years. Interviews were in all cases with the wife, though occasionally the husband was present. Information on the husband's occupation and duration of employment was first secured in every case where access was gained to the home; then if it was found that there were no children of school age, the interview was not continued. Furthermore, not all the data could be secured from every individual interview. Thus for the working class there are 182 families for whom data were given on the length of time the husband had worked for his present employer, 165 for whom data were given on unemployment, 124 going through the entire schedule, giving information on all or nearly all points, and 100 for whom income distribution was secured.

The working class families were secured by following up as a nucleus the addresses of all shop workers on the pay-rolls of three leading plants situated in three different sections of the city and engaged in three different representative types of production in glass and metal. Care was taken to keep the total of families interviewed in any one section of the city roughly proportional to the working class population of that section. The working class interviews were not arranged in advance. Each took from two to three hours and, despite this fact and the suspicion of strangers that had to be overcome, the three staff members who did the interviewing were able to carry through the interviews in something over four out of each five families in which the wife was actually at home and the family contained children of the appropriate age. With nine exceptions, all of the 124 interviews were made during the day. This had the disadvantage of cutting down the proportion of the families in which the wife was actually working away from home at the time and of increasing the proportion of families of the more prosperous workers, including foremen. This inadequacy in the sample is somewhat offset, however, by the fact that local industry was depressed at the time of the interviews, which tended to reduce the number of women employed in industry, and by the use of five-year records on such points as the wife's working for pay. When the names taken from the pay-rolls had been exhausted, the interviewers in some instances supplemented them by the random method of ringing the door-bells of other houses in the vicinity, where the woman just interviewed stated that there were children of school age.

These 124 working class families are believed to be a representative sample of those employed in the manufacturing and mechanical industries which dominate Middletown. The 100 working class men for whom income data were secured may be roughly classified by occupations as follows: machinists, 26; foremen, 9; molders, 7; sorters, inspectors, checkers, 5; skilled mechanics, 4; pattern-makers, 3; tool-makers, 3; filers and buffers, 3; engineers, 2; carpenters, 2; craters and packers, 2; furnacemen, 2; sheet metal workers, 2; plumbers, 1; electricians, 1; railway yardmen, 1; fancy spotters in a dye house, 1; and a miscellaneous group of little-skilled workers, including laborers, 26. The occupations of the additional twenty-four whose families were interviewed in detail were machinists, 7; foremen, 2; sorters, 2; skilled mechanics, 1; engineers, 1; hammermen, 1; and a miscellaneous group of little-skilled workers, including laborers, 10.

The forty business class interviews are not as representative of all levels of the business group. It was necessary to make appointments in advance for this group and to interview people who were willing to coöperate. In most cases the interviews took longer than those with the working class. None of the interviews of either class took place until the latter part of the field study. In the light of half a year's residence, of the exigencies of time, and of the major emphasis throughout the study upon dynamic trends in the local life, it was deemed wise to contrast to the base-line group of working class families a fairly homogeneous, though necessarily small, group whose ways represent in most things the direction of change for the rest of the city. The occupations of the forty business class men whose wives were interviewed were: manufacturers, 5; engineers, 4; lawyers, 4; retail merchants, 4; wholesale merchants, 2; bankers, 2; teachers, 2; religious workers, 2; physicians, 2; bookkeepers, 2; sales managers, 1; salesmen, 1; real estate dealers, 1; hotel managers, 1; newspaper managers, 1; public utility managers, 1; bank tellers, 1; purchasing agents, 1; writers, 1; small promoters, 1; undertakers, 1. The families of four very wealthy manufacturers were purposely omitted, and with the exception of three wealthy families the forty families are what Middletown would describe as "just good, substantial folks." They are the people who shape the policies of Rotary and Kiwanis, the Masons and the Elks, the Chamber of Commerce, and the Presbyterian and Methodist Churches.

The difficulties involved in the long interviews with these sample house-wives were particularly acute in the case of personal questions such as those touching religious beliefs. The interviewers accordingly had frequently to approach these questions indirectly rather than, as in the case of the others, as point-blank questions asked under standardized conditions; they were usually slipped in at varying points in the interview as an opportunity arose.

The value of different types of data secured through such interviews varies greatly. Information given by the women interviewed on the habits of their mothers' families, while concrete and specific in all cases where it is used, is obviously less reliable than information on current practices in their own families. And in any case, the groups are too small to allow heavy dependence on these data. Although members of the staff believe that the data secured from these families are in most cases widely representative of the two groups in Middletown, the claim is not made on the basis of these samples that "the business class" or "the working class" in Middle-town does thus and so. Figures based upon them are purposely given in numbers rather than percentages, and probable error is omitted, to avoid any fallacious appearance of accuracy. They are offered not as absolute proof but as what are believed to be significant indices to the behavior of this community.

5. *Questionnaires.* As extensions of the interview, questionnaires were used at some points in the study. A questionnaire on club membership and activities was sent to the more than 400 clubs in the city as they were located in the spring of 1924. A questionnaire, dealing with the life of the high school population, was given to all sophomore, junior, and senior English classes, which included roughly three-fourths of the high school population of these last three years, a total of between 700 and 800 boys and girls. A "true-false" questionnaire, given to all junior and senior social science classes, approximately 550 boys and girls, dealt largely with points of view on certain public questions. These high school questionnaires were given by staff members during school hours, throughout the same morning, each taking approximately fifty minutes; they were entirely anonymous and the students were urged to leave any questions blank which they did not care to answer seriously. Informal conversation with a num-ber of boys and girls afterward confirmed the impression of the staff members that the answers were for the most part genuine.

No answers from negroes were included in the tabulations. In both questionnaires students were asked to check the section of town in which they lived and to give their fathers' occupations (without stating the name of the firm or plant) and their religious and political affiliations. Different sub-groups in the community were thus roughly differentiated.

Since these questionnaires were not used to measure any general "atti-tudes," scores for the questionnaires as a whole were not compiled, nor were they tested for reliability or validity. The answers to individual ques-tions, representing as they do verbalizations, are, like the interviews, used not as proof but as suggestion of tendencies.

The schedules and questionnaires used are here omitted at the request of the sponsors of the study because of considerations of space, as are like-wise a number of the tables on which the text is based.

One other point should be noted. The population figure of 38,000 for 1924 used throughout the study represents a careful estimate. The U. S. Census estimate for the inter-censal year 1924 was 41,000, but this was obviously incorrect, since it was based on a prolongation of a misleading rate of growth between 1910 and 1920 involving the addition of adjoining suburbs in 1919. An attempt was made to use combined building permits for dwellings and the excess of births over deaths, but this, too, had to be rejected on account of irregularly kept records in the case both of building permits and of vital statistics. The fact that "at least 500 vacant houses" were reported in Middletown in the late fall of 1924 after several months of "bad times" made this attempt to use building permits still more unreliable. The elementary school enrollment figures indicated 5,311 in 1920, when the population was under 37,000, 5,644 in 1921, 5,364 in 1922, 5,529 in 1923, 5,651 in 1924—attendance being compulsory throughout. The school enumeration totals showed an actual decrease of 948, 8.6 per cent. between 1920 and 1924, though these figures are unreliable, especially in view of the fact that during this period the State Department of Education was active in squeezing water from enumeration figures in certain other cities, a condition likely to encourage conservative enumeration in Middletown. The estimate of 38,000 is based upon a careful weighing of many factors, including the trend in local industry since 1920.

TABLES

TABLE I: PER CENT. OF THE APPROXIMATELY 13,000 MALES AND 3,500 FEMALES GAINFULLY EMPLOYED IN BUSINESS CLASS AND IN WORKING CLASS ACTIVITIES, 1920 [1]

GROUP	TOTAL	MALES	FEMALES
	100.0%	100.0%	100.0 %
Business class (engaged in activities addressed primarily to persons) [2]	29.4	25.5	43.5 [3]
Working class (engaged in activities addressed primarily to things)..........	70.6	74.5	56.5

[1] These percentages are based on a careful redistribution of the 1920 Census classification of those gainfully employed in Middletown. They are based on exact numbers, though round numbers are used in the caption to avoid identifying Middletown. Such a classification is rough at best, involving difficult decisions in the case of certain ambivalent groups; the decisions depend upon months of familiarity with the working habits of Middletown and upon consultation with local business and working men. The general method of classification is revealed by the following treatment of the occupations grouped by the Census under the general heading, Manufacturing and Mechanical Industries. The entire group, both males and females, were placed in the working class, with the following exceptions: four men among the group of bakers were transferred to the business class as being more distinctly retailers, etc. than working men; all builders and building contractors, ten of the dressmakers and seamstresses, one male and fourteen of the female milliners and millinery dealers, eleven of the tailors, all of the managers and superintendents, and all of the group of manufacturers and officials were placed in the business class.

[2] Four of the twenty-nine in each 100 who constitute the business class, as explained in Ch. IV, N. 3, are professional technicians addressing their activities primarily to things rather than to persons but are here classed, for the reason given, with the business class.

[3] This group includes a larger number of ambivalent cases than any other, e.g., girls clerking in stores.

TABLE II: AGE DISTRIBUTION OF EMPLOYEES[1] IN THREE REPRESENTATIVE MANUFACTURING PLANTS[2] IN 1924 COMPARED WITH DISTRIBUTION OF POPULATION IN 1920

AGE	MALES				FEMALES			
	1920 Population 15 yrs. of age and over 13,797	Plant I 713 employees	Plant II 838 employees	Plant III 264 employees	1920 Population 15 yrs. of age and over 13,245	Plant I 238 employees	Plant II 22 employees	Plant III 96 employees
TOTAL	100.0%	100.0%	100.0%	100.0%	100.0%	100.0%	100.0%	100.0%
15 years	} 11.7	0.0	0.0	0.0	} 12.4	0.0	0.0	0.0
16 "		0.0	0.0	1.1		0.0	0.0	3.1
17 "		0.0	0.4	1.1		0.0	0.0	3.1
18–19 "		2.5	4.0	8.7		5.0	4.6	26.0
20–24 "	11.7	12.4	18.5	27.3	13.6	28.6	27.3	36.5
25–44 "	42.9	47.4	56.7	47.4	42.7	55.5	63.6	30.2
45–54 "	} 26.5	16.4	12.2	8.7	} 24.1	5.5	4.5	0.0
55–64 "		12.4	4.5	3.4		0.4	0.0	1.1
65–74 "	4.6	1.1	2.3		0.0	0.0	0.0
75 and over	} 6.8	1.1	0.0	0.0	} 7.1	0.0	0.0	0.0
Unknown	0.4	3.2	2.6	0.0	0.1	5.0	0.0	0.0

[1] Exclusive of office force.
[2] See Ch. V, N. 12 and accompanying text for description of these three plants.

TABLE III: UNEMPLOYMENT AMONG 165 WORKING MEN IN 1923 AND IN 9 MONTHS OF 1924[1]

YEAR	LOST NO TIME	LOST SOME TIME							
	Total No. Men	Total No. Men	Less than 1 Mo.	1–2 Mos.	2–3 Mos.	3–4 Mos.	4–5 Mos.	5–7 Mos.	7–10 Mos.
1923 (Jan.-Dec.)	119	46	25	1	8	7	2	2	1
1924 (Jan.-Sept.)	63	102	32	16	14	16	9	9	6

[1] See Appendix on Method for manner of selecting this sample and of interviewing. Information was secured from the wife and every effort was made to secure accuracy by checking over the time with her month by month. In case of any uncertainty the schedule was not used. Thus, while 182 women gave information on their husbands' employment, only 165 gave sufficiently definite information to be used on the actual number of months of employment during the preceding twenty-one months.

During 1923 and the first three months of 1924 times were "good" in Middletown; during the next six months of 1924 they were "bad." See the index of employment in Ch. VII.

TABLE IV: LENGTH OF TIME 182 WORKING MEN HAD WORKED FOR THEIR PRESENT EMPLOYERS [1]

PERIOD	NO. OF MEN	PERIOD	NO. OF MEN
Less than 6 months...	21	5–10 years.......	45
6 months to 1 year....	28	10–15 "	15
1–2 years............	20	15–20 "	9
2–3 "	16	20–25 "	8
3–4 "	11	25–30 "	4
4–5 "	4	39 "	1

[1] Figures are as of October 1, 1924. See note to Table III for employment conditions in 1923–24. See Appendix on Method for manner of sampling and interviewing. As pointed out in the text, this does not necessarily mean that these men had worked uninterruptedly for their employers during the period stated or that they had not even in some cases temporarily held other jobs for periods even of some months during lulls in their regular work.

TABLE V: DISTANCE EMPLOYEES [1] IN THREE REPRESENTATIVE MANUFACTURING PLANTS LIVE FROM PLACE OF WORK [2]

DISTANCE LIVED	TOTAL MEN AND WOMEN 2171	TOTAL MEN 1815	PLANT I 713 MEN	PLANT II 838 MEN	PLANT III 264 MEN	TOTAL WOMEN 356
TOTAL	100.0%	100.0%	100.0%	100.0%	100.0%	100.0%
Less than ½ mile.	27.9	27.4	34.2	19.1	35.2	30.3
½–1 mile	27.0	26.6	30.3	23.5	26.5	29.2
1–2 miles	22.7	22.8	18.0	27.2	21.6	22.5
2–3 "	3.2	3.1	5.6	1.2	2.3	3.9
3–10 "	13.1	13.4	10.9	15.9	12.5	11.2
10–15 "	3.3	3.5	0.7	6.8	0.8	2.3
15–20 "	1.5	1.7	0.3	3.1	0.8	0.6
20–25 "	0.6	0.6	0.0	1.3	0.3	0.0
25–45 "	0.7	0.9	0.0	1.9	0.0	0.0

[1] Exclusive of office force.

[2] See Ch. V, N. 12 and accompanying text for description of these three plants.

Addresses were secured from plant records. Distances under three miles were measured by drawing concentric circles of one-half, one, two, etc., mile radii about the factory door and locating each address within its circle. Since the radius of each circle represents an "as the crow flies" distance, a hypotenuse, whereas the worker must actually traverse the two sides of the right angle in all cases except where he lives directly down the street from the factory door, distances are somewhat underestimated, and many of those tabulated above as living at various distances under three miles probably belong in the group next more remote.

TABLE VI: EARNED INCOME, SURPLUS OR DEFICIT, AND EXPENDITURES FOR CERTAIN ITEMS OF 100 WORKING CLASS FAMILIES, OCTOBER 1, 1923–SEPTEMBER 30, 1924[1] (CARRIED TO NEAREST DOLLAR)

Family	Number in Family[2]	Total family income[3]	Father's income	Mother's income[4]	Surplus or Deficit[5]	Life insurance	Rent	House: investment in[6]	Furniture	Church, Sunday School, etc.	Charity	Lodges	Labor unions	Other clubs	Newspapers and periodicals[7]	Books (not school books)[8]	Music lessons	Sheet music, victrola and pianola records[9]	Concerts and lectures	Automobile: purchase, license, gas, and upkeep[10]	Vacation[11]	Other recreation[12]
1	8	345	345	0	−1200	53	0	48	0	2	1	0	0	0	10	0	0	0	0	0	0	26
2	7	461	425	36	−250	32	152	0	100	3	0	0	0	0	11	0	0	0	0	0	0	26
3	3	588	480	108	?[13]	15	216	0	36	0	0	0	0	0	23	3	0	0	0	45	0	26
4	4	660	660	0	−250	72	300	0	0	25	12	0	0	0	10	2	0	0	0	0	0	5
5	5	784	784	0	0	96	0	162	26	10	1	0	0	0	11	0	0	0	2	0	0	5
6	3	888	768	120	0	0	0	0	0	25	0	30	0	0	12	0	21	0	0	0	0	0
7	3	888	825	63	−200	16	180	0	60	1	0	0	0	0	10	0	0	0	0	0	0	10
8	4	927	900	173	0	27	0	0	0	1	0	0	0	1	11	0	0	2	0	62	0	5
9	5	960	754	60	0	32	0	83	0	3	0	13	0	0	7	3	43	0	0	0	0	5
10	5	968	968	0	−1000	65	0	772	0	30	0	0	0	0	14	0	0	0	0	0	0	0
11	6[17]	971	804	167	0	26	0	1098	0	8	0	0	0	21	12	0	0	0	0	348	0	55
12	3	975	923	0	−475	52	0	0	0	26	2	10	0	0	10	0	2	0	0	0	0	0
13	5	986	750	236	0	84	216	0	60	25	5	12	0	0	12	0	0	0	0	254	0	33
14	5	988	988	0	150	0	240	0	0	0	0	0	0	0	8	0	0	0	0	0	0	0
15	5	991	855	0	0	100	0	105	104	22	0	0	0	0	12	0	0	0	0	51	0	13
16	8	1003	833	0	+300	116	180	0	0	1	0	0	0	0	10	0	0	0	0	0	0	40
17	4	1065	1065	0	−50	59	364	0	0	2	3	0	0	0	10	0	0	0	0	97	0	3
18	5	1075	1075	0	?[13]	57	0	140	0	5	0	0	0	0	10	0	0	0	0	0	0	45
19	8	1075	1075	0	?[13]	18	0	56	0	3	0	22	0	0	14	0	0	3	0	0	0	0
20	6	1092	936	0	0	0	0	30	0	1	0	0	0	0	10	0	0	0	0	0	4	10
21	8	1101	1081	20	0	47	0	240	25	15	11	10	0	0	17	0	0	0	0	0	0	2
22		1150	1150	0	−250	60	0	370	0	5	102	0	0	0	8	0	0	3	0	0	0	4

26	52	0	0	0	7	0	14	10	0	0	0	0	59	0	0	204	63	0	150	1191/1035	3[17]/7	25
27	62	6	254	0	0	7	0	16	0	0	0	0	4	100	0	240	59	56	40	1196/1040	5	26
28	26	0	0	0	0	0	3	10	0	0	3	2	3	49	0	240	40	0	156	1206/1050	4	27
29	13	0	0	0	0	0	0	15	0	0	0	0	10	0	0	0	50	0	312	1217/905	3	28
30	0	0	0	0	0	0	0	8	0	0	20	0	1	0	279	210	54	35	0	1230	7	29
31	64	0	223	0	4	0	0	0	0	31	10	6	0	8	0	240	26	0	0	1230/1253	3	30
32	0	0	0	1	0	44	0	17	0	0	33	20	4	0	25	216	75	200	0	1253/1260	7[17]	31
33	25	0	0	0	0	0	0	10	0	0	10	13	62	0	0	0	61	400	75	1260/1264	5	32
34	12	0	0	0	0	0	0	22	0	0	0	0	60	18	256	132	18	0	0	1264/1196	4	33
35	31	0	103	0	3	0	0	10	0	0	3	0	0	200	0	180	34	0	85	1271/1274	7	34
36	48	30	25	0	1	0	0	12	0	0	0	0	1	0	0	0	32	0	0	1274/1190	4	35
37	52	0	52	0	1	27	27	10	0	18	12	0	3	100	218	0	26	0	0	1275/1278	4[17]	36
38	16	0	0	0	0	0	0	16	0	0	0	10	39	0	190	0	10	0	91	1278/1284	3[17]	37
39	58	10	0	2	0	33	0	20	0	0	29	5	8	0	354	0	20	0	215	1284	7	38
40	16	0	39[14]	0	0	0	5	10	1	0	0	0	63	202	0	0	55	0	506	1287/1196	7	39
41	16	30	166	0	0	26	0	22	0	0	0	30	31	8	42	0	70	0	0	1295/1080	7	40
42	83	0	67	0	0	0	1	12	0	0	20	10	3	0	65	0	22	0	0	1316/810	5	41
43	6	27	0	1	2	0	0	10	1	0	20	5	18	0	0	0	81	0	0	1319/1319	4	42
44	39	0	0	0	106[15]	0	0	10	0	0	10	0	12	0	48	0	44	0	0	1363/1363	4	43
45	0	20	0	0	0	0	0	17	0	0	18	0	15	55	0	315	78	50	738	1400/1400	7	44
46	43	20	29	0	3	0	1	17	0	0	0	15	5	0	0	216	40	0	0	1410/960	6	45
47	70	0	0	0	6	0	0	21	0	0	36	5	3	0	0	0	121	100	0	1427/1427	6	46
48	26	0	0	0	0	0	0	18	0	0	6	0	6	130	13	182	73	0	0	1429/691	4[17]	47
49	16	13	0	0	0	0	0	23	0	0	0	3	30	0	0	0	43	100	0	1441/1441	6	48
50	41	0	0	5	0	0	0	12	0	0	0	0	39	45	20	240	66	50	0	1442/1456	5	49
51	1	3	60	0	0	0	5	10	0	52	0	25	100	50	0	240	90	0	0	1456/1268	3	50
52	20	0	128	0	0	0	0	11	0	31	0	5	7	0	21	0	130	0	65	1495/1495	5	51
53	20	3	0	0	0	104	0	23	0	0	0	29	175	200	100	0	81	100	0	1499/1229	3[17]	52
54	36	0	90	0	0	0	4	12	0	0	0	0	5	40	77	216	31	0	624	1512/1512	8	53
55	21	0	0	0	0	45	0	17	0	0	15	5	9	19	0	0	52	?13	304	1521/1456	8	54
56	82	0	0	2	0	0	0	12	0	0	10	5	29	0	350	192	55	?13	350	1610/1610	3	55
57	0	50	474	0	0	52	0	10	0	0	0	5	50	78	0	180	0	100	170	1630	7	56
58	26	0	118	5	12	0	0	19	1	0	15	5	0	0	1000	0	115	0	0	1635/900	3[17]	57
59	52	0	492	0	0	42	0	17	0	0	10	2	10	0	40	468	122	1000	0	1637/1287	8	58
60	38	175	0	0	5	0	0	12	1	0	11	0	16	0	0	0	31	60	0	1638/1468	8	59
61	59	0		0	0		0	14	0	0	0	0	0	0	0	0	38	1520	0	1680/1680	3	60
62	62	0	0	0	5	0	0	17	0	31	0	5	2	1000	152	0	36	?13	0	1680/1680	7	61
																	10			1680/1680		62

EXPENDITURES FOR CERTAIN ITEMS

Family	Number in Family[2]	Total family income[3]	Father's income	Mother's income[4]	Surplus or Deficit[5]	Life insurance	Rent	House: investment in[6]	Furniture	Church, Sunday School, etc.	Charity	Lodges	Labor unions	Other clubs	Newspapers and periodicals[7]	Books (not school books)[8]	Music lessons	Sheet music, victrola and pianola records[9]	Concerts and lectures	Automobile: purchase, license, gas, and upkeep[10]	Vacation[11]	Other recreation[12]
63	3[17]	1698	1248	450	0	10	0	195	100	33	24	20	0	0	10	5	0	2	0	53	0	23
64	4	1718	1380	338	?[13]	0	0	35	0	34	0	0	0	0	12	0	0	0	1	0	0	10
65	4	1728	1045	260	−1000	70	240	0	0	78	38	0	0	0	14	5	8	0	5	0	0	58
66	3	1732	1474	258	0	47	0	1009	0	2	80	0	25	1	11	0	13	0	0	60	0	56
67	3	1748	1639	109	+50	81	180	0	200	5	5	14	0	0	17	4	38	3	0	142	25	78
68	4	1820	1820		+260	62	240	0	90	35	10	3	0	3	10	0	0	0	0	0	7	0
69	7	1824	1824		0	36	360	0	60	8	1	0	0	0	10	0	0	0	0	451	0	15
70	6[13]	1860	1326	257	+200	36	0	0	45	21	0	0	31	10	25	0	4	39	0	0	0	166
71	7[13]	1871	1838	388	0	114	300	223	0	8	6	10	0	0	10	0	0	0	0	103	4	8
72	4	1873	1616		500	143	0	0	125	0	8	8	0	10	21	0	0	25	0	0	0	189
73	7	1892	1504	56	+40	109	216	48	0	56	0	0	0	0	17	3	0	0	0	0	0	10
74	4	1900	1900		0	21	0	0	0	22	100	0	0	3	14	0	0	2	0	125	9	104
75	6	1932	1876		+100	30	0	0	188	4	0	12	0	0	22	0	0	1	6	54	50	41
76	4	2080	1976		+500	10	0	0	0	36	6	10	0	0	10	0	0	0	5	90	10	10
77	3	2080	2080	182	0	48	144	449	0	16	10	12	0	0	23	0	41	0	0	78	0	21
78	4	2080	2080		+100	20	180	0	0	5	0	4	0	0	13	32	0	7	0	58	1	26
79	5	2080	2080	1092	0	100	0	0	139	52	13	0	60	3	8	0	52	4	0	89	15	33
80	5	2100	2100		+500	57	300	600	0	16	12	26	0	0	10	1	112	0	0	132	50	27
81	7	2181	1949	34	+40	40	240	0	75	50	5	0	0	0	29	3	0	0	8	170	4	40
82	5	2184	2184		0	13	182	0	132	27	14	0	0	0	13	0	0	0	0	133	0	78
83	9	2209	1117	1092	+100	14	0	0	200	10	0	10	31	1	10	27	0	0	0	78	0	0
84	7	2218	2210		+100	6	0	0	0	13	39	0	18	0	44	9	4	5	0	135	30	23
85	3	2218	2184	34	0	350	0	100	15	27	5	0	0	0	10	10	27	0	0	0	0	16
86	4[18]	2246	2246		+500	2	0	300	0	50	0	0	0	0	20	3	0	0	11	97	20	65
87		2246	2246		0	200	0	0	0	10	7	30	0	0	15	0	50	10	0	0	10	151
88		2470	2236		+100	0	0	400	43	50	0	0	0	0	19	0	0	0	0	51	0	38

	Income[3]	Expenditure	Saved (±)[5]
93 [5]	2860	2860	0 +
94 [4][17]	2876	2600	276
95 [3][17]	3061	2736	325
96 [4][18]	3120	2600	0
97 [7][20]	3198	1820	−1500
98 [4][17]	3356	3200	0
99 [10]	3396	2040	0
100 [5]	3460	2860	600 +
Median	1495	1303	0
Q1	1194	1048	
Q3	2006	1857	115

[1] See Appendix on Method for manner of sampling and interviewing and for occupational distribution. There are nine foremen's families among these 100 families.

Each item of expenditure listed here represents a careful check with the housewife and in many cases is a composite of various smaller items, each of which was checked separately. Medians and quartiles are figured from the actual expenditures, not from amounts carried to nearest dollar.

[2] Includes all those dependent upon the family income, but not self-supporting children or other relatives or boarders who take their meals with the family but are not dependent upon it. Such additional persons are indicated in each case.

[3] Includes income from father, mother, children, or other contributors. Amounts not listed as father's or mother's income are, in cases where other persons taking meals with the family are indicated, contributed by self-supporting children (income from other boarders is included under mother's income); where such persons are not indicated these amounts are contributed by dependent children who work only part time. Experience in other studies has shown that wives frequently underestimate family income through ignorance of how much the husband really earns; some of this possible source of error has been drained off here through the elimination from the total families interviewed of 24 cases in which the wife was vague as to income details.

[4] Does not include money from children, but includes money from other boarders and all other amounts mother earns either at home or outside.

[5] Includes money saved (+) and withdrawn from reserve or borrowed (−).

[6] Includes purchase, interest on mortgage, and major repairs or alterations, but not taxes or insurance.

[7] The morning paper, delivered by carrier, costs $10.40 and the evening paper $7.80.

[8] All but one family had spent money on school books, the median amount being $8 with first and third quartiles at $5 and $15 respectively. In addition family No. 22 spent $35 on tuition; No. 81, $40; No. 65, $45; No. 39, $60.

[9] In cases where the only sheet music purchased was music for the children's lessons the amount spent was included with the amount for music lessons. Six other families had purchased sheet music, twenty-three families victrola records, and three pianola records.

[10] There is considerable chance for error here in such matters as the housewife's estimate of the amount spent on gas.

[11] Any holiday trip of more than two days.

[12] Effort was made to cover all forms of recreation not separately listed. The largest items are expense for motion pictures and extra expense for guests. The number of guests per week in winter and in summer months was secured and twenty cents was taken as the additional cost for one guest for a single meal.

[13] Withdrew from reserve or borrowed but not able to specify amount.

[14] Do not own a car, but spent money for gasolene while driving with friends.

[15] Includes $100 for purchase of a victrola.

[16] Includes $150 for purchase of a radio.

[17] One other person takes meals with this family.

[18] Two other persons take meals with this family.

[19] Three other persons take meals with this family.

[20] Five other persons take meals with this family.

TABLE VII: MINIMUM COST OF LIVING FOR A FAMILY OF FIVE, 1924 [1]

Food... $627.08

Clothing
 Husband............................... 87.14
 Wife.................................. 105.70
 Boy (12)............................... 73.03
 Girl (6)................................ 41.71
 Boy (2)................................ 31.38

  Total................................. 338.96

Furniture (annual replacement)........................ 60.09
Cleaning supplies.................................... 40.13
Rent... 300.00
Fuel and light...................................... 80.69

Miscellaneous
 Insurance
  Life: $7,500 (straight life)........ 113.25
  Fire: on $700 (furniture)......... 1.87

  Total 115.12

Carfare
 Husband......................... 30.00
 All others........................ 20.80

  Total........................... 50.80

Help—one day a week (or laundry) 145.60
Maintenance of health..................... 67.00
Amusements.............................. 25.00
Newspapers (daily and Sunday)............. 10.40
Church.................................. 15.00
Labor organizations....................... 24.00
Telephone, postage, tobacco, etc............. 20.00
One good magazine 1.00

  Total.................................. 473.92

  Grand total....................................... $1,920.87

[1] Based on the "Minimum Quantity Budget Necessary to Maintain a Worker's Family of Five in Health and Decency," U. S. Bureau of Labor (*Monthly Labor Review*, Vol. X, No. 6, June, 1920).

Prices are in all cases Middletown prices in 1924, determined by records kept over a period of weeks by a small group of coöperating Middletown working class housewives and by consultation with retailers.

TABLE VIII: COST OF LIVING INDEX FOR A WORKING CLASS FAMILY OF FIVE IN 1924 (1891=100) [1]

| To-TAL | FOOD | CLOTHING | | | | | HOUSE FUR-NISH-INGS | RENT | FUEL AND LIGHT | MIS-CELLA-NEOUS |
		Total	Men's	Wom-en's	Boys'	Girls'				
216.5	218.5	214.0	201.3	236.5	193.1	226.3	250.2	240.0	533.3	137.8

[1] The method of calculating this index follows that of the cost of living index of the United States Bureau of Labor Statistics except that every effort was made to make the weights by which the prices were multiplied correspond to the amounts consumed annually by a Middletown family of five in 1891.

Most of the food weights for the Middletown index were procured from the food consumption records of 104 families in this state and the state adjoining it on the east, collected in 1891 in the course of a Federal cost of living survey and published in the 1892 report of the Labor Commissioner. In obtaining average consumption the amounts consumed by American families only were used, and only for those families with one to five children. The families whose records were utilized averaged 2.97 adult male equivalents, according to the Bureau of Labor scale. (The standard family equals 3.35 adult males.) The average meat consumption figures obtained from these records are very high and the figure for milk low, as compared with the figures obtained in the 1918 cost of living survey. The difference seems to represent a real change in food consumption from 189- to 1918, particularly since the 1891 figures from Middletown's state and its neighbor to the east are nearer to the food consumption figures obtained in the Federal Cost of Living Survey in 1903 than to the 1918 figures. Where food consumption figures were not available either in quantity or cost units in the 1891 survey and the article was known to have been commonly purchased in that year, the detailed figures obtained in the Labor Bureau's 1918 survey were used.

The 1891 survey did not record details as to clothing and it was necessary here to take the Labor Bureau's figures for 1918 consumption and revise them in consultation with the older Middletown merchants familiar with the buying habits of the early nineties. This resulted in substituting material by the yard for many items of ready-to-wear clothing and in excluding a few items entirely.

In the case of rent the higher range of rents for 1891 and the lower range for 1924 were used in order to allow for the inprovement occasioned by the introduction of running water. The cost of heat and light for a five-room house was procured for the fuels used in each period.

All prices were verified from at least two reliable sources, and quality and quantity in both periods were considered, even where the style had changed.

TABLE IX: HOUSING MOBILITY OF BUSINESS CLASS AND
WORKING CLASS FAMILIES 1920–24 AND 1893–98 [1]

NO. OF TIMES MOVED	BUSINESS CLASS		WORKING CLASS	
	Jan. 1, 1920–Oct. 1, 1924	1893–98	Jan. 1, 1920–Oct. 1, 1924	1893–98
Total answering...............	40	38	124	106
Did not move...............	22	25	49	68
Moved once...............	14	10	41	16
Moved more than once........	4	3	34	22
Moved twice............	4	1	13	9
" 3 times...........	0	0	7	10
" 4 "	0	1	5	0
" 5 "	0	1	7	2
" more than 5 times	0	0	2	1

[1] Answers were secured from the housewives for themselves and for their mothers. The discrepancy in numbers between the two generations is accounted for by the fact that, while in every case the woman interviewed could tell how many times her family had moved during the preceding five years, a few could not give precise information regarding their mothers' families. The latter were omitted.

The period from the summer of 1893 to the spring of 1898 was used since the Chicago World's Fair and the Spanish War were convenient terminal periods in asking this question.

The smallness of the groups and the factor of memory make these data only roughly significant.

TABLE X: SIZE OF FORTY BUSINESS CLASS AND 124 WORKING
CLASS FAMILIES INTERVIEWED [1]

NUMBER IN FAMILY	BUSINESS CLASS	WORKING CLASS
3	6	17
4	13	30
5	12	29
6	6	15
7	2	13
8	1	12
9	0	4
10	0	3
11	0	0
12	0	1

[1] The number given in this table for each family is the number for whom the housewife cooks. It excludes children living away from home but includes self-supporting children living at home, relatives, and others living with the family.

TABLE XI: RATIO OF DIVORCES GRANTED TO MARRIAGE LI-CENSES ISSUED IN MIDDLETOWN'S COUNTY, 1889–1924 [1]

YEAR	No. OF DIVORCES GRANTED	No. OF MARRIAGE LICENSES ISSUED	PER CENT. DIVORCES CONSTITUTE OF LICENSES
1889	26	303	8.6
1890	30	283	10.6
1891	51	336	15.2
1892	47	333	14.1
1893	69	360	19.2
1894	41	376	10.9
1895	84	458	18.3
1900	117	631	18.5
1905	135	581	23.2
1910	147	557	26.4
1915	158	515	30.7
1916	150	579	25.9
1917	196	643	30.5
1918	248	463	53.5
1919	264	672	39.3
1920	261	798	32.7
1921	345	625	55.2
1922	279	692	40.0
1923	272 [2]	731	37.2
1924	273	644	42.3

[1] All figures are taken from the annual and biennial reports of the State with the exception of marriages for 1889–94 inclusive and for 1905. No figures on marriages for 1889–94 are available except in the *Special Reports of the United States Census Office: Marriage and Divorce, 1867–1906*. The number of marriages performed in 1905, according to the State Bureau of Statistics, was 817; the likelihood of error in this sudden leap in the total suggested the wisdom of using the smaller figure given in the *Special Reports* of the Census.

[2] The figure 272 given in the State report is used instead of 253, the figure given in the Federal Census Report for 1923. This choice rests upon the possibility, suggested by the State official responsible for these records, that the Federal figures do not include returns from both the circuit and superior courts of the county whose clerks make separate reports.

TABLE XII: STATED CAUSES OF DIVORCE IN MIDDLETOWN'S COUNTY 1919–22 AND 1889–92 [1]

YEARS	Total	Abandon-ment	Adultery	Impotency	Criminal conviction	Cruel treatment	Drunken-ness	Failure to provide	Other causes
1919–1922 (1,149 divorces)	100.0%	14.9	14.5	0.0	0.4	51.7	1.2	11.7	5.6
1889–1892 (154 divorces)	100.0%	29.6	23.7	0.0	0.0	30.3	4.6	11.8	0.0

[1] Percentages for the early period are based upon the State *Biennial Reports* for 1889–90 and 1891–2. Those for the current period are based upon the County records.

TABLE XIII: SOURCES OF DISAGREEMENT BETWEEN 348 BOYS
AND 382 GIRLS AND THEIR PARENTS[1]

SOURCE OF DISAGREEMENT	BOYS CHECKING		GIRLS CHECKING	
	No.	Per cent.	No.	Per cent.
1. Use of the automobile...............	124	35.6	113	29.6
2. The boys or girls you choose as friends.	87	25.0	103	27.0
3. Your spending money...............	130	37.4	110	28.8
4. Number of times you go out on school nights during the week	157	45.1	182	47.6
5. Grades at school...................	140	40.2	119	31.2
6. The hour you get in at night.........	158	45.4	163	42.7
7. Home duties (tending furnace, cooking, etc.)	66	19.0	101	26.4
8. Clubs or societies you belong to......	19	5.5	40	10.5
9. Church and Sunday School attendance.	66	19.0	71	18.6
10. Sunday observance, aside from just going to church and Sunday School..	50	14.4	53	13.9
11. The way you dress.................	55	15.8	94	24.6
12. Going to unchaperoned parties.......	53	15.2	105	27.5
13. Any other sources of disagreement[2]...	33	9.5	32	8.4
"Do not disagree"[3]................	7	2.0	8	2.1

[1] This is one question in the questionnaire given to all the English classes in the three upper years of the high school. (See Appendix on Method.) It read: "Check the things listed below about which you and your parents disagree. State any other causes of disagreement." The items are given here in the order in which they were presented in the questionnaire. Since no limit was placed on the number of items to be checked and most children checked more than one item, the percentages add to more than 100. The boys averaged 3.3 and the girls 3.4 checks each.

[2] Among other sources of disagreement listed by the boys were: "Spending all my time on athletics," "Smoking," "Drinking," "How much I should work," "Having a rifle."

Among those listed by the girls were: "Cigarettes," "Boys," "Petting Parties," "Bobbed hair," "Playing cards," "Reading too many books," "Dancing," "Machine riding to other towns at night with dates," "Evolution."

[3] This item was not on the questionnaire. The answers here so classified were volunteered by the children and probably do not include all those who "do not disagree." Fifty-seven boys and eighty-two girls answering the questionnaire did not check this list.

TRAIT	Business Class 37 mothers: 1924				Business Class 34 mothers: 1890				Working Class 104 mothers: 1924				Working Class 67 mothers: 1890			
	No. marking A	No. marking B	No. marking C	No. marking Zero	No. marking A	No. marking B	No. marking C	No. marking Zero	No. marking A	No. marking B	No. marking C	No. marking Zero	No. marking A	No. marking B	No. marking C	No. marking Zero
Frankness in dealing with others	16	7	14	0	13	11	10	0	22	48	33	1	12	32	23	0
Desire to make a name in the world	0	3	26	8	1	12	16	5	7	40	50	7	4	24	34	5
Concentration	7	22	8	0	2	9	22	1	6	11	87	0	2	11	54	0
Social-mindedness	8	11	17	1	3	20	8	3	10	27	67	0	4	24	39	0
Strict obedience	16	10	9	2	24	8	2	0	48	40	15	1	41	22	4	0
Appreciation of art, music, and poetry	5	12	20	0	3	10	21	0	8	34	60	2	2	13	41	11
Economy in money matters	3	15	19	0	2	20	11	1	32	51	20	1	20	40	7	0
Loyalty to the Church	13	12	10	2	23	6	3	2	58	32	12	2	47	17	3	0
Knowledge of sex hygiene	6	19	12	0	0	5	7	22	15	33	50	6	2	12	21	32
Tolerance	4	17	16	0	2	10	19	3	4	17	80	3	3	12	49	3
Curiosity	0	3	29	5	1	1	25	7	1	7	67	29	0	2	46	19
Patriotism	8	13	16	0	4	15	13	2	21	47	34	2	13	34	19	1
Good manners	7	20	10	0	13	19	13	0	36	51	17	0	28	29	10	0
Independence	17	12	8	0	6	9	17	2	18	37	49	0	10	18	38	1
Getting very good grades in school	1	9	25	2	4	16	14	0	26	46	32	0	11	33	22	1

[1] The women were instructed to mark the three traits they considered of greatest importance "A," the five of next importance "B," any of third importance "C," and any of no importance "Zero." See Ch. XI, N, 21 for method of marking. Questions are in the same order here as on the questionnaire.

It is probable that differences between generations are considerably underestimated, as even with the greatest efforts to mark carefully there was apparently a tendency to minimize differences. Where a woman was vague or uncertain as to her mother's emphasis or where she said, "Oh, I guess she did just about as I do," and marked the list accordingly, the ratings for her mother were not used. In cases, however, where she considered the list carefully and said, "I try to do just as nearly as I can what my mother did," the ratings were used.

In the case of some items, notably "Loyalty to the Church," there was some tendency for a woman to rate according to what she thought she *ought* to emphasize.

Data of this type, particularly when based upon such small groups, must, of course, be regarded as suggestive only. It should be noted that the mothers of 1890 are here regarded as belonging to business class or working class according as their daughters today belong to one group or the other.

TABLE XV: RATINGS BY HIGH SCHOOL BOYS AND GIRLS OF TRAITS MOST DESIRABLE IN A FATHER AND IN A MOTHER [1]

In a Father

Trait	369 Boys		415 Girls	
	No.	Per cent.	No.	Per cent.
Total answers	738	200.0	830	200.0
1. Being a college graduate	86	23.3	55	13.3
2. Spending time with his children, reading, talking, playing with them, etc.	227	61.5	276	66.5
3. Making plenty of money	45	12.2	47	11.3
4. Being an active church member	99	26.8	126	30.3
5. Owning a good-looking car	11	3.0	7	1.7
6. Being prominent in social life	18	4.9	9	2.2
7. Never nagging his children about what they do	45	12.2	50	12.0
8. Being well dressed	23	6.2	21	5.1
9. Having a love of music and poetry	16	4.3	26	6.3
10. Respecting his children's opinions	125	33.9	176	42.4
(Second trait not marked)[2]	43	11.7	37	8.9

In a Mother

Trait	369 Boys		423 Girls	
	No.	Per cent.	No.	Per cent.
Total answers	738	200.0	846	200.0
1. Being a good cook and housekeeper	212	57.5	221	52.2
2. Being prominent in social life	13	3.5	16	3.8
3. Respecting her children's opinions	89	24.1	94	22.2
4. Being well dressed	28	7.6	12	2.9
5. Always having time to read, talk, go on picnics or play with her children	126	34.1	172	40.7
6. Having a love of music and poetry	22	6.0	14	3.3
7. Being a good hostess	14	3.8	17	4.0
8. Being an active church member	94	25.5	105	24.8
9. Being a college graduate	26	7.0	16	3.8
10. Never losing her temper or nagging	85	23.0	143	33.8
(Second trait not marked)	29	7.9	36	8.5

[1] Directions regarding this question given to the three upper years of the high school (see Appendix on Method) read: "Place a capital letter A before the 2 qualities below that you consider most desirable in a father." The similar question on traits most desirable in a mother occurred on a different page of the questionnaire. The questions are presented here in the same order in which they appeared on the questionnaire.

[2] This number of children in each group marked only one trait. These omitted markings are included in the totals on which percentages are based in order to make the totals 200 per cent.

TABLE XVI: DISTRIBUTION OF SUBJECTS TAUGHT IN THE HIGH SCHOOL IN THE FIRST SEMESTER OF 1923-24[1]

SUBJECT	NO. OF CLASSES	NO. OF HOURS PER WEEK PER CLASS	TOTAL NO. OF STUDENTS ALL CLASSES	STUDENT-HOURS PER WEEK	
				No.	Per cent.
Total.....................	262			31,291	100.0
Mathematics...............	39	5	934	4,670	14.9
English...................	53	5	1,380	6,900	22.0
Social science..............			974	4,870	15.6
Civics.................	3	5	81	405	1.3
Sociology..............	1	5	24	120	0.4
History................	30	5	869	4,345	13.9
Science...................	18	5	422	2,110	6.7
Language..................			875	4,375	14.0
Latin.................	28	5	653	3,265	10.4
French................	5	5	124	620	2.0
Spanish................	5	5	98	490	1.6
Art......................	3	5	60	300	1.0
Gymnasium................	14	2	413	826	2.6
Music....................			436	1,778	5.7
2 hour courses..........	5	2	104	208	0.7
3 hour courses..........	2	3	45	135	0.4
5 hour courses..........	8	5	287	1,435	4.6
Domestic science...........			211	971	3.1
Food courses...........	6	4	84	336	1.1
Other courses..........	8	5	127	635	2.0
Commercial dep't...........	19	5	544	2,720	8.7
Manual arts...............			140	661	2.1
Manual arts proper.....	2	4	39	156	0.5
Other courses..........	7	5	101	505	1.6
Vocational................	6	10	111	1,110	3.6

[1] A similar distribution of subjects for the second semester showed no marked differences.

TABLE XVII: BOOKS BORROWED FROM THE ADULT DEPART-
MENT OF THE PUBLIC LIBRARY IN 1923 AND IN 1903[1]

SUBJECT	1923		1903	
	No.	Per cent.	No.	Per cent.
Total...............................	92,618	100.00	22,265	100.00
Encyclopedias, General Works, etc....	1,941	2.10	701	3.15
Philosophy, Psychology, etc..........	1,097	1.18	42	0.19
Religion.............................	836	0.90	73	0.33
Sociology............................	1,219	1.32	129	0.58
Philology............................	207	0.22	0	0.00
Science..............................	697	0.75	112	0.50
Useful Arts.........................	1,546	1.67	25	0.11
Fine Arts............................	1,777	1.92	64	0.29
Literature...........................	2,557	2.76	362	1.63
History..............................	1,755	1.90	228	1.02
Travel...............................	1,140	1.23	174	0.78
Biography...........................	1,116	1.20	194	0.87
Fiction..............................	76,666	82.78	20,161	90.55
Unclassified.........................	64	0.07	0	0.00

[1] 1903 is the first year for which circulation figures are available.

TABLE XVIII: NUMBER AND KIND OF CLUBS IN 1924 AND IN 1890 AND MEMBERSHIP IN 1924[1]

KIND OF CLUB	NO. OF CLUBS 1924	NO. OF CLUBS 1890	No. of clubs for which membership secured	Total no. of members	MEMBERSHIP OF CLUBS: 1924 No. of male clubs	No. of members of male clubs	No. of female clubs	No. of members of female clubs	No. of mixed clubs	No. of male members of mixed clubs	No. of female members of mixed clubs	No. of unclassified members of mixed clubs
Total adult clubs	363	86	290	23,963	67	13,132	179	6,131	44	1,714	1,445	1,541
Athletic (not including teams)	3	3	3	404	1	50			2			354
Benevolent	50	27	36	10,868	17	8,681	10	709	9	316	674	488
Auxiliary	10	0	9	704			8	589	1	14	101	
Business and professional	9	1	9	408	8	368			1	36		
Church and other religious	101	8	77	4,371	3	177	57	3,150	17	339	585	120
Religious other than church	4	0	4	585			4	585				
Civic	11	1	10	1,399	6	342	2	711	2			
Literary, musical, and study	24	12	20	710			17	680[2]	3			
Military and patriotic	13	2	13	2,627	8	2,361	5	266				
Social	129	21	101	1,722	5	263	88[3]	1,255	8	918	68	15
Trade unions	19[4]	8	19	879	18	815[5]			1	12	18	64
Miscellaneous	4	3	2	575	1	75			1	93	96	500
Total juvenile clubs	95	6	81	2,836	23	577	37	1,363	21	319	502	75

[1] A club is defined as an organized group having at least one meeting a month which is entirely or partly social. A Sunday School class, e.g., even though it is organized and holds an annual picnic, is not included unless it has at least a monthly social meeting. Only white groups are included. See Ch. XIX, N. 19 for the method by which these figures were collected.

[2] The clubs constituting the Federated Club of Clubs are classified as Literary, Musical, and Study Clubs.

[3] Including thirteen "savings clubs," with a total membership of 171.

[4] All trade unions except the Women's Union Label League.

[5] Including twenty-six Negroes among the members of the Hodcarriers' and Building Laborers' Union.

TABLE XIX: NUMBER AND KIND OF CLUB AFFILIATIONS OF BUSINESS CLASS AND WORKING CLASS MEN AND WOMEN IN 1924[1]

| | WOMEN | | MEN[2] | |
AFFILIATIONS	Business class	Working class	Business class	Working class
Total no. answering	39	123	39	123
No. not belonging to any organization	3	79	1	53
No. belonging to 1 or more organizations.......................	36	44	38	70
1 organization..................	3	23	8	46
2 organizations.................	10	13	8	14
3 "	5	3	12	6
4 "	9	0	5	3
5 "	6	2	4	1
6 "	2	2	1	0
7 or more organizations.........	1	1	0	0
No. belonging to lodges............	0	20	34	60
1 lodge.......................	0	13	24	42
2 lodges......................	0	3	7	11
3 "	0	2	2	4
4 "	0	2	1	2
5 "	0	0	0	1
No. belonging to church social clubs..	23	17	7	0
1 church club.................	10	10	7	0
2 church clubs................	9	7	0	0
3 " "	2	0	0	0
4 " "	2	0	0	0
No. belonging to labor unions........	0	2	0	17
1 union......................	0	2	0	15
2 unions.....................	0	0	0	2
No. belonging to other clubs........	35	18	30	1
1 club.......................	10	13	13	1
2 clubs......................	10	1	13	0
3 "	9	2	4	0
4 "	6	1	0	0
5 "	0	1	0	0

[1] For definition of a "club" see Table XVIII, N. 1.
[2] Answers for men were for the most part secured from their wives.

TABLE XX: AVERAGE WEEKLY ATTENDANCE AT RELIGIOUS SERVICES BY WHITE POPULATION IN NOVEMBER, 1924[1]

ATTENDANCE	TOTAL		MALE		FEMALE	
	No.	Per. cent.	No.	Per cent.	No.	Per cent.
Total white population—1924[2]	35,872	100.0	18,187	100.0	17,685	100.0
Average total weekly attendance at all services[3]...................	20,632					
Average at Sunday A.M. service[4]	5,157	14.3	2,057	11.3	3,100	17.5
" " " P.M. " [5]	4,317	12.0	1,857	10.2	2,460	13.9
" " Prayer Meeting[6].....	1,603	4.5	622	3.4	981	5.5
" " Sunday School[7]......	6,624	18.4				
" " Young People's Meeting[8]	977	2.7	404	2.2	573	3.2
" " Other services (except Catholic)	152	0.4				
" " 6:30 and 8:30 A.M. week-day Mass[9]...	1,802	5.0	767	4.2	1,035	5.9

[1] Blanks for recording attendance, accompanied by return envelopes, were sent to every minister in Middletown preceding each of the four Sundays in November, 1924. In some cases precise counts of attendance were made by some one in the choir or in the back of the church; in some of the smaller churches the minister made the count; in a few cases only an estimate was secured from the minister. In the majority of cases figures were secured for all four Sundays; in certain others for only two or three; while in a few of the very small struggling churches attendance was secured for one Sunday only. An average attendance for each service for each church was figured. The totals here given are the sums of these averages. The four weeks selected are believed to be a good sample, including both good and bad weather; attendance probably fluctuates slightly above these figures in mid-winter and below them in summer. Attendance figures secured in the manner described above were checked wherever possible by counts made by staff members at the various services.

[2] Population figures were arrived at by reducing the estimated population of 38,000 in 1924 by the percentage of negro males and females given in the 1920 Census.

[3] This involves many duplications. A number of ministers, e.g., estimated that half their Sunday evening congregation was present at the morning service.

[4] Five churches do not have a Sunday morning service. The Catholic 6:30 Mass is included with the Sunday morning services.

[5] Seven churches do not have Sunday evening service.

[6] Six churches do not have Prayer Meeting.

[7] Two churches do not have Sunday School. Sunday School totals were not distributed according to sex.

[8] Fifteen churches do not have Young People's Meetings. A total equivalent to 9.7 per cent. of the white population 5–21 years (8.1 per cent. of the male population 5–21 and 11.2 per cent. of the females) attended Young People s Meetings.

[9] Figures for week-day Mass are the total for six days.

TABLE XXI: ATTENDANCE AT RELIGIOUS SERVICES BY BUSINESS CLASS AND WORKING CLASS FAMILIES INTERVIEWED IN 1924 AND BY THE WIVES' PARENTS' FAMILIES IN THE NINETIES [1]

ATTENDANCE	SUNDAY MORNING SERVICE				SUNDAY EVENING SERVICE				SUNDAY SCHOOL			
	Business class		Working class		Business class		Working class		Business class		Working class	
	1924	1890	1924	1890	1924	1890	1924	1890	1924	1890	1924	1890
Entire family Total no. of families..........	40	40	123	119	40	40	123	119	40	40	123	119
Attending 4 times a month........	11	24	14	50	0	8	14	29	5	14	15	44
" 3 " "	4	0	3	3	0	0	4	2	2	0	5	3
" 2 " "	2	1	4	9	0	1	4	6	0	0	1	1
" 1 time "	2	0	2	1	0	0	4	2	0	0	1	0
" less than 1 time a month.....	0	0	0	0	0	0	0	3	0	0	0	0
Never attending..............	21	15	100	56	40	31	97	77	33	26	101	71
Father [2] Total no. of fathers...........	40	39	123	110	40	39	123	110	40	39	123	110
Attending 4 times a month........	16	27	26	49	0	12	22	30	11	15	22	42
" 3 " "	6	0	3	6	1	0	2	2	3	0	5	3
" 2 " "	4	1	3	7	2	2	7	7	0	0	1	1
" 1 time "	4	0	4	3	1	0	7	1	0	0	1	0
" less than 1 time a month.....	0	0	2	1	0	0	3	4	0	0	1	0
Never attending...............	10	11	85	44	36	25	82	66	26	24	93	64

Mother[2]												
Total no. of mothers	40	40	123	119	40	40	123	119	40	40	123	119
Attending 4 times a month	16	31	24	63	0	14	18	37	6	16	25	52
" 3 " " "	6	1	7	6	1	0	5	4	4	0	6	3
" 2 " " "	5	0	6	6	1	2	8	9	0	0	4	1
" 1 time " "	2	0	4	1	1	0	9	3	0	0	1	0
" less than 1 time a month	0	0	2	3	0	0	7	5	0	0	0	0
Never attending	11	8	80	40	37	24	76	61	30	24	87	63

Children[3]												
Total no. of families with children	40	40	123	119	40	40	123	119	40	40	123	119
Attending 4 times a month	15	30	32	73	0	12	22	39	32	32	77	94
" 3 " " "	3	1	5	5	0	0	4	4	4	2	17	5
" 2 " " "	2	1	9	8	0	2	8	7	2	0	10	5
" 1 time " "	2	0	2	1	0	0	7	2	0	0	3	0
" less than 1 time a month	0	0	0	0	0	0	0	4	0	0	1	0
Never attending	18	8	75	32	40	26	82	63	2	6	15	15

[1] These figures were secured from the housewives interviewed. In each case their estimates of the church-going habits of the family "during a usual month" were checked by the last four weeks. Every effort was made to secure careful reporting for the earlier period, and it is probable that in a routinized, repeated activity like church-going memory would tend to be reasonably accurate. The discrepancy in numbers between the two generations is accounted for by the fact that in some cases father or mother was dead and in others reports were too vague to be used. A month is here taken as having four Sundays.

[2] With or without other members of the family. Included, also, under "Entire Family."

TABLE XXII: ADVERTISING IN THE LEADING PAPER IN OCTOBER, 1923, AND IN OCTOBER, 1890[1]

KIND OF ADVERTISING	NO. OF LINES		PER CENT.	
	1923	1890	1923	1890
Total lineage..........................	604,292	101,448	100.00	100.00
Amusements........................	27,749	3,940	4.59	3.88
Musical instruments......................	16,131	113	2.67	0.11
Autos, accessories.......................	35,648	0	5.90	0.00
Livery stable, carriage equipment..........	0	2,155	0.00	2.12
Department stores and dry goods..........	66,474	22,586	11.00	22.26
Women's wear...........................	55,277	863	9.15	0.85
Men's wear.............................	25,289	7,523	4.19	7.42
Shoes..................................	18,426	6,366	3.05	6.27
Furniture...............................	41,420	0	6.86	0.00
Hardware and household appliances.........	13,168	3,079	2.18	3.04
Foods, groceries........................	44,947	9,055	7.43	8.93
Tobacco................................	6,904	3,121	1.14	3.08
Drug stores	17,757	3,226	2.94	3.18
Medical advertising (including patent medicines)................................	45,451	25,004	7.52	24.65
Jewelers, opticians......................	6,968	0	1.15	0.00
Financial advertising.....................	9,381	3,795	1.55	3.74
Legal advertising........................	4,375	1,063	0.72	1.05
Real estate and building..................	6,142	984	1.02	0.97
Agriculture.............................	22,084	0	3.66	0.00
Hotels, resorts..........................	0	84	0.00	0.08
Railroads, steamships....................	3,282	3,037	0.54	2.99
Classified advertisements.................	90,413	1,519	14.96	1.50
Miscellaneous..........................	29,984	3,294	4.96	3.25
"House copy" (the newspaper's own copy)....	17,022	641	2.82	0.63

[1] 1923 includes seven issues a week and 1890 six issues, as there was no Sunday paper in 1890. The 1923 paper is a morning paper; the 1890 paper was an afternoon paper, there being no morning paper. A similar check for the month of May in both periods yielded similar results.

TABLE XXIII: GEOGRAPHICAL SCOPE OF NEWSPAPER READ-
ING MATTER[1] DURING ONE WEEK IN 1923 AND IN 1890[2]

	1923		1890	
SCOPE	Morning paper (7 days)	Evening paper (6 days)	Evening paper No. 1 (6 days)	Evening paper No. 2 (6 days)
Total inches of news space.......	6,900	6,560	2,057	1,286
Per cent. of news space.........	100.0	100.0	100.0	100.0
Local.........................	17.6	19.7	24.1	45.6
County........................	7.0	6.1	5.7	4.5
State.........................	13.5	6.7	11.8	8.4
National......................	39.4	25.9	31.2	25.0
International..................	9.0	11.1	8.9	5.2
Non-geographical..............	13.5	30.5	18.3	11.3

[1] Including editorials.
[2] The count was made for the first week in March in both periods.
    Figures derived from counts for such a short period must, of course, be regarded as sug-
gestive only.

TABLE XXIV: DISTRIBUTION BY SUBJECT OF NEWSPAPER READING MATTER [1] DURING ONE WEEK IN 1923 AND IN 1890 [2]

SUBJECT OF NEWS	1923		1890	
	Morning paper (7 days)	Evening paper (6 days)	Evening paper No. 1 (6 days)	Evening paper No.2 (6 days)
Total inches news space.........	6,900	6,560	2,057	1,286
Per cent. of news space..........	100.0	100.0	100.0	100.0
Public affairs [3]	18.2	13.2	13.0	5.1
Politics [4]	1.5	0.8	21.6	13.0
Business.......................	9.7	3.5	3.9	2.9
Legal [5]	2.3	1.3	1.2	1.5
Labor..........................	0.3	0.2	0.9	0.8
Agriculture [6]	0.9	1.3	6.5	2.1
Police..........................	6.8	8.1	4.8	7.6
Divorce........................	0.6	1.0	0.0	0.0
Accidents.......................	1.8	2.1	6.4	4.4
Personal.......................	8.9	12.0	8.3	19.1
Social..........................	5.6	3.9	3.1	4.3
Women's news [7]................	2.3	4.5	1.0	0.0
Health.........................	1.3	0.5	0.1	0.2
Charity.........................	0.3	0.3	0.2	0.5
Religion........................	2.5	2.1	1.8	9.4
Education.......................	1.3	0.7	1.9	2.5
Science.........................	1.1	0.9	2.7	1.3
Radio..........................	0.2	0.3	0.0	0.0
Sports..........................	15.9	10.4	4.4	3.1
Lectures........................	0.4	0.2	0.1	0.4
Arts [8]	0.7	0.9	1.7	2.4
Theater.........................	2.2	1.7	0.0	0.2
Travel..........................	1.2	1.8	1.7	5.3
Cartoons........................	7.6	21.6	0.0	0.4
Fiction.........................	1.8	3.0	3.8	0.0
Jokes—jingles	1.4	1.3	2.9	7.9
Weather........................	0.5	0.6	1.1	1.4
Miscellaneous..................	2.7	1.8	6.9	4.2

[1] Including editorials.

[2] See Table XXIII, N. 2 for week counted. Checks for a single week can be only roughly significant.

[3] Legislative action, building of a new sidewalk on Main Street, etc. Paper No. 1 in 1890 carried a regular column summarizing the work of the preceding day in Congress and the Federal departments. No such systematic summary of governmental proceedings appears in Middletown papers today.

[4] Election news, candidates, etc. The political news in 1890 was somewhat influenced by an approaching election—though in reading Middletown's papers of a generation ago one gets a strong impression that Middletown was always either looking forward to or recovering from an election.

[5] Real estate transfers, suits filed, decrees and judgments in other than divorce and criminal cases.

[6] Exclusive of market quotations which are included under Business.

[7] Dorothy Dix, styles, dressmaking, recipes, household hints.

[8] Exclusive of clubs devoted to music and art which are included under Social news.

INDEX

INDEX

School houses, used as community centers, 485 n.
Secularization, 500; and control of parenthood, 123; of books on science, 237 n.; of charity, 462, 468; of church programs, 401; of divorce, 121; of health matters, 457; of lectures and lecturers, 229; of marriage ceremony, 112; of "Sabbath," 339; differential rate of, 26 n., 123, 203, 338 n., 343
Self-criticism, stifling of, 222
Sermons, 345, 372 f.; excerpts from, quoted, 374 f.
Servants, 169 f.; and labor-saving devices, 171
"Service," 304
Services, church, see "Churches"
Sewing, 164 f.
Sex, and magazines, 241; and motion pictures, 267 f.; and revivals, 380 n.
Sex hygiene, 145
Sexual relations between unmarried people, 112, 115 n., 123
Sexes, different activities of, 25, 28, 116, 123, 126, 140, 148, 164, 167, 270, 277
Shelter, 62
"Shut-downs," 55 f.; see also "Unemployment"
Sickness, 439 n.; and wife's working, 28; defined, 454
Sin, 122 n., 361, 374
Skilled worker, 74, 81
Sloan, George W., Fifty Years in Pharmacy, 157 n.
Smith, J. Russell, North America, 7, 225 n.
Snyder, Raymond H., An Analysis of the Content of Elementary High School History Texts, 199 n.
Social change, and automobile, 253; and proximity to centers of cultural diffusion, 5; and "social problems," 496 f.; music, 247; ornamentation and, 95 n., 166; process of, 8, 35 n., 95 n., 98, 294; resistance to, 38, 42 n., 142, 150 n., 157, 160, 413; reversal of trend, 174
Social entertainment, see "Parties"
Social illiteracy, new types of, 166, 222
Social Organization, Charles Horton Cooley, 494 n.

Social Organization, W. H. R. Rivers, 4 n., 132 n.
"Social problem," 3, 34 n., 35 n., 53, 59, 63, 70 n., 105 n., 107, 129
Social problems, and social change, 496; masked by "boosting," 222; masked by emotion, 178; masked by formulas, 178; masked by slogans, see "Slogans"; of care of health, 445; of crime, 431; of high cost of living, 84; solution, 426 n.
Social Service Bureau, 295, 433, 438, 458, 462
Social studies, 196 f.
Social Trend, E. A. Ross, 38 n.
Socialism, 228 n.
Socialization of accident hazard, 70
Solidarity, pressure for, 278; see also "Conformity," "Standardization"
Sophistication of children, 140
Specialization, 40, 45, 75, 249, 409; see also "Standardization," "Conformity"
Speeches, 228; political campaign, 419; topics, 228; see also "Lectures," "Leisure"
Spending money, sources of children's, 141
"Spring sickness," 157; see also "Health"
"Stability," 414
Standard of living, 83, 99 ff., 161, 167, 171; and number of children, 131, 243 n.; and wife's working, 28, 131 n.
Standardization, 106, 161, 491; art, 249 n.; of leisure time pursuits, 309; relation of credit to, 47; see also "Credit," "Conformity," "Advertising"
State Statistician, quoted, 74; see also "Reports, state"
Statistics, compilation of, in regard to study, 507
Style show, 82 n.
Success, 65
Sunday observance, 339 f.; and the automobile, 258 f.; recreation, 341 f.; sports, 340 f.
Sunday School, the, 341, 360 f., 383 f., 391, 393; changes in, 391 f.; direction of, 383; graded lessons, 391; services, 383, 387; teachers, 383 f.; teaching, 385